IEE TELECOMMUNICATIONS SERIES 47

Series Editors: Professor J. O'Reilly
Professor G. White

Local Access Network Technologies

Other volumes in this series:

Local access network technologies

Edited by
Paul France

The Institution of Electrical Engineers

Published by: The Institution of Electrical Engineers, London,
United Kingdom

The Institution of Electrical Engineers,
Michael Faraday House,
Six Hills Way, Stevenage,
Herts., SG1 2AY, United Kingdom

www.iee.org

British Library Cataloguing in Publication Data

Local access network technologies. – (Telecommunication series ; 47)
1. Computer networks
I. France, P. II. Institution of Electrical Engineers
004.6

ISBN 0 85296 176 6

Typeset in India by Newgen Imaging Systems (P) Ltd., Chennai
Printed in the UK by MPG Books, Bodmin, Cornwall

Contents

6   VDSL – The story so far   117
D. Clarke

7   Implementing local loop unbundling – an account of the key
challenges   129
A. Cameron, D. Milham, R. Mistry, G. Williamson, K. Cobb
and K. O'Neill

M. Begley and A. Sago

Preface to Local Access Network Technologies

The access network, or local loop, is a vital consideration for any network engineer. It is used to provide the connection through to the end customer and, since there are now billions of customers around the world, there are billions of access connections. However, this world is changing rapidly, driven by the liberalisation of telecommunication markets and the rapid development of new technologies. Networks are currently undergoing a fundamental structural change, moving from narrowband TDM networks supporting the delivery of PSTN or POTS, to broadband packet-based networks supporting a range of advanced services and applications. At the heart of this transformation lies the access network. For more than 100 years based on the simple but effective twisted copper pair, the access network is now being replaced or augmented by ADSL, optical fibre, radio, hybrid fibre-coaxial cable, wireless LAN, GSM or 3G, satellite, and many other access technologies. This bewildering choice of local access technologies is an important consideration for any network engineer, since the cost of the access network can typically amount to some 80 per cent of overall telecommunication networks.

The work presented in this book reviews the major new access technologies that are finding their way into significant use. The book is a compilation of chapters written by several contributors, all of whom have extensive experience in the field and are internationally recognised for their work. By and large, the authors are based or associated with BT's development facilities, now known as BT Exact, at Adastral Park, near Ipswich in the UK.

The first chapter is a general introduction to access networks, and overviews the major new technologies and their use by network operators. Chapter 2 examines market trends for narrowband and broadband applications, gives some indication of bandwidths required, and describes the access bottleneck. Chapters 3, 4, 5, 6 and 7 then discuss aspects of copper technologies including ADSL and VDSL, but also including 'local loop unbundling' and 'frequency planning'. In particular, Chapter 5 describes how ADSL has been used in the broadband programme and discusses many of the operational issues such as increasing coverage into rural areas. Chapters 8, 9 and 10 then cover optical fibre technologies, outlining the different topologies in use today, discussing both PONs and point-to-point technologies in some depth, as well as the physical infrastructure issues of fibre-to-the-home. Chapters 11, 12, 13 and 14 move on to radio technologies, from wireless LANs and fixed radio technologies through to satellite and UMTS systems. The large scale deployment of cable systems warrants a specific chapter on this topic, Chapter 15, which describes not only cable technology and means for delivering TV and voice, but also covers the DOCSIS standard for data (and Internet) delivery using cable modems. Chapter 16 discusses the increasing use of SDK in the access network. Finally, Chapter 17 discusses management of the access network, based on TMN architecture, and presents many of the issues concerned with network management of broadband access.

The book is aimed at engineers working not only in the field of access networks but also with telecommunication systems in general, since many of the issues are concerned with local access. However it may also be useful for students both at graduate and post graduate level.

Finally, I would like to acknowledge the help and support of all those who have played a part in putting this book together, in particular:

- all the contributors, who have put a great deal of effort into their respective chapters;
- the IEE for their encouragement and support, and in particular Diana Levy, Production Manager, and Gerry White, Series Editor;
- Debbie Hawkley and Adrian Silvertown for their help in co-ordinating the manuscripts and producing the Glossary found at the back of the book. Adrian has developed his own tool for extracting acronyms that other authors and editors may also find useful.

Finally I would like to acknowledge the support of my own organisation BT Exact, based at Adastral Park near Ipswich. This world-renowned division of BT continues to be a driving force behind telecommunication technologies and much of the content of this book is a by-product of that work.

Paul France
November 2003

Contributors

Dave Austin
BT Wholesale
Technology & Support Programmes
pp G09
Oberon House (B67-MH)
Adastral Park
Martlesham Heath
Ipswich
Suffolk, IP5 3RE

Maurice Begley
BTExact
Technical Team Leader
pp RSB10 7
Aquarius Building (B65-MH)
Adastral Park
Martlesham Heath
Ipswich
Suffolk, IP5 3RE

Martin Booth
BTExact
pp OP3
Polaris House (B29-MH)
Adastral Park
Martlesham Heath
Ipswich
Suffolk, IP5 3RE

David Bryant
BTExact
Senior Spectrum Management Engineer

pp RSB G10/7
Aquarius Building (B65-MH)
Adastral Park
Martlesham Heath
Ipswich
Suffolk, IP5 3RE

Alan Cameron
BTExact
Access Solution Designer
pp 102
Polaris House (B29-MH)
Adastral Park
Martlesham Heath
Ipswich
Suffolk, IP5 3RE

Spyros Christou
BTExact
Lead IP & Data Solution Designer
pp 5 MLB2
Orion Building (B62-MH)
Adastral Park
Martlesham Heath
Ipswich
Suffolk, IP5 3RE

Don E A Clarke
BT Wholesale
Access Network Technologist
pp T1F
Telephone House

West Stockwell Street
Colchester
Essex, CO1 1BA

Ken Cobb
BTExact
Senior Professional – Broadband
Network Engineering
pp G34
Polaris House (B29-MH)
Adastral Park
Martlesham Heath
Ipswich
Suffolk, IP5 3RE

John W Cook
BTExact
Engineering Adviser – Copper
Access Systems
pp G21
Polaris House (B29-MH)
Adastral Park
Martlesham Heath
Ipswich
Suffolk, IP5 3RE

David W Faulkner
BTExact
Venture Leader in Access Technologies
pp G11 pp G20
Polaris House (B29-MH)
Adastral Park
Martlesham Heath
Ipswich
Suffolk, IP5 3RE

Andy Fidler
BTExact
Project/Technology Manager
pp RSB10G/pp 6A
Aquarius Building (B65-MH)
Adastral Park
Martlesham Heath
Ipswich
Suffolk, IP5 3RE

Simon Fisher
BTExact
Broadband Network Engineering
pp G38
Polaris House (B29-MH)
Adastral Park
Martlesham Heath
Ipswich
Suffolk, IP5 3RE

Kevin T Foster
BTExact
Unit Manager – Advanced Copper
Technologies
pp G20
Polaris House (B29-MH)
Adastral Park
Martlesham Heath
Ipswich
Suffolk, IP5 3RE

Paul France
BTExact
Centre Manager – Internet & Solution
Design
pp ORION 5/2
Orion Building (B62-MH)
Adastral Park
Martlesham Heath
Ipswich
Suffolk, IP5 3RE

Guillem Hernandez
BTExact
Project/Technology Manager
pp 6 RSB-G10
Aquarius Building (B65-MH)
Adastral Park
Martlesham Heath
Ipswich
Suffolk, IP5 3RE

Keith James
BTExact
Access Solution Manager

pp B29 OP11
Polaris House (B29-MH)
Adastral Park
Martlesham Heath
Ipswich
Suffolk, IP5 3RE

Andrew Kerrison
BTExact
Broadband Design Manager
pp 5 2nd Floor
Orion Building (B62-MH)
Adastral Park
Martlesham Heath
Ipswich
Suffolk, IP5 3RE

Rob H Kirkby
BTExact
Senior Professional – Spectral
Compatibility
pp HWD397
Virtual Postbox (HOM-NZ)
PO Box 200
London, N18 1ZF

Andi Mayhew
BTExact
Broadband Network Engineering
pp G31
Polaris House (B29-MH)
Adastral Park
Martlesham Heath
Ipswich
Suffolk, IP5 3RE

Dave Milham
BTExact
OSS Industrial Collaborations
pp B36 Room 7
B36 Adastral Park (B36-MH)
Martlesham Heath
Ipswich
Suffolk, IP5 3RE

Raj Mistry
BT Global Services
Business Process Transformation
Consultant
pp B67 220
Oberon House (B67-MH)
Adastral Park
Martlesham Heath
Ipswich
Suffolk, IP5 3RE

Reza Mostafavi
BTExact
Radio Systems Engineer
pp RSB10 6C
Aquarius Building (B65-MH)
Adastral Park
Martlesham Heath
Ipswich
Suffolk, IP5 3RE

Kevin O'Neill
BTExact
Access Network Designer
pp B29/102B
Polaris House (B29-MH)
Adastral Park
Martlesham Heath
Ipswich
Suffolk, IP5 3RE

Steven Page
BTExact
Broadband Network Engineering
pp G31
Polaris House (B29-MH)
Adastral Park
Martlesham Heath
Ipswich
Suffolk, IP5 3RE

David B Payne
BTExact
Manager – Broadband Architectures and
Optical Networks Unit

pp B29 G16
Polaris House (B29-MH)
Adastral Park
Martlesham Heath
Ipswich
Suffolk, IP5 3RE

Tim Pell
BTExact
Senior Satellite Systems Engineer
pp RSB10 pp 6
Aquarius Building (B65-MH)
Adastral Park
Martlesham Heath
Ipswich
Suffolk, IP5 3RE

Ian Rose
BTExact
Technical Area Leader – Satellite
Systems
pp RSB G10/6
Aquarius Building (B65-MH)
Adastral Park
Martlesham Heath
Ipswich
Suffolk, IP5 3RE

Simon J Rees
Director DSL Solutions
Fujitsu Telecommunications Europe Ltd

Andy Sago
BTExact
Wireless Solutions Designer
pp RSB10 7
Aquarius Building (B65-MH)
Adastral Park
Martlesham Heath
Ipswich
Suffolk, IP5 3RE

Dave Spirit
BT Global Services
Commercial Propositions Manager
pp 5

Adhara Building (ADH-IP)
Adastral Park
Martlesham Heath
Ipswich
Suffolk, IP5 3RE

Jerome Tassel
BTExact
Broadband IP Solution Designer
pp 5th Floor pp 2
Orion Building (B62-MH)
Adastral Park
Martlesham Heath
Ipswich
Suffolk, IP5 3RE

Andrew Walker
BTExact
Broadband Network Engineering
pp G31
Polaris House (B29-MH)
Adastral Park
Martlesham Heath
Ipswich
Suffolk, IP5 3RE

Gary Williamson
BTExact
IP and Data Platforms Integration
engineer
pp RM123C
Gemini Buildings (SST-MH)
Adastral Park
Martlesham Heath
Ipswich
Suffolk, IP5 3RE

Ben Willis
BTExact
Fixed Wireless Access
pp RSB10 7
Aquarius Building (B65-MH)
Adastral Park
Martlesham Heath

Ipswich
Suffolk, IP5 3RE

Peter N Woolnough
BT Wholesale
Wholesale Market Analysis Manager

pp B83 230/1
Columba House (B83-MH)
Adastral Park
Martlesham Heath
Ipswich
Suffolk, IP5 3RE

Chapter 1

An introduction to the access network

P. W. France and D. M. Spirit

1.1   Introduction

Fixed access networks have always been a critical component of telecommunication networks. Essentially they provide the final connection through to the customer and at the same time they are usually the most expensive component in terms of capital investment and ongoing cost of maintenance and repair. Typically they may form some 70% of the total investment required. Competition in the access and local telecommunications market has become fierce in recent times, as major players seek to form new mergers and acquisitions, and cable TV companies and new entrants compete head-on with traditional Telcos (PTT). This competition is being fuelled by regulatory changes as markets become more liberalised, and indeed by new technology which increasingly threatens to make existing operators and their networks obsolete. Already there has been a massive investment into fixed access networks around the globe where some 700 million access lines have now been installed – by far the majority using twisted copper pairs. Increasingly, new access technologies are being developed and deployed with a view to delivering new services and to reducing capital and operating costs. However, with such a large installed base of twisted pairs, a key focus of the new technologies has also been to enhance the capability of existing copper.

There are, therefore, a considerable number of technology options available to both existing and new operators, and the optimum choice will be dependent on a number of different factors. This chapter presents an overview of current and forthcoming access network technologies and how they are likely to be deployed in access networks around the globe. These technologies are discussed in the light of commercial considerations, such as the current market structure, the impact of competition and regulation, as well as the demand for new broadband and Internet services.

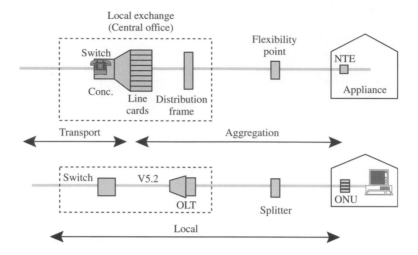

Figure 1.1   Definition of access network

1.1.1   Definition of access

Figure 1.1 shows the key elements of access networks as used today. Traditionally the access network connected the customer through to the local exchange (or central office) and did not include exchange equipment. The only part of exchange equipment included was the main distribution frame – the first point of flexibility in the network. More recent definitions, however, separate out the aggregation business (access) from the transport business (core). Aggregation equipment would include the fixed access connection to the customer's premises (copper, fibre or radio) as well as any aggregation devices such as multiplexors. In particular, it also includes exchange line cards (digital line interface into switch). The new definition takes into account the impact of new fibre access technologies, which can shift the position of line cards out to the customers' premises or into the street, as well as providing concentration, although this can still be categorised as aggregation.

The market in the USA is categorised slightly differently into local and long distance, with local networks in this case including the local switch and providing an end-to-end call connection between customers.

1.1.2   UK market structure

The liberalisation of the market in recent years has led to some 200 new operating licences being granted in the UK, offering a range of different services. Figure 1.2 is an attempt to classify some of the key operators in terms of their services and customers they are targeting.

The traditional Telco is shown offering telephony services to all customers. The diagram gives a view of the number of niche operators that have been formed in recent years, some competing with the PTTs and others, such as the cable TV companies,

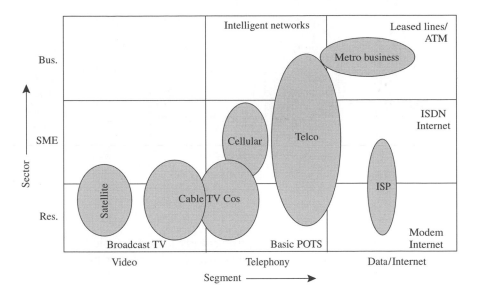

Figure 1.2   UK market structure

Table 1.1   Service/network options

Services	Access network
Bulk telephony/Private circuits	Copper pairs/Fibre
Mobile/Niche telephony	GSM/Fixed radio
Metropolitan/Business services	Fibre access/SDH
Broadcast TV/Telephony	HFC/Copper pairs
Broadcast TV only	Satellite
Long distance/Internet	Indirect access
Value-added service providers	Leased lines

also offering new services. By and large different companies have selected different access technologies and Table 1.1 gives a view of technologies used today. These technologies are summarised in Section 1.2.

1.2   Current networks

1.2.1   Copper access network

An outline structure of BT's telephony network is given in Figure 1.3. Over many decades BT's access network has grown to encompass access nodes in over 5600

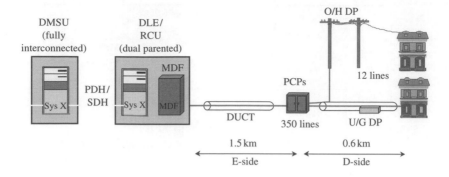

Figure 1.3   Copper access network

buildings and an overhead network comprising nearly 1.5 million poles. This network reaches all corners of the UK and is capable of delivering services to the vast majority of dwellings and businesses.

In the main, access is provided by copper pairs with some 29 m working pairs in the UK today terminating on a two tier switching network, comprising a small number of trunk switches (DMSUs) used to interconnect a large number of local switches (DLEs). The MDF is the main point of physical flexibility in the exchange allowing copper pairs to be flexibly connected into the switch. E-side (exchange) cables might typically consist of 1000 pairs and fan out from exchanges to PCPs – the main flexibility points in the network. A PCP typically serves some 350 customers and allows better utilisation of E-side cables. D-side (distribution) cables might typically consist of some 100 pairs and fan out from PCPs to DPs. DPs are the final flexibility points in the network and typically serve about 12 customers. Although E-side cables are usually installed underground, D-side cable exists as both underground and overhead. Generally underground cable is installed in ducts, within the UK, to allow for additional cable installation. In continental Europe access cables are frequently directly buried.

Because of the uncertainty in demand for service, and the desire to keep connection times down to a minimum, a considerable number of spare pairs has to be installed, based on Telco planning rules. Directly buried networks generally need to have more spare capacity installed than ducted cable. Typical distances are given in Figure 1.3, and are characteristic for Europe where population densities tend to be high.

Not surprisingly, the cost of such a network is dominated by fixed costs resulting from depreciation and overheads such as accommodation. However, as a result of many initiatives in recent years direct operating costs, although substantial, are now amongst the best in the world.

Access networks traditionally have been vulnerable to the effects of weather; e.g. storm damage or water ingress. But as a result of continuous improvement, the average customer will now see just one fault in eight years. There is still scope for improving the weather resilience of the access network since the unpredictability of severe weather leads to less than optimum deployment of labour resources. Detailed

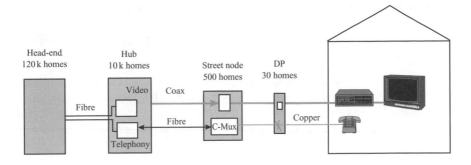

Figure 1.4   Cable TV network

studies show that modern access plant is inherently reliable and that it is interventions and work practice that drive fault rates.

1.2.2   Cable TV network

Regulatory freedom in the UK has encouraged cable TV operators to roll out networks offering both cable TV and telephony services. A typical network is shown in Figure 1.4 which indicates how the two services are offered. Although technology now exists to offer both TV and telephony services over a single integrated coaxial cable network, at the time these networks were installed a safer option was to install both coaxial cable and copper pairs for the final connection. Optical fibre was used as the E-side cable. Essentially there are two parallel networks side by side: hybrid fibre/coaxial cable for the TV network and hybrid fibre/copper pairs for telephony. A primary multiplexor (C-mux) is installed in the street node so that signals on copper pairs can be digitally multiplexed onto optical fibre. The networks are integrated at the physical cable and duct level which is where most of the cost saving can be found. Local authorities generally require cable TV networks to be installed entirely underground using duct. Analogue TV is distributed using AM-VSB, although two-way digital upgrades are now extensively being installed across the UK. A detailed view of digital cable networks is given in Chapter 8.

1.2.3   Other current networks

Other niche operators are now offering services and frequently have specialised in the type of access technology they use. These are discussed in Section 1.4 but generally:

- Mobile operators use GSM or other PCN technology. In addition, a number of companies provide fixed access using point-to-multipoint radio equipment over limited geographic areas.
- Direct fibre through to customers is now increasingly used both by BT and other operators, although because of the high cost associated with civil works this can only really be justified for high bandwidth connections to large business customers. These point-to-point fibre systems generally use either PDH or

increasingly SDH multiplexors and can provide bandwidths from 2 to about 155 Mbit/s.
- Satellite technology can be used effectively to distribute broadcast TV and satellite companies have a large number of connections in the UK.

At this stage it is also worth mentioning that the lack of an access network does not preclude operators from offering services to customers. Indirect access, though the use of another operator's network, allows long distance operators to provide service, and indeed this is the norm in the USA market. Indirect access is also used by Internet Service Providers to allow dial-up connections into their Internet services.

1.3 Market dynamics

The choice of access technology is crucial to the survival of operators since access is such a dominant part of overall costs. That choice will, however, be dependent on a number of factors, best summarised through the use of the Porter diagram of competitive forces in Figure 1.5. New entrants have been discussed in Section 1.2.2, with many new niche operators selecting specific technology to support their overall strategy. The new technologies available are discussed in Section 1.4.

1.3.1 Regulation

Without doubt regulation makes a significant contribution to the choice of network. Competition was first introduced in the UK in 1984 when a duopoly was established allowing both BT and then MCL (Mercury Communications Ltd) to offer services. This followed liberalisation within the USA and divestiture of AT&T in 1984, as well as the modified final judgement giving RBOCs the exclusive right to offer local services and allowing competing long distance companies. Following the duopoly

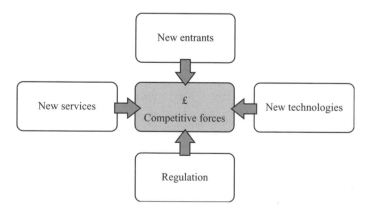

Figure 1.5   Competitive forces

review in 1991, the UK market was opened to much more competition, and the cable TV companies allowed to offer full telephony services.

In the early years the regulatory policy in the UK followed that of 'managed competition' whereby new entrants were encouraged into the market place through the implementation of favourable rules. For example, BT was prohibited from offering broadcast TV services and from the general use of fixed radio technologies. Regulation was used to encourage new access infrastructure build within the UK and the policy was largely successful. This was based on the assumption that economies of competition (through competing networks) outweighed economies of scope – whereby a single network may have been used to offer competing services.

As the market has matured regulatory policy has shifted towards 'behavioural regulation' whereby established players are discouraged from anti-competitive behaviour through the use of penalties. In addition operators are encouraged to move towards more open networks, and network interconnect established on fair terms. One example is 'copper unbundling' whereby operators may lease the use of copper pairs from another operator to offer their own services. Inevitably this entails 'collocation' of equipment since interconnection at the MDF requires additional equipment in the exchange. One contentious area is 'infrastructure sharing' whereby operators may share the use of physical assets such as duct and cable. Although not aggressively pursued in the UK yet, it seems likely that, as we move towards a more mature market with full open competition, operators will inevitably co-operate as well as compete in order to prosper.

1.3.2   New services

A key factor in determining the choice of technology is of course the services that will be offered over the network. Undoubtedly, they will broaden out from the three key services of today of telephony, cable TV and mobility to include:

- Teleworking
- Video-on-Demand
- Home Shopping
- Tele-education
- Videoconferencing
- LAN Interconnect
- Intranet
- High Speed Internet

Indeed these new services will begin to dominate over a period of time in terms of overall revenues as margins on commodity services reduce. Figure 1.6 is one view of how the overall market mix for services may change within the UK over a period of ten years. The figure has been compiled from a variety of sources and is meant for illustrative purposes only.

Although to a certain extent new services will be offered at low bandwidths (9.6 – 56 kbit/s) feasible with V32 modems over conventional copper pairs, the requirement for video services means that customers will be better served by higher speed digital

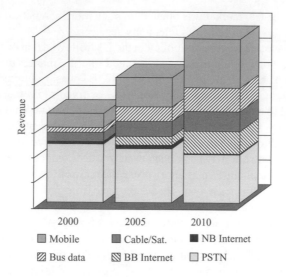

Figure 1.6   The market mix is changing

access connections. Some outline requirements in terms of bandwidths for these new services are given in Table 1.2.

Table 1.2   Requirements

Corporate	Lan interconnect, Intranet (34,155 Mbs symmetric)
Teleworking	Remote LAN access (0.5/0.5 Mbs, 2.0/0.5 Mbs)
SME	Fast Internet access (0.5/0.5 Mbs, 2.0/0.5 Mbit/s)
Residential	Fast Internet access (0.5/0.5 Mbs)
	Entertainment (VoD 2.0/0.5 Mbs)

1.4   Emerging access technologies

The delivery of innovative services requires the presentation of new and increasingly digital interfaces at the network termination. In many cases, the capacity required is in excess of those required for single line PCM speech channels (64 kbit/s). This has led to a range of developments in access technologies that give network operators a great variety of options with which customers may be provided with services. The selection of technology depends not just on the service requirements of any one customer, but also on the type of network provider and the access infrastructure already in place (see Table 1.3).

A schematic of a selection of the access transport options available to current network operators is shown in Figure 1.7. In this example, the access network is

Table 1.3   Access options

Type of operator	Example of current access infrastructure	Most appropriate new technologies
Incumbent operator in developed country	Twisted pair copper dominates (high penetration), but substantial optical fibre	xDSL on copper, expansion of fibre (PON + point-to-point) in existing duct
Cable TV operator	Coaxial cable, some optical fibre	Cable modems (if possible)
Incumbent operator in developing country	Twisted pair copper (but low penetration)	Wireless local loop, expansion of fibre (PON + point-to-point)
New entrant in developed country	None	Mixture of fibre, point-to-point radio, wireless local loop, leased lines from other operators and indirect access
New entrant in developing country	None	Wireless local loop, satellite systems for quick rollout, leased lines and indirect access

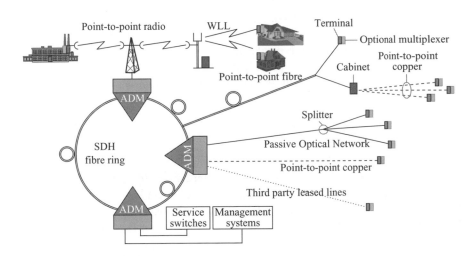

Figure 1.7   Modern access networks

based on an SDH fibre ring, which provides aggregation of traffic onto protected paths and a high degree of resilience between the service switches/routers and the access add-drop multiplexors (ADMs). From the ADMs, connections to customer sites are provided over star or tree-and-branch networks.

In certain cases, technological innovations are being matched by an increasing level of network management functionality in the access network (particularly where customer SDH/SONET is provided). This can allow the management of core and access networks from a common network control layer, removing, to a certain extent, the boundary between core and access networks.

The choice of technology with which to serve a particular customer or group of customers depends on a number of parameters, the key issue being the aggregate bandwidth to be delivered to/from the customer site. Where a network operator already possesses access infrastructure that enters the customer premises, the approach should be to provide services over existing infrastructure if at all possible. The focus is on the enhancement of capabilities by changing only the equipment at either end of the access network, rather than making wholesale changes to the access infrastructure. This approach enables network operators to reduce both the time and the cost of responding to customer requirements.

New network operators may choose to obtain customer access using incumbent (or other competitor) facilities such as interconnect, leased lines, or 'dark copper', where regulation permits. All these options require a substantial outflow of current account payments to competitors. Consequently, there are a growing number of new operators who are, to some degree, deploying their own access infrastructure. In addition to bandwidth, the choice of technology to serve customers depends upon the proximity to other customers or potential customers: a concentrated customer cluster could be most efficiently served by optical fibre infrastructure for high bandwidths or wireless local loop (WLL) for telephony. For incumbent and new network operators, efficient planning systems and forecasting techniques are essential to minimise the impact of changed or new customer requirements on the access network infrastructure.

The sub-sections that follow provide a summary of key access technologies. A more detailed overview of how these technologies are currently deployed is given in subsequent chapters.

1.4.1   Wireline technologies

1.4.1.1   Twisted pair copper cables

The rapid growth in dial-up access to the Internet has led to customers using 'high speed' voiceband data modems which are capable of operation at data rates of up to 56 kbit/s over analogue copper lines. However, this technology is reaching the fundamental limits set by Shannon's Law for transmission using these types of protocol over installed twisted pair cables.

Further enhancements to data rates on copper cables requires the use of different formats of line coding, implemented in two main classes: Integrated Services Digital Network (ISDN) basic rate at 128 + 16 kbit/s and the range of standard methods known generically as Digital Subscriber Line (DSL). There are a range of DSL systems operating at a variety of data rates which are often referred to as xDSL (Figure 1.8). The first to be deployed was High speed DSL (HDSL), which provides 2 Mbit/s transmission over distances of up to several km. Transmission spans may be increased by using multiple pairs. More recently asymmetric DSL (ADSL: up to

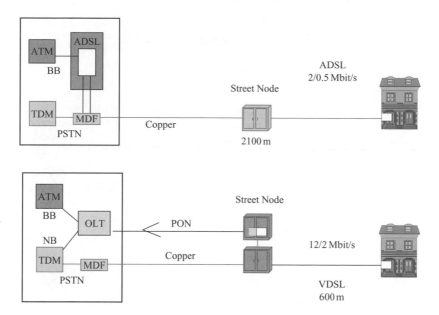

Figure 1.8   DSL technologies

6 Mbit/s towards the customer) has become fairly widespread in the consumer market as Telcos have moved into high speed Internet services. Very high speed DSL (VDSL) allows speed up to and in excess of 15 Mbit/s, at the expense of reductions in the transmission distance. It is still largely an emerging technology, although it has been used to deliver entertainment TV services.

1.4.1.2   Coaxial copper cables

The dominant usage of coaxial copper cables in access has been to provide the bearer for analogue cable TV networks. Where the existing allocation of RF spectrum to the broadcast TV signals permits, unused spectrum may be used for data transmission between terminals known as 'cable modems' (640 kbit/s is typical). New cable TV network architectures, particularly hybrid fibre-coax (HFC: see Figure 1.9) are being deployed. The final customer connection is over a much shorter coaxial cable than the original all-coax network, implying that end-to-end transmission at a higher data rate will be possible with this new architecture.

1.4.1.3   Electrical mains power cables

Telecommunications deregulation in many countries has led to the emergence of network operators with a wide range of owners. In particular, electricity distribution companies have diversified into telecommunications, in many cases using their regional or national grids as the basis of a core network. To enable extensions into the access area, new technology has recently become available to carry digital signals (up to a shared 1 Mbit/s) over the electrical mains distribution network. This technology

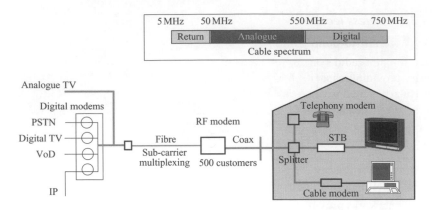

Figure 1.9   Digital HFC network

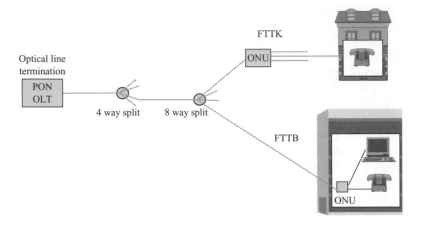

Figure 1.10   Telephony over passive optical network (TPON)

is being trialled within the UK, although there are key concerns regarding both the capacity available to a customer and radio frequency interference.

1.4.1.4   Optical fibre cables

Optical fibre cable systems are being deployed in access networks to carry aggregated traffic in one of two configurations: point-to-point or 'passive optical networks' (PONs). Point-to-point capacities of 4×2 Mbit/s are common, although 34 Mbit/s and 155 Mbit/s are also deployed. Figure 1.10 shows an example configuration of TPON systems that are in established commercial use for fibre-to-the-kerb (FTTK) and Fibre-To-The-Building (FTTB) applications. TPON systems can provide capacities of up to 16×2 Mbit/s and offer a cost-effective alternative to point-to-point configurations for medium sized sources of traffic. PONs offering ATM-format transmission capacities up to 622 Mbit/s have also now been developed, and have been deployed by a number of network operators world-wide.

1.4.2 Wireless technologies

1.4.2.1 Point-to-point radio

Point-to-point radio systems are commercially available in a wide range of spectral bands, providing capacity between 2 Mbit/s and multiples of 155 Mbit/s. Radio is particularly suited to applications where customers with bandwidths of 2 Mbit/s or more are situated an appreciable distance from cable networks. The reach of a radio system will be determined by the characteristics of the transmitter/receiver pair and radio propagation in the spectral band(s) available to the network operator.

1.4.2.2 Point-to-multipoint terrestrial radio

Cellular networks provide the most common form of mobile access, with radio channels being assigned to establish connections to customers requiring service. Whilst the service capability of current mobile networks is extensive, the maximum bandwidth available to an individual customer remains somewhat limited, although increasing with the introduction of 3G technologies.

Wireless local loop (WLL) is a radio alternative to twisted pair copper networks. These systems provide point-to-multipoint radio transmission from a base station to a fixed network termination. In current commercial products, an $n \times 32$ kbit/s link is provided to each customer on demand (n is 1 or 2). Speech paths are provided over 32 kbit/s digital compressed speech channels (adaptive differential PCM, or ADPCM). Two main product groupings currently exist: the Digital European standard for Cordless Telephony (DECT) in the 800 MHz band, where the cell size is typically up to 1 km, and proprietary systems operating at high frequencies and hence larger cell radii (several km is typical). Evolution of these two classes of systems is occurring in different directions: DECT towards some limited mobility (within the same base station area), and the larger cell systems towards ISDN basic rate capability. Convergence of DECT and cellular technology (WLL for 'home' use, GSM for roaming away from home) is already taking place, with dual mode DECT/GSM handsets in commercial production.

A limited number of manufacturers are producing broadband (multiple 2 Mbit/s per customer) point-to-multipoint radio which is being deployed in a number of locations around the world. Just as for WLL systems, bandwidth is assigned to customers on demand, even if the capacity required changes during a period of communication over the link. However, this is a separate technology to WLL and is not an evolutionary path from it.

1.4.2.3 Satellite

Satellite technologies are particularly suitable for delivering services over very wide geographic areas. Services that are most appropriately delivered over satellite include:

- subscription-based and pay-per-view broadcast TV, for both residential and business applications;

- contribution quality TV links, particularly from transportable earth stations for occasional use;
- data, particularly using very small aperture terminals (VSAT) in areas where terrestrial access is not available;
- speech, fax and voice-band data with global reach (currently provided via geostationary satellites, but soon to be augmented by several constellations of low and medium earth orbit satellites under construction).

All of these application areas are undergoing major changes introduced by technological advances, including increased bandwidth/decreased cost per bit, low earth orbit satellite mobile systems and digital broadcast TV. New services are becoming available, particularly those which are based on the Internet protocol (e.g. Internet push/pull) and are generally not particularly delay sensitive. Satellite remains an expensive technology so there is growing interest in high altitude platforms.

1.4.3   Infrastructure-independent technologies

To complete the portfolio of access options available to network operators, it is necessary to introduce other options which are particularly (though not exclusively) applicable to new operators. Around the world, operators and bandwidth suppliers lease capacity that can provide access links. To increase the range of services that can be carried over a single leased circuit, equipment suppliers produce a range of proprietary multiplexors to aggregate traffic from different services, originally for the corporate networks market. Historically, these have been based on time division multiplexing (TDM), although cell relay, asynchronous transfer mode (ATM) and Ethernet based equipment is also in increasing commercial use. Multiplexor manufacturers include voice compression capability within their product ranges to ensure that the capability offered by a single leased line is maximised.

In countries where public voice telephony has been deregulated, access to alternative service providers can obtained via the incumbent's access network. An appropriate prefix is added to the dialled digits to signal to the incumbent voice switches that the call should be routed to an alternative service provider's network ('indirect access'). This prefix can be entered manually by the customer or automatically by a PBX or 'smartbox' using a pre-assigned allocation of a prefix to classes of dialled digits.

1.5   The immediate future – Telco

In order to address the current issues affecting a typical Telco access network, a range of commercial and technical solutions will be used to continue to drive down costs whilst simultaneously increasing the earning potential of the access business.

Costs: Cost reduction will be achieved by progressive reduction in capital investment in mature service delivery mechanisms. As competition increases so too will churn between one operator and another, so new solutions will be deployed specifically to control the cost of churn and the impact that churn traditionally has on network quality. The impact of overhead costs will be lessened by initiatives that spread the cost across an increasing volume of new services carried by the access network or by selective write-offs of obsolete networks.

Volumes: Volumes for mature delivery mechanisms (such as narrowband copper for POTS) will remain static in the face of growing retail competition; but there may be scope for growth in the wholesale market. Stock levels and just-in-time delivery processes will be optimised especially when providing connectivity to new green-field sites. There will be a shift from a mainly reactive build process in the access network to a focused pro-active 'build & sell' strategy.

The datawave is producing a growing demand for digital narrowband and broadband capacity both from existing customers and the emerging Service Provider market. Mobile networks are competing more with the traditional hardwired markets to provide narrowband speech service and there will be an increasingly blurred boundary between fixed and mobile services and network infrastructure.

Capacity management: Access will continue to be predominantly via a hardwired network. Increasingly the copper network will be reserved for the provision of higher bandwidth services using DSL technology and continuing use will be made of pair gain devices with up to six channels to provide additional channels. Fibre will be used where it is cost effective to do so and where it enables high bandwidth services. A significant proportion of provision will be by radio to manage churn and to reduce the need to install new copper or fibre cables.

Structural changes: Certain customers and market segments will require levels of reliability significantly better than average. Investment will be selectively applied to improve infrastructure in a targeted manner. In other cases where speed of provision becomes a crucial competitive advantage, pre-provision allowing high levels of 'met from stock' may well be justified. Less capital intensive solutions will be needed for cases where the commercial value is less or where significant market share is unlikely.

The evolution of broadband services demands that the design approach to the first concentration point in the access network is improved. Solutions are being developed to provide for short term needs but with sufficient flexibility to evolve and support the data-wave thereby avoiding the risk of a proliferation of 'stovepipe' solutions. A similar approach to standardisation and simplification of NTE (network terminating equipment) for broadband and business products is being developed.

Operational costs: More of the network will be managed remotely by reducing the need for reactive interventions to provide service. Increasingly, flexibility will be provided by using radio and pair gain devices. Improved reliability combined with a substantially lower requirement for flexibility will change the management of the access network from 'hands-on' to 'hands-off'.

Network health and reliability: Systems will be used to monitor the health of the network, adopt optimum stock levels, anticipate component failure and direct engineers to network components that require work. Network health scores will also become a key management performance measure. Maintaining accurate data and records is absolutely crucial to informed decision making and business success.

People and culture: The work force will be organised in customer service teams trained, focused and 'incentivised' to provide service at a cost and quality that will meet the commercial requirements of the business and provide exceptional customer service. They will function as separate businesses to commercially exploit the local access network. This will also enable the development of more pride and ownership of the 'patch' which should be reflected in improved quality and customer satisfaction.

1.6   Telco evolution to broadband

There are then a bewildering number of choices for an existing Telco to consider when upgrading the access network. One key element will be capital cost since this will have a major bearing on the commercial case for upgrade. Figure 1.11 compares the cost of potential broadband upgrade technologies as a function of penetration expressed as a percentage of homes passed. The figure assumes that these are costs to an existing PTT with a copper access network already installed. Not surprisingly therefore the DSL technologies come out lowest cost since they exploit the presence of copper pairs. FTTE represents ADSL technology in a fibre to the exchange architecture. FTTCab represents VDSL in a fibre to the cabinet architecture (Figure 1.8). HFC assumes the use of some existing BT duct but otherwise demands new infrastructure. FTTK and FTTH represent fibre to the kerb and home respectively. Figure 1.11 clearly shows that ADSL is the lowest risk way of offering higher bandwidths since it is by far the lowest cost technology, particularly at the low penetrations experienced during the early phases of the market. FTTCab would similarly be the lowest risk way

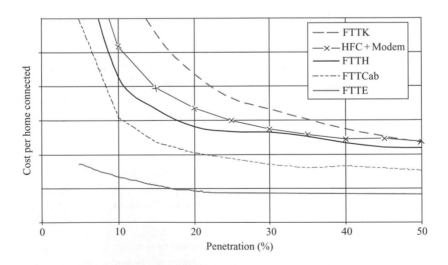

Figure 1.11   Cost versus penetration

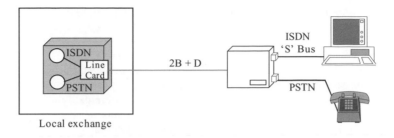

Figure 1.12   Lowband–ISDN Home Highway

of offering higher bandwidths required for a full range of new interactive services, including broadcast TV.

However, it should be recognised that it will take some time to roll-out any broadband technology in high volume, and migration steps are indicated in Table 1.4. Lowband is essentially ISDN technology and offers a digital service with 3–4 times the data capacity available to users of current voice band modems. Midband is the current step offering the first big leap in bandwidths. Highband could then be introduced at a later date when demand for even greater bandwidth emerges and when the cost of the requisite technology becomes commensurate with potential network revenues.

Lowband is currently offered by many Telcos as standard ISDN. BT also introduced a lowband product in the form of Home Highway (Figure 1.12) which allows both PSTN and ISDN connectivity at the NTE.

Midband is being deployed around the world and is the subject of Chapter 5. Similarly, there is much activity on highband around the world largely in the form of the Full Service Access Network (FSAN) initiative.

Table 1.4   Broadband evolution steps

Lowband	ISDN2/Home Highway	64–128 kbit/s
Midband	ADSL	500 kbit/s–6 Mbit/s
Highband	FTTCab/VDSL	up to 25 Mbit/s

1.7   Choice for other operators

The upgrade path for other existing operators will largely depend on their initial choice of installed technology. Table 1.5 shows some possible ways forward based on the information first given in Table 1.1.

The mobile and fixed radio operators will offer narrowband data connections and in the longer term may turn to UMTS (the third generation mobile standard). The MAN

Table 1.5   Upgrade options

Business	Possible upgrade
Mobile/Niche telephony	GPRS/UMTS
Metropolitan/Business services	Geographical coverage
Cable TV/Telephony	Digital/Cable modem
Satellite TV	Interactive back channel
Long distance	Direct access (owned and leased)

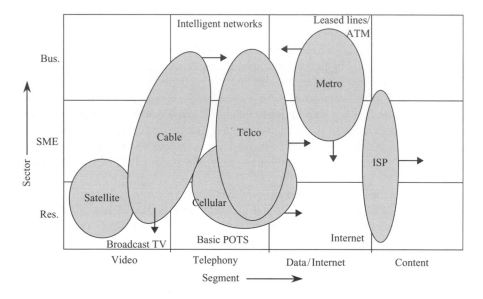

Figure 1.13   Future market structure

operators already offer broadband access through the use of direct fibre connection and will opt for increased geographical coverage. The cable TV companies are now offering digital services and install cable modems for Internet services. Satellite TV is offering interactive back-channels using PSTN to enhance their services. Finally, the long distance companies will move into the local market and direct access either through their own or leased infrastructure.

1.8   Future market

As we move into the future then, the market structure will change as regulation allows more liberalisation and operators jostle for position. Figure 1.13 outlines the direction the market is moving in. One key difference shows ISPs moving more towards content rather than network as regulatory changes allow Telcos to provide IP networks as part of the regulated businesses. Telcos are shown having moved extensively into the data market and offering broadband access. Long distance companies have moved into local services through the merger with cable companies. Although this gives a predominantly UK view, there are also signs that the US market is following similar trends.

1.9   Conclusions

In conclusion we have seen that the telecommunications market is very dynamic at the moment with new and existing operators jostling for position. Factors affecting the market include regulation, which is encouraging new entrants into the market, as well as the advent of new interactive services, which is placing new requirements on the network. In addition there is now a large choice of new access technologies becoming available, and this choice is a critical part of each operator's strategy. Access forms a dominant part of the investment required as well as determining the level of service offered.

The choice may clearly be different between new entrant operators and established companies who may already have an installed base of one particular network. In particular a suitable choice for a Telco would be the use of DSL technologies since they exploit the existing copper network. In addition the best way forward would be through low, mid and highband phases as the market develops over a period of time.

In a world where telecommunications continues to change at an increasing pace, the access business will need to re-invent itself. There are no easy or cheap options available for access and no single initiative will instantly transform an incumbent access network. Although the access network dates back to the earliest dates of telecommunications it remains critical in the delivery of even the most advanced and modern service to the user. Thus the future of access is assured but the structure of the access business and the underlying access technology will inevitably continue to evolve.

Chapter 2

Bandwidth drivers for future networks

D. B. Payne and P. N. Woolnough

2.1 Introduction

There has been considerable hype over the bandwidth growth of telecommunications networks driven by the phenomenal growth of the Internet. Overall the Internet has been growing at about 100% per annum for the last few years (except during 1995/6 when the World Wide Web appeared and much higher growths occurred) [1]. User volumes rather than user bandwidth demands drove the largest portion of Internet growth (most users were using dial modems with average session bandwidths in the 5–10 kb/s range).

Along with the euphoria that existed for everything Internet and communications generally was an assumption that if the bandwidth was delivered, services would materialise and generate new revenues. These additional revenues would pay for the provision of the bandwidth so producing a virtuous circle.

There is no doubt that the Internet, data and video services provide a huge potential demand for bandwidth. The problem is how to deliver that bandwidth in an affordable way. Bandwidth demand can easily outstrip the revenue growth that is needed to pay for the network investment and, not surprisingly, the access network provides the greatest challenges both in the provision of bandwidth and the necessity to reduce cost and expenditure.

Access is still the major bottleneck for broadband services. The vast majority of customer sites are connected via copper twisted pair, the majority of which have not yet been enhanced with DSL (digital subscriber loop) technology or fibre. A major decision for incumbent operators with large embedded assets in copper technology will be how far to upgrade using DSL technologies and when to replace it with fibre.

2.2 Service scenario model

To assess the potential for bandwidth growth, both core and access bandwidth, a bottom up or customer driven model for the UK was used. This model was based on

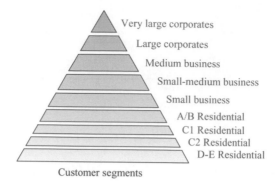

Figure 2.1   Customer site segmentation scheme used in services and traffic scenario

a site segmentation scheme whereby the customer base was divided into a number of segments. A site was defined to be either an individual business site or a single residential household. Although UK based, the model has also been used for other European countries and could be adopted by the US market.

Each segment represents a strata within the customer site population (see Figure 2.1). The parameters chosen for classification of a customer site into a particular segment was to some extent arbitrary but were largely related to site size (in terms of telecommunications services) and revenues.

The business market was divided into five site segments. The smaller sites represents a one-line business site, moving through small to medium enterprises and on to large corporate and multi-national company sites. The top segment of sites includes the largest individual locations in the UK measured in terms of their aggregate telecommunications requirements. This group would include the principal sites of top companies, major financial institutions, large R&D centres, central government departments, large ISPs, content hosting sites, and major OLO switch and interconnect sites.

The residential market was divided into four groups based on the propensity of a household to adopt and use various voice, data and entertainment services based on telecommunications. The propensity ratings were assumed to be closely linked to the 'standard' UK socio-economic groupings A&B, C1, C2, D&E. It must be acknowledged that this classification is becoming less popular because of new schemes, but for the purposes of the study it was considered adequate.

2.3 Modelling methodology

Given a segmentation scheme a range of generic services were considered such as high speed Internet, VoD, corporate data services, etc.

For modelling traffic, the key attributes associated with the different customer site segments fell into four basic parameter types – service or application related

Table 2.1   Parameters used to model traffic

Class of parameter	Main traffic parameters
Application related	Data/session bit rate requirements
	Peak bit rate requirements
	Peak-to-mean (burst) characteristics
Traffic routing	Fraction of source traffic that impacts the core network
	Average hop count of traffic in core network
Customer usage	Frequency of use
	Aggregate duration of use
	Time-of-day usage characteristics
	Fraction of simultaneous users
Market	Market size

parameters, traffic routing parameters, customer usage parameters, and market parameters. The main parameters in each of the four groups are listed in Table 2.1.

For each of the parameter types, the key underlying drivers or limitations are considered. For example, for the application related parameters particular attention is given to understanding the impact of new technologies such as DSL. For the customer usage parameters, the model takes into account the behaviour and likely usage patterns of different types of user for a particular type of service.

Before considering how much traffic may be expected in the future, it is important to gain an understanding of the various parameters shown in Table 2.1 in relation to today's customer base, services and usage patterns. This stage calibrates the model and establishes a solid basis for estimating future traffic growth.

Following the calibration step, the same parameters were determined for the different customer site segments for future years. After parameter values had been determined, traffic could be estimated on a per-household or per-business site basis for different services for each of the scenarios to be modelled.

The procedure was performed for each year of the analysis period, or in some cases, for simplicity, for chosen spot years. The resulting figures were multiplied by a figure corresponding to the total size of each customer segment to yield figures for traffic volumes for the market as a whole.

2.4   Overall pattern of traffic growth

Using the methodology outlined above a number of different future service and traffic scenarios can be considered. The modelled results for a 'high bandwidth' scenario are shown in Figure 2.2. The major assumptions for this scenario in 2006 are:

- 7 million broadband Internet customers;
- 5 million video/VoD customers;

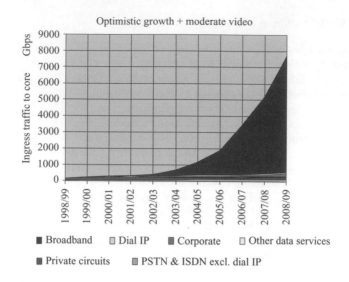

Figure 2.2   UK traffic – high bandwidth

- 80 mins/day average Internet session time (4 × today);
- 500 kb/s average session bandwidth (50 × today);
- 22 mins/day average video/VoD session time;
- 3 Mb/s average video/VoD session rate (assumes mix of VDSL/fibre and ADSL rates).

2.4.1   Voice traffic

Total fixed voice traffic growth is expected to be static or a few percent at most; revenues may decline as tariffs continue to fall.

2.4.2   Private circuit traffic

With private circuits, the full bandwidth is guaranteed to all end-users at all times. There is no statistical gain derived from customers' actual usage behaviour as there is in the case of services operating over shared networks such as the PSTN or public Internet. In this study therefore, aggregate circuit capacity rather than aggregate traffic has been modelled. It can be argued that private circuit platforms are significantly over-dimensioned for the traffic they carry. Indeed it has been estimated [1] that, on average, private circuits in the US are only used at a few percent of their capacity. Clearly, some private circuits will be heavily utilised, in which case any statistical gain from migration to a shared network such as a VPN would be negligible. However, the evidence indicates that for the majority of circuits, this is not the case and for large scale migration of private circuits to shared networks, bandwidth growth, in the early years, will be offset by the statistical multiplexing gain afforded by use of VPNs.

2.4.3 Dial IP traffic

It was estimated that dial IP traffic will grow substantially over the next few years as a result of a growing customer base, increasing average session duration, and increasing average session bit rate. However, the growth of broadband traffic (assumed to be xDSL and cable modem delivered in the above scenarios) will rapidly dominate dial IP traffic.

The assumptions for dial IP traffic acknowledges the increased usage that is likely to occur with the introduction of unmetered Internet services.

2.4.4 Broadband (xDSL) traffic

From the scenario shown in Figure 2.2 it can be seen that in the latter years of the analysis the overall traffic is increasingly dominated by broadband traffic. It must be acknowledged, however, that there is substantial uncertainty due to the difficulty in gauging how demand for mass-market broadband services will develop. The following factors will have a significant impact on the rate of growth of broadband traffic.

- Speed of roll-out of broadband connections.
- Mix of business and residential end-users.
- Mix of 'fast internet' and video customers.
- Average session bit rates.
- Contention and overbooking ratios.
- Adoption rates of local loop unbundling.

Clearly many of the factors listed above are interdependent. Nevertheless it should be apparent from the number of factors and the uncertainties associated with them that widely differing scenarios in terms of traffic volume are possible. The bounds of the uncertainties were explored in the different scenarios modelled.

It was found that the aggregate traffic from broadband services for 2006 could vary by well over an order of magnitude – from less than 100 Gbit/s in a very conservative scenario to over 2000 Gbit/s in a very high scenario.

2.4.5 Corporate data traffic

The corporate traffic shown in Figure 2.2 is assumed to be comprised of two main sources: 'new' traffic associated with increased usage of data services from this sector of the market and traffic previously carried over private circuits. The data portion (Internet/intranet) of this private circuit traffic is assumed to migrate over time, onto VPNs or other shared networks. Corporate traffic was varied between the scenarios modelled using a simple structure corresponding to 'normal', 'high', and 'very high' rates of growth respectively, as shown in Figure 2.3.

Growth projections for each of the corporate traffic scenarios were based on knowledge of the size of the customer base, to which were then added assumptions for rates of bandwidth growth.

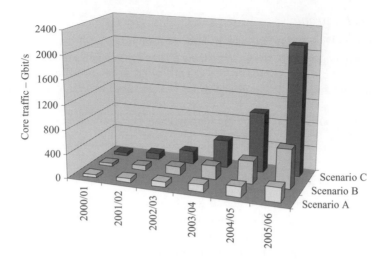

Figure 2.3   Three growth scenarios for corporate data traffic

Broadly, scenario B corresponds to the high growth scenario shown in Figure 2.2. The results include the private circuit element of traffic that migrates across to data networks and VPNs. Indeed in the first two or three years of the study period, the overall corporate traffic will be dominated by private circuit traffic. In subsequent years however, for all scenarios it was assumed that an increasing fraction of private circuit traffic would be carried over public or private data networks and VPNs.

2.4.6   Mobile traffic

One of the findings from the analysis was that the volume of traffic associated with mobile data services remained very small compared with traffic generated by fixed network users. This result arises as a result of the restricted bandwidth on cellular mobile networks – a situation that will persist even with the emergence of new 3G technologies.

It follows that mobile data and Internet services will consist principally of lower bandwidth applications than their fixed network counterparts. This view is supported in studies carried out by Ovum [4].

2.5   Different traffic scenarios

It is acknowledged that traffic modelling is subject to significant uncertainty and for this reason, it is useful to consider a number of scenarios. Figure 2.4 shows four other modelled scenarios with different assumptions for market size, growth and usage of services.

In this analysis the traffic modelled in Figure 2.4 has been assigned to one of three basic types rather than to specific service types as used in the analysis accompanying

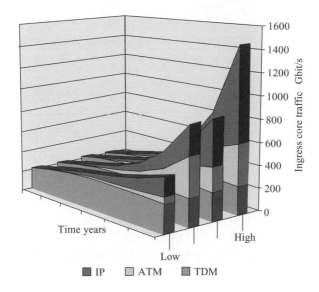

Figure 2.4   Set of four modelled scenarios for core traffic growth

Figure 2.2. This traffic was determined to be comprised of either TDM, IP or ATM according to a set of simple apportionment rules.

It was found that much of the variation in traffic across the different scenarios is due to uncertainty in the level of broadband traffic. The low bandwidth scenario in Figure 2.4 assumed three million broadband customers, with an average session bandwidth of 100 kbit/s and one hour per day usage. It was assumed that 1 million of the connections were used for some type of streamed video service, in addition to the fast Internet usage.

The high bandwidth scenario in Figure 2.4 assumed seven million broadband connections and much higher session bit rates. The assumptions are similar to those used for Figure 2.2 but with less entertainment video content. The two middle scenarios were based on volumes and usage characteristics between the high and low scenarios. These scenarios had a similar overall traffic envelope but a different mix of IP and ATM traffic.

2.6   The cost issue

The major problem associated with the higher bandwidth scenarios shown in Figures 2.2 and 2.4 is the capital expenditure required to build the network capacity to meet the predicted growth. Figure 2.5 illustrates the problem. This shows a growth curve fitted to the bandwidth scenario shown in Figure 2.2 and corresponds to 50% per annum growth, about half the growth of the Internet over the past few years. Also shown in Figure 2.5 is a cost curve based on the traditional rule of thumb for bandwidth provision of electronic transmission systems, i.e. quadruple the bandwidth for

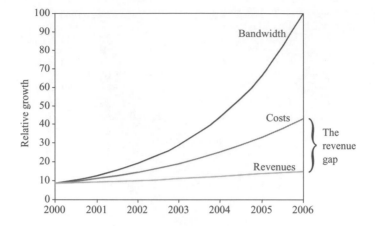

Figure 2.5   Growth curves for bandwidth costs and revenues for high growth scenario

2.5 times price increase. If this rule is applied to a price learning curve where bandwidth is substituted for volume it corresponds to an 80% learning curve, i.e. double the volume and the price reduces to 80% of the previous price. It is an interesting coincidence that products from the electronics industry have traditionally followed an 80% price learning curve! Of course the capital cost of bandwidth provision is only part of the costs to a telecommunications operator and in practice the cost curve would be even steeper than shown here.

The other curve in Figure 2.5 is a revenue curve. This shows a 7% per annum growth in revenue. This is quite optimistic for year-on-year growth for the industry as a whole although some service revenues and company revenues can grow much faster than this. Generally for all services taken together over the longer term a few percent per annum is a sustainable rate. Revenue growth is typically a few percent per year and it is this revenue that ultimately must pay for network investment.

Clearly with these bandwidth growth figures there will be a decline in margins unless new solutions are found to bandwidth provision, that can yield faster unit cost decline than electronic solutions have traditionally produced.

2.6.1   Core price declines

Another way of looking at this problem is to consider the revenue per unit of bandwidth delivered and compare it with the cost of providing a unit of bandwidth in the core and access. Because total revenue generation is to a first approximation independent of bandwidth there is an almost inverse relationship between the network capacity growth and the earnings per unit of bandwidth carried. New services generally do not generate entirely new revenue streams, nearly always there is a significant substitutional element – although there can be a delay between cause and effect. Traditionally network growth and the 80% price learning curve have kept in reasonable step (20–30% per annum growth can be supported). However, the potential explosion of growth bought about by the data-wave, and the Internet in particular, will break that comfortable cycle. To meet the growth curve shown in Figure 2.2 price learning curves of 50% to 60% will be required.

One possible way that this may be achieved in the core network is to remove the expensive opto-electronic conversion elements at each interface within the core network elements and move to a more 'transparent' or photonic layer. This is sometimes referred to as 'all optical networking'. This may be possible in the future but some significant technical problems will need to solved. Not the least of these is the issue of management of a network where analogue-like impairments can accumulate in different ways as the network is reconfigured via the 'transparent' optical cross connects. This approach would also move intelligence, traffic consolidation and grooming functions towards the logical edge of the network, producing a high capacity, simpler and possibly much lower cost but relatively 'dumb' core.

2.6.2   Access price declines

Putting the price decline issues in very rapid growth situations to one side, overhauling and providing capacity in the core is relatively straightforward. Traffic is consolidated, costs are shared over a relatively large user base and capacity can be added in a piecemeal manner to satisfy demand as it arises. The access network poses significantly different sets of problems.

Interestingly, the price decline per unit of bandwidth delivered can be quite dramatic as new access technologies are deployed and 60–70% learning curves look possible (again using a bandwidth parameter rather than product volume). To illustrate this, Figure 2.6 shows some extrapolated costs for a range of access technologies assuming adequate volumes for each technology (volumes are biased towards the lower bandwidth technologies). These were normalised and the cost per unit bandwidth was plotted against peak customer bandwidth. For the PON systems, dynamic bandwidth assignment was assumed to be implemented and an over-booking ratio of

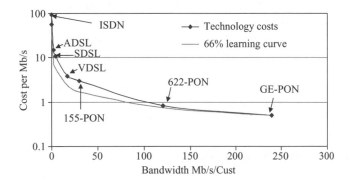

Figure 2.6   Access technologies – bandwidth cost decline

three was used. Also plotted is a price decline curve of 66%. Although the VDSL and 155 Mbit/s PON sit above this curve it is a reasonable fit when the higher bandwidth PONs are included. Although this was an approximate calculation with simple cost models it does show that unit bandwidth costs for access technologies can produce a bandwidth learning curve closer to the required decline rates.

The greater problem for access is the sheer level of investment required. This is compounded by the logistics of targeting the desired markets and achieving sufficient volume deployment to realise the equipment cost reductions. With insufficient volumes, equipment costs will not fall fast enough to enable a return on the investment. This 'chicken and egg' cycle has plagued access deployments for at least the last two decades and pushed incumbent operators to an incremental approach in the deployment of new access technologies.

2.7   The access bottleneck

A further barrier to bandwidth growth is the legacy access network. The ubiquitous nature of the deployed base of copper twisted pairs for existing operators and the need to re-use that investment is the main limiting factor to high bandwidth provision. Even after deployment of advanced enhancement technologies such as ADSL or SDSL, access bandwidth restrictions will remain the limitation to very high bandwidth services and broadband multi-service packages.

The advent of the Internet is leading to services that demand higher bandwidths and is increasing the pressure to upgrade the access network. The first stage of this, for incumbent operators, is the deployment of ADSL technology. This will significantly enhance the Internet service experience providing significantly higher speed access and enabling a much richer set of services to be offered.

For the incumbent operator, ADSL can be deployed with minimal physical infrastructure investment and is a continuation of the incremental approach. Only exchange equipment needs to be deployed to gain coverage, which means coverage can be rapid with much of the expenditure deferred until service take up occurs.

The take up of ADSL and cable modem technology may stimulate demand for yet greater bandwidths which, when coupled with high quality video services, could exceed the bandwidth capacity of these technologies. The issue is what should the next upgrade be. The next incremental step for the incumbent operator would be VDSL (very high speed DSL) technology which could deliver new high bandwidth services or service bundles. A more radical step could be to deploy fibre to the home/office which would have the potential to deliver much higher bandwidths with almost indefinite future upgrade capability. The choice of the next access technology to follow ADSL will be one of the major decisions for operators in the coming years.

2.8   Access bandwidth contention

When operating in a multi-service environment a consequence of restricted bandwidth on access pipes is service contention at the customer site, even if core bandwidth exists

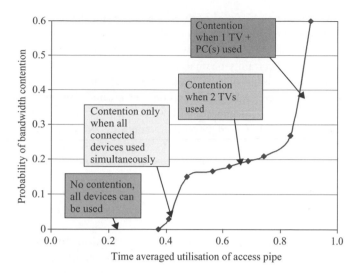

*Figure 2.7   Access bandwidth contention example for a household with two TVs
and two PCs and 5 Mbit/s access bandwidth*

to deliver the services. Contention for bandwidth within a customer site can arise if
there are a number of devices at the site that can request services with aggregate
bandwidth greater than can be delivered over the access connection. The results from
a simple model are shown in Figure 2.7, they illustrate the increasing contention
a customer will experience as service bandwidths increase but access bandwidths
remain constant. The results are for a specific example – a household with a 5 Mbit/s
access connection and four output devices, two TVs and two PCs. The x axis is a
measure of the service bandwidth increasing and is defined as the average utilisation
of the bandwidth on the access pipe.

Figure 2.7 can be interpreted by considering the different regions of the curve.
There is a region where the total bandwidth demand from all devices is less than the
bandwidth of the access connection and no contention occurs. This corresponds to
relatively low average utilisation as the probability of all devices being used simultaneously for long periods is quite low (in the example used this occurred for average
utilisations less than 37%). In the example for this 'uncontended' region, the TV and
video services operate at 2.3 Mbit/s and the bandwidth required for other services,
available via the two PCs (voice/videophone, fast Internet and streamed multimedia),
is collectively less than 0.4 Mbit/s.

As the bandwidth of individual services is increased, the average utilisation of
the access pipe increases and the probability of bandwidth contention starts to rise.
The shape of this part of the curve will depend on the mix, bandwidth and usage of
services but it can be distinctly non-linear as indicated in the example shown.

Figure 2.7 indicates that even at moderate average utilisations (~50–70%)
significant bandwidth contention can be experienced. The final part of the curve

corresponds to the utilisation of the access connection exceeding ~85%. In this operating region, for the example shown, contention arises when just one TV and either one or both PCs are used. The probability of bandwidth contention in this region of operation would almost certainly be unacceptable.

This also raises the issue of 'soft' contention and 'hard' contention. The former may be defined as contention that can be 'got around' by one or more of the output devices reducing its throughput without undue degradation on service. Hard contention may be defined as contention that cannot be avoided in this way, either because the contending services are CBR, or because reducing throughput is not possible without significant degradation to one or more services. In this case a call acceptance control (CAC) mechanism would be required. Soft contention is very similar to the way diffserv. would work in such an environment and could be problematic for contending CBR services (e.g. high quality video) unless CAC is also implemented.

The most important conclusion from Figure 2.7 is that uncontended bandwidth in the network does not guarantee or imply that there will be no contention/blocking from multiple devices in the home/office requesting simultaneous access. The other conclusion is that with multiple devices and a typical mix of services, the average utilisation of the access pipe in the busiest period probably cannot exceed around 85% before severe bandwidth contention and/or blocking is experienced.

In summary, access bandwidth contention has implications for content selection, traffic management, customer-perceived quality of service, and for access strategy.

2.9   Optical access solutions

One way of avoiding contention at the customer site is to provide an access connection with sufficient bandwidth for simultaneous use of all the customer devices and services. Optical access solutions offer the promise of providing very high bandwidths and also dynamic bandwidth assignment that will allow high peak bandwidths to be allocated to customers and also increase the utilisation of the access transport system.

Drivers that could push operators to upgrade the access network beyond ADSL technologies are shown in Table 2.2.

DVD quality video services will require 4 to 6 Mbit/s and although this could be delivered by ADSL technology to some customers, there would be severe limitations in reach over the copper network. This would have major restrictions on service deployment and raises many marketing issues. As bundled service packages are offered, the bandwidth contention issue discussed above also becomes a significant problem and could lead to customer dissatisfaction. These bundled service bandwidth demands could be provided via VDSL technology.

If higher bandwidth services such as HDTV emerge and they are also offered within bundle packages, then bandwidth demands could exceed the capability of VDSL systems. Fibre to the home/office then becomes an increasingly attractive technical solution. For services that demand greater symmetry of up and downstream bandwidth, the attraction of fibre solutions increases still further. Finally, when very

Table 2.2   Bandwidth drivers towards optical access networks

Services requiring bandwidths greater than ~2 Mbit/s	Minimum access bandwidth
Broadcast or DVD quality video	~4–6 Mbit/s
Bundled delivery of multiple video channels and high speed Internet services	~5–15+ Mbit/s
HDTV services	~14 Mbit/s
HDTV + other services simultaneously	>20 Mbit/s
Greater demand for return bandwidth e.g. tele-workers, video telephony, video conferencing etc.	
'Wavelength' services to medium and larger business customers*	

* Large and multinational customers will already be connected via fibre; for this table it is assumed that 'wavelength' services penetrate to smaller customer sites that are currently serviced via copper pairs.

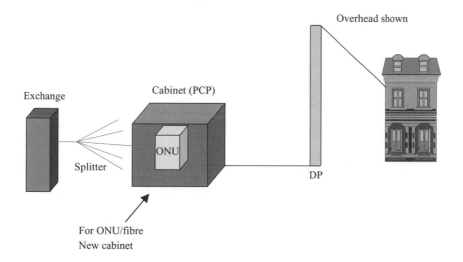

Figure 2.8   Fibre to the cabinet (VDSL)

high capacity systems are required, including so called 'wavelength' services, then fibre becomes the only option.

The VDSL architecture relies on fibre feeders to the street cabinet located about 1 km from the customer. This feeder network could be a PON (passive optical network) solution as shown in Figure 2.8 or point-to-point fibre. For fibre to the home/office (illustrated in Figure 2.9) PON is the much more serious contender as the mass market solution. The decision for operators needing to install network technology with capacity much greater than ADSL will be to choose between the VDSL and FTTH options for this upgrade.

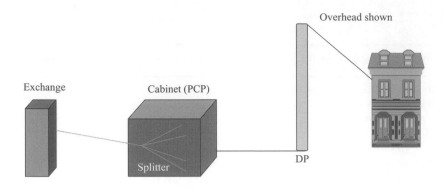

Figure 2.9   Fibre to the home/office

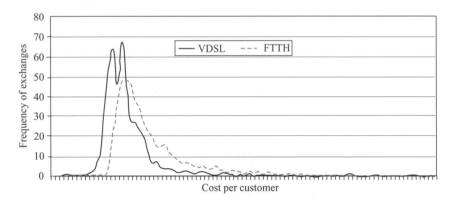

Figure 2.10   Cost comparison of FTTH versus VDSL (FTTCabinet) (all UK exchanges ranked by residential population)

The VDSL option is available here and now and if the decision had to be made in the near future then this may well be the option chosen. If, however, the decision can be delayed for a few years and ADSL meets demand for that period, then the FTTH option becomes a real contender. The extra time would enable the suppliers to develop the mass market optical solution and get cost down to be competitive with the VDSL solution. VDSL is likely to remain slightly lower cost than FTTH for incumbent operators because of the re-use of the last kilometre of copper cable. However, FTTH will be capable of providing much higher bandwidths into the future and studies have shown that for high volumes the deployment costs of the two solutions can be quite close even for incumbent operators.

This close cost parity is illustrated in Figure 2.10 where VDSL was compared with FTTH for a 5 Mbit/s service offering. The model used took into account the geographical distribution of customers and assumed an average take up across the high end residential customers and the small business customer segments of about 20%. The average cost for the FTTH solution was only 16% greater than the VDSL solution! It should be pointed out that the costs used in the model were projected volume costs for both technologies and are not available today. They were based on an incumbent operator with an installed base of access infrastructure and copper cables.

Although the VDSL solution was overall the lower cost, the cost difference is sufficiently close, given the uncertainties in the overall assumptions, to make the fibre solution worthy of serious consideration and study – particularly when the FTTH solution would be the last infrastructure upgrade the operator should need to make. All further upgrades should be implemented be terminal equipment changes only.

2.10 Conclusions

Traffic forecasting is a notoriously fickle business, the only certainty is that forecasts will be wrong. However by use of rational argument and an acceptance of 'social and economic inertia' combined with sensible linkage to base numbers, e.g. number of customers, then bounds can be found for likely traffic growth. All growth curves are in reality 'S' curves and it is important to avoid the simple extrapolation of the early exponential regions too far into the future.

The results shown in this chapter are based on bottom up models and consideration of likely customer behaviour patterns for usage. Even the most bullish of the scenarios modelled only generate a few terabits of traffic for the UK. Network capacity will be some multiple of this traffic and will depend on the network utilisation that can ultimately be achieved. It is reasonable to assume that network capacity will be at least a factor of ten greater than the ingress traffic into the core from the access network. However, this traffic will be spread over a number of core network nodes and these nodes will be interconnected with multiple fibres. Terabit plus fibre transmission systems can be bought today and optical cross connects with multi-terabit capacity are also becoming available. Bandwidth capacity in the core is not the problem – reducing cost is the major issue!

The capacity of the access network is the ultimate barrier to bandwidth growth and while copper dominates the last mile, the upper bounds can be relatively easily determined. The really disruptive technology is fibre to the home and this has not yet been deployed in any quantity that seriously perturbs these bounds. However, if incumbent and other large scale CLECs start to mass deploy fibre to customers then future network bandwidths could become truly enormous!

2.11 References

1 COFFMAN, K. G. and ODLYZKO, A. M.: 'Internet growth: Is there a "Moore's Law" for data traffic?', in ABELLO, J., PARDALOS, P. M. and RESENDE, M. G. C. (Eds): 'Handbook of massive data sets' (Kluwer, 2001)
2 ODLYZKO, A. M.: 'The low utilization and high cost of data networks', http://www.research.att.com/~amo/doc/networks.html
3 Ovum Report, 'Third Generation Mobile: Market Strategies', September 1999
4 Ovum Report, 'Global Mobile Markets 2001–2005', February 2001

Chapter 3

Realising the potential of access networks using DSL

K. T. Foster, J. W. Cook, D. E. A. Clarke, M. G. Booth and R. H. Kirkby

3.1   Introduction

There are around 700 million copper pairs in telephony access networks that provide for extensive connectivity of the world's population. A combination of this existing copper infrastructure and new transmission technologies mean that a new era of universal broadband access can now begin at a fraction of the cost, and in a fraction of the time required for optical access networks.

However, there is a threat to this vision of the future – it is relatively straight-forward to design transmission systems that work well in computer simulations and in a few specific laboratory tests, but more difficult to deliver useful capacity when subjected to the hostile environment of the real network [1]. Also, the uncontrolled deployment of such advanced transmission systems in multi-pair cables may result in mutual interference (i.e. crosstalk). Such degradation can occur even when the systems are quite thoroughly specified, since they can be inadvertently operated so as to cause problems. Even a low rate of such instances would significantly pollute the copper network. It is therefore vital to understand the crosstalk environment and spectral compatibility issues for different DSLs [2,3] if the broadband potential of the existing copper access network is to be fully realised.

Most Telcos have a broadband access strategy that involves deployment of ISDN, ADSL, SDSL and VDSL transmission technologies for a new generation of services, and the use of HDSL/SDSL for reducing costs and connection time for existing services. BT is no exception. It has invested considerable effort in understanding copper transmission technologies. This is exemplified by high profile involvement in standards bodies such as ETSI, ANSI, ITU-T, the DSL Forum and the Full Services

Access Network (FSAN) initiative. The work has also involved extensive trials, and development of world class test facilities and test methodologies [4].

The issues associated with the spectral compatibility of the various digital transmission systems that have been or will be deployed in the world's copper access networks are technically complex. This chapter provides an overview of digital subscriber line (DSL) technology and describes some of the key technical issues facing DSL engineers. It examines the theoretical information capacity of the copper access network, and explains how measurements have been made on live BT cables to provide realistic performance models. The critical areas of frequency planning and spectral compatibility, which are essential if the broadband potential of this valuable national asset is to be realised, are then considered in Chapter 5.

3.2   Technology overview

Originally, owing to the use of FDM channel multiplexers in the PSTN trunk network, customers were constrained to using the 4 kHz voiceband to convey data over the copper access twisted pair and onwards through the network. Voice-band modems initially worked at just a few tens of bit/s, before progressing through to 2.4 and 4.8 kbit/s then developing rapidly to pack even more information into the 4 kHz bandwidth. Families of modems operating at 9.6, 19.2, 28.8 and 33 kbit/s subsequently appeared in the market.

With the advent of PCM digital transmission in the core network and digital switching, there was no longer a need to constrain the signals transmitted in the copper access network to a 4 kHz bandwidth. In principle, a copper transmission system could now use any bandwidth as long as the information transported could be conveyed through the 64 kbit/s narrowband digital PSTN.

Data-over-voice pair-gain systems and private circuit systems were among the first to make use of additional bandwidth on the copper pairs by using frequencies above the voice-band. Subsequently, basic-rate ISDN systems operating at 160 kbit/s used more modern modulation schemes (2B1Q, 3B2T and 4B3T), together with advanced digital signal processing, such as non-linear echo cancellation, to exploit more efficiently the available capacity by improving the spectral efficiency. The transmitted signals for these ISDN systems are mostly constrained to frequencies below 80 kHz.

Next in the evolution of digital transmission systems came high bit-rate DSL (HDSL) that was originally a 'scaled-up' version of basic rate ISDN. HDSL uses the same line code as the ISDN standard (2B1Q), but operates at 784 kbit/s to 1 Mbit/s on each pair, the bulk of transmitted power being below 400 kHz. It uses two or three copper pairs to deliver multi-megabit services in the UK, such as BT's ISDN30 and 2 Mbit/s private circuits. Single-pair or symmetric DSL (SDSL) is now emerging which is based on older HDSL but uses trellis coded pulse amplitude modulation (TC-PAM) to achieve greater spectral efficiency whilst preserving the ability to deliver symmetric bit rates. It utilises a single copper pair for service delivery, which makes it an attractive technology for mass deployment. SDSL will initially find application

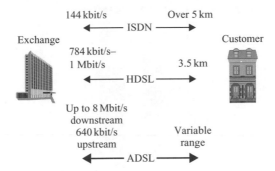

Figure 3.1   Classes of copper access transmission systems

in the business or SME market sector and has the ability to transport delay sensitive voice traffic (VoDSL) as well as normal data because the latency is bounded.

Asymmetric DSL (ADSL) is the most recent DSL system to enter full commercial deployment. ADSL employs asymmetric transmission in terms of both capacity and bandwidth in order to improve the crosstalk environment, and hence improve the capacity in the direction of exchange to customer compared to HDSL or the emerging single pair SDSL. Unlike ISDN, HDSL and SDSL which use baseband transmission techniques, ADSL uses passband modulation (between 25 kHz and 1.1 MHz) to keep the 4 kHz voice-band free for simultaneous analogue telephony. Figure 3.1 illustrates these three basic classes of modern copper transmission systems.

All of these systems were designed to operate between the customer's premises and their local exchange. In order to achieve further improvement in information capacity, it is necessary to make the copper network shorter (less signal attenuation). This is achieved by taking fibre deeper into the access network. Very high-speed DSL (VDSL) is then used for transmission over the remaining D-side and drop-wire copper pairs.

VDSL further increases the bandwidth used on the copper pair, with standardised systems using frequencies up to 12 MHz (see Figure 3.2). The use of these high frequencies for transmission over copper pairs requires greater attention to electromagnetic compatibility (EMC) and radio frequency interference (RFI) compatibility during the design [5] because of diminishing cable balance with increasing frequency.

Note that it is desirable to use the same line for both the new high rate connections and the original telephony service. This will normally be achieved by frequency division, using analogue band splitter filters located at each end of the line as shown in Figure 3.3.

With the multi-megabit access capability of ADSL and VDSL comes the need to upgrade the remainder of the network to avoid bottle-necks arising elsewhere, e.g. by deploying ATM switches together with SDH and WDM transmission in the core.

Development of new types of DSL modem technology has not halted improvement and development of the existing systems. For example, voice-band modems have evolved to deliver 56 kbit/s to customers by exploiting the reduction in quantisation

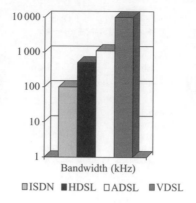

Bandwidth (kHz)

☐ISDN ■HDSL ☐ADSL ■VDSL

Figure 3.2   Growth in copper access bandwidth usage (kHz)

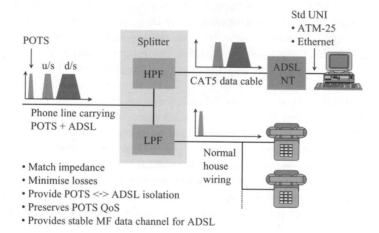

Figure 3.3   Purpose of a splitter filter

Note: Normally the high pass filter (HPF) resides in the same box as the ADSL network termination (NT) transceiver and is only shown here collocated with the LPF to illustrate the signal separation.

noise when the analogue/digital conversion process at the POTS line-card is bypassed on the service provider side. Also, narrowband ISDN line cards with much improved range have become available.

HDSL is no longer seen as just an expedient way of providing existing E1 or T1 legacy services using multiple wire-pairs. These systems have become more flexible, e.g. rate-configurable, trading range for bit rate. The latter type of system is known as single-pair or symmetric DSL (SDSL). This potentially opens up new markets for DSL, such as campus LANs using a single copper pair.

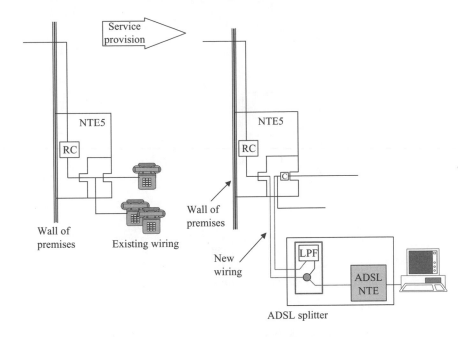

Figure 3.4   Normal ADSL installation

ADSL has also become rate-adaptive, and the systems are no longer simple 'bit pumps' with clock and data interfaces [6]. ADSL systems now exist that have integrated ATM functionality, and even integrated bridging or routing.

Originally the ADSL modem was envisaged as being installed close to the master telephone socket, as shown in Figure 3.4. An alternative proposal is to omit the splitter at the customer end. The idea is to remove the need for new internal wiring by allowing the customer ADSL unit to be directly plugged into any existing telephone socket as shown in Figure 3.5. It would also save the cost of an engineering visit or truck roll. This is commonly called 'splitterless' ADSL or DSL Lite.

It should be noted that BT originally chose to implement an LPF filter into a new (patented) front plate of the NTE5 master socket thereby achieving cost savings and improving the aesthetics of the installation. The new wiring is normally to CAT5 standard and simply connects the ADSL NTE to the master socket. This is now termed as a 'classical' installation. Mass deployment in the UK is moving towards the use of distributed microfilters at each piece of CPE including the ADSL modem.

For splitterless ADSL to operate successfully, modifications are needed to the ADSL modem design to compensate for the increased mutual interference between POTS and ADSL services on the same wire-pair. This interference results in additional noise for telephony, and reduced capacity for the ADSL. Moreover, the performance of splitterless ADSL will be very dependent on the type and number of telephones and other telephony CPE that are installed. It may be that, unlike classical ADSL, simultaneous operation of telephony and splitterless ADSL will not be possible for

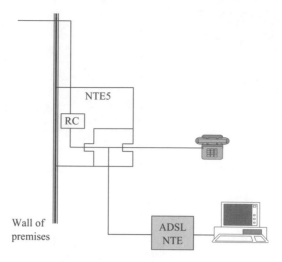

Figure 3.5   'Splitterless' ADSL installation

some customers without the addition of a second line (or of course a splitter or one or more distributed low-pass microfilters). The options for splitterless installation are varied and encouraging developments are being made in distributed micro filter technology.

VDSL developments, arising since conception of the fibre to the cabinet (FTTCab) architecture, include increased interest in video-centric multiple services for the consumer and SME markets, and the use of VDSL in campus environments to extend the reach of Ethernet based LAN systems, both on data cable, and on existing telephony wiring.

All DSL technologies have benefited from the continuing rapid advances in electronics (e.g. increasing processing power, and reduction in size and power consumption). Apart from improvements in functionality, there is modest scope for DSL modems to access more of the intrinsic information capacity of the copper pairs. This relies on more sophisticated modulation and coding techniques, and making use of improvements in silicon integration to generate improvements at acceptable cost, and reduced power consumption. Table 3.1 (reproduced from Chen [7]) compares the capacity actually obtained by some existing systems against the theoretical capacity of their channels. (For system details and assumed channel characteristics, see Chen [7].)

The signal processing used in voice-band modems operates at relatively slow rates, enabling sophisticated modulation and coding algorithms, such as multi-dimensional trellis coding which can be implemented at low cost. Voice-band modems are close to the limit in realising the capacity available in the voice band.

Similar techniques were not practical when ISDN and HDSL were originally conceived and standardised, so these systems are less efficient. By the time ADSL appeared, signal processing power and integration had improved and ADSL began

Table 3.1   Practical capacity as percentage of theoretical

Attribute	V.34 modem	ISDN	HDSL	ADSL
Channel capacity (kbit/s)	34.88	700	1700	10 000
Throughput (kbit/s)	28.8*	160	800	7 000
Transmission efficiency (%)	83	23	47	70

*On the basis of equivalent assumptions, a 33 kbit/s modem operates at 95% transmission efficiency. 56 kbit/s modems effectively operate over a different channel with higher capacity owing to the removal of a source of quantisation noise.

to make better use of available capacity than its DSL predecessors by exploiting techniques such as multi-tone modulation, Reed-Solomon error-correction coding and trellis coding.

The increased availability of fast signal processing power is now being used for new improved variants of the earlier DSL systems. HDSL-2 is a new variant of HDSL that seeks to deliver 1.5 Mbit/s T1 services over a single copper pair with similar range to that currently achieved with 2-pair T1 HDSL systems. To achieve this increased transmission efficiency, HDSL2 uses advanced modulation techniques [8] that are more sophisticated than the existing 2B1Q line code. It also employs coding techniques.

SDSL is very similar to HDSL-2 in that both acronyms are often used to cover a new generation of HDSL technology that is characterised by: operation on a single pair, symmetric bit-rate, low latency, and rate selectability (to trade range for bit-rate). There are very exciting opportunities for SDSL technology because of its inherent spectral compatibility with ADSL (and other heritage xDSL) and also the inherent flexibility to deliver a wide range of business as well as residential services.

In addition to improvements in DSL technology, any further improvement in capacity depends on maintaining the available network capacity by judicious control of the crosstalk environment, and careful spectrum management. For example, HDSL-2 has been defined for standardisation in the T1 HDSL market (primarily North America). However, HDSL-2 may not be able to deliver 2 Mbit/s E1 services over a single copper pair without causing interference to ADSL systems operating in the same cable. Preserving spectral compatibility between different DSL technologies will be vital if the inherent information carrying capacity of access networks is to be realised.

3.3   Impairments for DSL

Copper access transmission systems face a variety of impairments that present barriers to their operation. These can be broadly classified as *intrinsic* or *extrinsic* to the cable environment.

3.3.1 Noise

Examples of intrinsic noise impairments are thermal noise, echoes and reflections, attenuation and crosstalk. There are also other components that reside in the cable infrastructure that can impair the operation of DSL systems. These include surge protectors, RFI filters, and in some networks, bridged taps and loading coils.

Another intrinsic impairment is the condition of the cable infrastructure which may exhibit faults such as split pairs, leakage to ground, low insulation resistance, battery or earth contacts, and high-resistance joints. All these impairments reduce DSL performance.

Examples of extrinsic impairments are impulsive noise originating from lightning strikes, electric fences, power lines, rotating machinery, electric arc welders, switches, fluorescent lighting, etc. There is also radio interference from a variety of sources, such as AM broadcasting and SW/amateur radio transmitters.

The noise sources mentioned above can alternatively be classified as capacity limiting or performance limiting. Capacity-limiting noise is usually slow changing, such as thermal noise and crosstalk. These noise levels are often predictable and relatively easy to take into account when a Telco creates well-engineered deployment-planning rules.

Performance limiting noise, such as impulses and RFI, is intermittent in nature. It is geographically variable and unpredictable, and so is usually accounted for in planning rules by using a safety margin. DSL systems seek to use additional signal processing such as error correction with interleaving, and adaptive line codes, to mitigate such sources of noise.

3.3.2 Crosstalk

Crosstalk causes by far the largest contribution to capacity limiting noise for DSL systems, so it is worth examining it in a little more detail. There are two very different types of crosstalk in multi-pair access network cables, near-end crosstalk (NEXT) and far-end crosstalk (FEXT) as shown in Figures 3.6 and 3.7.

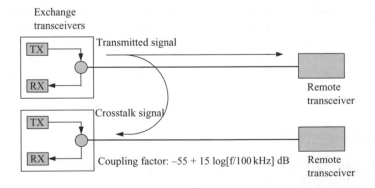

Figure 3.6   NEXT

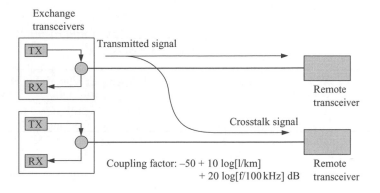

Figure 3.7   FEXT

NEXT is interference that appears on an adjacent pair located at the same end of the cable as the source of the interference. Its level is substantially independent of the length of the cable.

FEXT on the other hand is interference that appears on another pair at the opposite or far end of the cable to the source of the interference. Its level is attenuated at least as much as the signal itself if both have travelled the same distance.

NEXT affects any systems that transmit in both directions at once (e.g. echo-cancelling systems), and where it occurs it invariably dominates over FEXT.

NEXT can in principle be eliminated by not transmitting in both directions in the same band at the same time, separating the two directions of transmission either into non-overlapping intervals in time or into non-overlapping frequency bands. This converts duplex transmission into independent simplex transmissions, avoiding NEXT at the cost of a reduced bandwidth in each direction. At high enough frequencies the advantage of transmitting against FEXT noise rather than NEXT noise becomes so great that it can outweigh the disadvantage of reduced bandwidth.

As a rule of thumb the transition between these two modes of transmission occurs at a reach-dependent frequency given by:

$$F_{\text{critical}} = \{2.5/(\text{reach/km}) - 0.45\}\,\text{MHz}$$

This compares with heritage systems as shown in Table 3.2.

It can be seen that BT's heritage systems are, in the main, optimal according to this rule of thumb.

3.3.3   EMC

DSL transmission systems are required to operate on access wire-pairs that exist in a harsh physical and electromagnetic environment. As they cannot be screened and are often hung from poles, they have the capacity to act as very inefficient unintentional antennae. This means that they can pick up radio broadcast signals that may become sources of interference to DSL systems, and, equally, there is the potential for DSL

Table 3.2   System reach

System reach	Rule of thumb recommendation	Actual systems
5 km	Multiplexed above 50 kHz	ISDN echo cancelling < 40 kHz ADSL echo cancelling < 138 kHz Multiplexed > 138 kHz
3 km	Multiplexed above 380 kHz	HDSL echo cancelling < 292 kHz
1.5 km	Multiplexed above 1.1 MHz	Fibre to the cabinet VDSL, multiplexed ~ 1.1–12 MHz

signals to leak out of the cables and cause interference to radio systems. Obviously it is vital that both of these possibilities are understood and their impact carefully controlled.

3.3.3.1   Emissions

Most DSL line signals up to and including ADSL use frequencies and signal levels that are so low that emissions are very unlikely to cause problems. Access network cables transmit DSL signals in a balanced mode (equal and opposite voltages on each wire) that tends to cancel out potential emissions. Any signal that does find its way into the unbalanced or common mode is likely to be poorly radiated because the wavelength is so long that the antenna efficiency of the cables is very low. This has been confirmed by measurements both by BT [5] and independent authorities.

At VDSL frequencies the picture changes somewhat. Although the signal levels are still very low, the frequencies are much higher to the extent that the degrading balance of the cables allows more of it to enter the (radiative) common mode. Once there, the shorter wavelength raises the antenna efficiency so that emission becomes more likely. By making some careful design choices, the problems can be reduced to manageable proportions, as has been confirmed by extensive practical and theoretical measurements [9–11].

3.3.3.2   Susceptibility

Antennas work reciprocally. Cables that can radiate emissions can also pick them up from external sources. Normal radio signals do not pose much of a threat to DSL signals because the sophisticated receivers used have intrinsic abilities to eliminate or ignore them. RFI pick-up only really becomes an issue when there are strong nearby transmitters. Again VDSL systems are most likely to be affected because of the increased antenna efficiency of network cables at these frequencies. Some special methods can be used to mitigate these effects, such as RFI cancellation. In very severe RFI environments, such as close proximity to a strong AM broadcast transmitter, DSL systems operating on aerial cabling may not be workable and substitute technology, such as direct fibre, or direct buried cabling carrying DSL may be the only viable alternative.

3.4　Measuring FEXT

FEXT is a well-understood problem for telephony cables at low frequencies. It is, however, not well characterised for higher frequencies in a real network. In a real network, FEXT is not just a function of the crosstalk in the cables, but also a function of joints, gauge changes, and other physical properties of the installed transmission plant. In order to understand the implications for VDSL, equipment has been developed to measure FEXT in a live network (one carrying existing narrowband telephony or ISDN).

Measuring FEXT is both time consuming even in a controlled laboratory environment; in the access network it is even more problematic. In a live network not only is it more difficult to make any measurement, but also the disruption to the network must be kept to an absolute minimum.

In the laboratory, though cumbersome, it is straightforward to measure the FEXT in a cable. Typically the cable is either new or well-maintained with few faults; also the relative arrangement of the pairs in the cable is known. FEXT between two pairs drops off markedly with their separation in a cable, and the couplings between two adjacent pairs will be similar from pair to pair. Therefore the FEXT signal-to-noise ratios (SNRs) do not have to be measured for all pair combinations to have sufficient confidence in the characterisation of the cable. This, though, is not the case in the access network, and to adequately characterise a cable here, it is necessary to measure as large a set of the pair-to-pair combinations as possible.

Figure 3.8 is a schematic of part of a distribution network. There are two major differences from the simple cable case:

- there are joints (represented by 'X' in Figure 3.8);
- sets of pairs terminate at distribution points (DPs), resulting in different lengths over which crosstalk can affect the signal.

In order to measure FEXT in a live network, VDSL splitters are spliced into all 100 pairs at the cabinet; similarly VDSL splitters are spliced into all the pairs at a given DP (see Figure 3.9). This reduces the disruption to the customer's POTS service to the time it takes to splice in the splitter – a few seconds – and not the time it takes for the measurement, which is in the order of an hour.

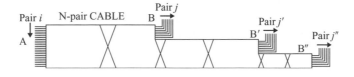

Figure 3.8　Schematic of a distribution network

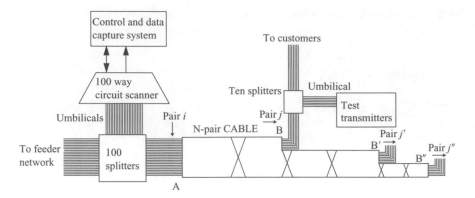

Figure 3.9   FEXT test arrangements for pairs at DP No B

A test signal generator is constructed that generates ten separate, distinct and broadband output signals. These signals are connected to up to ten of the VDSL connections at the DP. A scanning measurement then samples the signals on each of the up to 100 VDSL connections at the cabinet sequentially. As it measures each pair it is able to separate contributions to that pair from each of the up to ten connected distinct transmitters, so measuring up to ten FEXT couplings at up to 100 frequency points.

This is repeated at all the DPs present, thereby measuring all the required FEXT couplings to adequately characterise that cable.

The data is then analysed to extract the complete coupling matrix for the cable being measured of size n by m by l where n is the number of pairs scanned, m is the number of those pairs that had one of the transmitters connected at the DP at some stage during the testing, and l is the number of different frequency points measured. The cable attenuations are accounted for, and the matrix can then be used to make predictions for FEXT.

Figure 3.10 shows over-plotted the 100 FEXT SNRs for the 100 pairs of a calibration check cable. Each SNR represents the difference between a particular insertion loss and the power sum of the 99 FEXT couplings that contribute to FEXT interference in that pair.

The straight line in Figure 3.10 shows the 1% worst-case theoretical model that the work on FEXT is intended to either verify or modify. Figure 3.11 shows further analysis on this data and clearly illustrates the variation in FEXT as the number of like disturbers in the cable, n, increases. The normally accepted Werner FEXT model assigns this a dependence of $n^{0.6}$ and this is borne out by the measurements – the surface in Figure 3.11 has a dependence of $n^{0.6}$.

The outcome of the analysis is that the Werner model is correct except for one point for the BT access network. The value of the constant is not -55.8 (at a frequency of 100 kHz, a length of 1 km and 49 like disturbers) but better characterised by a constant of -50.0. The difference is accounted for by noting that the previously quoted constant was taken from measurements upon cables in a laboratory with a

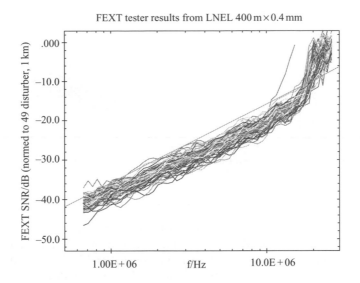

Figure 3.10   100 FEXT SNRs for 400 m of 0.4 mm cable

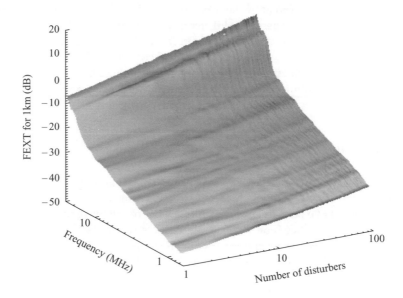

Figure 3.11   FEXT dependence upon frequency and n for the LNEL 400 m cable

shorter lay length of the pairs than is common in the BT access network. References [12,13] for example state that FEXT is reduced by having a shorter lay on the pairs in a cable and that the variance of the lay is important in the worst case. Intuitively this is correct since a long lay means that the wires lay nearly flat as opposed to

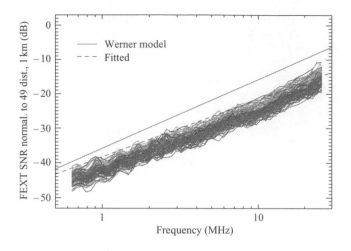

Figure 3.12   FEXT SNRs for cable from which Werner model was originally derived

keeping their shape in a tight spiral. This increases the likelihood of deviation from an ideal helix and so increases FEXT. Figure 3.12 shows the FEXT SNRs for a grade of cable from which the Werner model was originally derived. When this is compared to Figure 3.10 there is an appreciable difference in the measured FEXT and this is attributed to the fact that the cable measured in Figure 3.10 has a far longer lay length of that in Figure 3.12.

The importance of the world-leading innovative BT work on FEXT, and its implications for DSL spectrum management in the local loop unbundling environment, resulted in BT being awarded the IEE Measurement Prize in 1998.

3.5   Theoretical capacity

The Hartley-Shannon theorem [14] states that any channel has an intrinsic information capacity, C, for carrying data without error (this is a mathematical limit. Modern DSL modems achieve performance within a few dB of the limit, but become more complicated and introduce more delay as the limit is approached. At the limit they would be infinitely expensive and never deliver any data because of the processing latency). For a white band-limited Gaussian channel [15] it is given by:

$$C = B \cdot \log_2 \cdot \left(1 + \frac{S}{N}\right) \qquad \text{bit/s}$$

where B is the bandwidth, and S/N is the SNR at the receiving station.

For the more general case where signal and noise vary with frequency this can be simply re-arranged thus:

$$C = \int_{f_{\min}}^{f_{\max}} \log_2 \cdot \left(1 + \frac{S(f)}{N(f)}\right) df \qquad \text{bit/s}$$

This expression is classical, dating from the 1940s. However, it is worth re-examining its consequences for DSL applications with the aid of a modern model of our channel [16].

3.5.1   NEXT-limited capacity

For self-NEXT-limited capacity the Hartley-Shannon integral becomes:

$$C = \int_{f_{\min}}^{f_{\max}} \log_2 \cdot \left(1 + \frac{P_t(f) \cdot |H(f)|^2}{P_t(f) \cdot X_n(f)} \right) df \qquad \text{bit/s}$$

where $H(f)$ is the voltage transfer function of the channel, $X_n(f)$ is the NEXT power coupling function and $P_t(f)$ is the power spectral density of the transmitter. In computing the self-NEXT-limited capacity some key assumptions must be made.

First, the source of noise in the channel is assumed to be generated by NEXT from adjacent and like systems (hence the term self-NEXT, and throughout this chapter, NEXT limited capacity will imply self-NEXT-limited capacity). The term 'like system' is used to indicate that the noise sources have the same spectral occupancy or power spectral density (PSD) as the system being interfered with. This is most likely to cause worst-case interference because all the coupled signals will be in-band for the receiver of the interfered system. Also, the worst case will occur when 49 other disturbers in a 50 pair unit are active at one time (Large cables are constructed from groups of smaller units, typically of 50 pairs. The UK term is *unit*, but the American term *binder group* is more common in world literature). Within the unit or group the individual pairs are usually ordered and twisted in a particular fashion. When a wire-pair suffers interference from other wire-pairs in the same unit it is the nearest 6–8 interferers which cause most of the noise power or crosstalk. However, it cannot be predicted which ones these will be unless *a priori* knowledge of the physical location and spectrum of all the possible interferers in the cable is available. Hence, a worst-case assumption is usually made in which it is assumed that all 49 possible disturbers are active and that 1% worst-case coupling is achieved. This is rather pessimistic but yields a planning rule that is conservative and offers margin to cater for the effects of other (less tangible) noise sources.

Second, it must be assumed that the receiver is perfect in all other aspects, namely:

- no non-linear behaviour in the channel, transmitter or receiver;
- no internal sources of noise (in particular no ADC quantisation noise);
- equalisation cancels all inter-symbol interference (ISI) perfectly;
- echo cancellation removes own-system echoes perfectly.

With the caveats above the expression for the NEXT-limited capacity reduces to:

$$C = \int_{f_{\min}}^{f_{\max}} \log_2 \cdot \left(1 + \frac{|H(f)|^2}{X_n(f)} \right) df \qquad \text{bit/s}$$

The expression for NEXT-limited capacity is independent of the transmitter power spectral density and depends only on the channel and crosstalk power coupling characteristics.

In BT's access network, which is not dissimilar from many other traditional Telcos, two types of metallic twisted pair predominate – 0.4 mm and 0.5 mm diameter copper. Computer simulations of the information capacity have been made of lines consisting of one homogeneous length of either 0.4 mm or 0.5 mm cables. Figures 3.13 and 3.14 show their NEXT-limited information capacity. With reference to Figure 3.13, a 3-D plot of NEXT-limited capacity for 0.4 mm conductor diameter copper cable is shown on the left, whereas a contour plot of the same data is shown on the right.

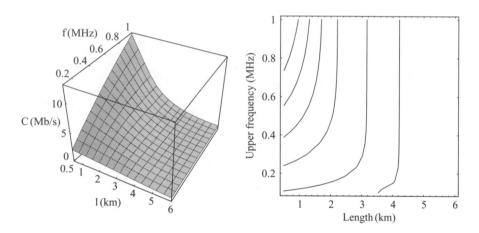

Figure 3.13   NEXT-limited capacity for 0.4 mm copper cable

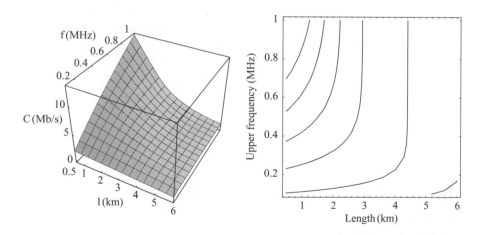

Figure 3.14   NEXT-limited capacity for 0.5 mm copper cable

The 3-D plot shows the surface of maximum capacity (C in Mbit/s) with variation in cable length (l in km) and bandwidth in use from DC to (f_{max} in MHz). Clearly it can be seen that as length increases the capacity decreases. At long cable lengths (~2–5 km), increasing the frequency of transmission has little or no effect on capacity. Only over relatively short lengths is it advantageous to use higher frequencies. Intuitively, this is to be expected from the band-limited nature of metallic access channels. This is illustrated more clearly by the 2-D contour plot shown on the right of the figure. Contours of fixed capacity are plotted for the following capacities – 10, 8, 6, 4, 2 and 1 Mbit/s. The 10 Mbit/s contour is shown at the top left of the plot, whereas the 1 Mbit/s contour is on the right. The 1 Mbit/s contour shows quite clearly that there is no advantage to using frequencies higher than about 200 kHz on long loops (This assumes a NEXT-limited environment. ADSL avoids this limit by reserving the higher frequencies (up to 1.1 MHz) for one direction only). However, the potential for high-speed transmission over short distances, using higher frequencies, would seem to exist.

Figures 3.13 and 3.14 show the increase in NEXT-limited capacity with increase in cable gauge. These plots show the remarkably high capacities on offer, and the potential for further exploitation of the local loop by systems occupying bandwidths from DC up to approximately 200 kHz.

In summary, the following conclusions may be drawn from the theoretical work on self-NEXT capacity. More detailed treatment is given in Foster [17]:

- capacity decreases with increasing cable length;
- for typical access network connections (2–5 km from serving exchange to customer termination), there is little point using frequencies above approximately 200 kHz in a NEXT-limited regime because the increase in capacity is very small;
- there is potential for very high speed transmission (e.g. 25–50 Mbit/s) over relatively short distances – copper tails on the end of passive optical networks may be viable utilising bandwidths in the copper network of up to 10–20 MHz.

3.5.2 FEXT-limited capacity

For FEXT-limited capacity the Hartley-Shannon capacity integral becomes:

$$C = \int_{f_{min}}^{f_{max}} \log_2 \cdot \left(1 + \frac{P_t(f) \cdot |H(f)|^2}{P_t(f) \cdot X_f(f)}\right) df \qquad \text{bit/s}$$

where $H(f)$ is the voltage transfer function of the channel, $X_f(f)$ is the FEXT power coupling function and $P_t(f)$ is the power spectral density of the transmitter.

Assuming that the sources of FEXT are owing to the FEXT from 49 other disturbers in the 50-pair binder group, and that the receiver is otherwise perfect (as assumed in the self-NEXT-limited analysis), then the above expression reduces to:

$$C = \int_{f_{min}}^{f_{max}} \log_2 \cdot \left(1 + \frac{|H(f)|^2}{X_f(f)}\right) df \qquad \text{bit/s}$$

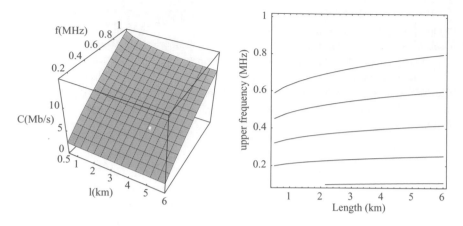

Figure 3.15   FEXT-limited capacity

Noting that $X_f(f)$ is the generally accepted Werner FEXT coupling function [18], this further reduces to:

$$C = \int_{f_{\min}}^{f_{\max}} \log_2 \cdot \left(1 + \frac{1}{\psi \cdot l \cdot f^2}\right) df \qquad \text{bit/s}$$

where l is the physical length of the channel (in feet), f is frequency (in Hz) and the constant ψ is determined empirically. This is the expression for FEXT-limited capacity. Notice that the expression is independent of transmitted power spectral density, and is frequency and length dependent. It is somewhat surprising that the only dependency on the channel is length, and therefore the FEXT limited capacity will not be dependent on cable type.

Figure 3.15 shows the FEXT-limited capacity of the access network. On the left is a 3-D plot of the FEXT-limited capacity whereas on the right is a 2-D contour plot of the same data.

Contours of fixed capacity are plotted for the following capacities – 10, 8, 6, 4 and 2 Mbit/s. The 10 Mbit/s contour is shown at the top of the plot whereas the 2 Mbit/s contour is shown at the bottom. It is instructive to compare this figure with a NEXT-limited equivalent (e.g. Figure 3.13 or Figure 3.14). Clearly, it can be seen that the contours of constant FEXT capacity are almost orthogonal to the contours of NEXT capacity. Also FEXT would not appear to limit the information-carrying capacity of the channel as much as NEXT does.

In contrast to the NEXT-limited scenario there does appear to be an advantage in increasing the frequency components of transmission (i.e. using greater bandwidths will release greater capacity in the FEXT-limited case, as exploited by ADSL at higher rates, and VDSL).

3.6 Conclusions

The DSL technologies outlined in this chapter are still evolving. Each new generation brings improvements in functionality, performance and levels of integration. This trend of technology development and innovation looks set to continue, in the same way that voiceband modems evolved to exploit the capacity of the voice channel more efficiently. The fact that VDSL occupies frequencies up to 12 MHz, implies that further innovation in DSL transmission will arise in large part by exploiting multi-pair transmission, and by reducing the distance between the customer and the network provider. VDSL is the first DSL technology to be designed from the outset to be operated remotely from the local exchange to provide high throughput transmission.

Since the key limitation to DSL capacity is crosstalk, BT has invested considerable effort in understanding the crosstalk environment of its network. This measurement and modelling effort is enabling the real-world performance of new DSL systems to be quantified, which in turn helps a Telco like BT to work with its suppliers to optimise system designs. It also facilitates more accurate performance predictions and equipment deployment rules.

In summary, the Telco-installed copper-pair network presents a challenging and hostile environment for high-speed, multi-megabit/s transmission. However, realisation of the megabit capacity of the existing network is critically dependent on understanding and controlling the crosstalk environment to ensure spectral compatibility for new and legacy xDSL transmission systems.

3.7 Acknowledgements

The authors would like to acknowledge the contributions and work of other members of BTExact's Advanced Copper Technologies Unit based at Adastral Park in Suffolk. Furthermore, the authors wish to acknowledge the wide-ranging discussions with other colleagues in BT and those from the Telco and vendor communities who represent their organisations at international DSL standards meetings. In particular, those involved in ETSI TM6, ANSI T1E1.4, ITU-T SG15 Q.4, the DSL Forum, FSAN initiative and the FS-VDSL committee.

3.8 References

1 COX, S. A. and ADAMS, P. F.: 'An analysis of digital transmission techniques for the local network', *BT Technol J*, 1985, **3**(3), pp. 73–85
2 YOUNG, G., FOSTER, K. T. and COOK, J. W.: 'Broadband multimedia delivery over copper', *BT Technol J*, 1995, **13**(4), pp. 78–96
3 'High speed access technology and services, including video-on-demand', IEE Colloquium, London (October 1994)
4 CLARKE, D. *et al*: 'Realising VDSL in Telco networks', International Conference on *VDSL*, London, UK (April 1998)

5 FOSTER, K. T.: 'The radio frequency environment for high speed metallic access systems', IEEE Globecom '96 VDSL Workshop (invited paper), London (November 1996)

6 DICKIE, A. and MACKENZIE, J.: 'A solution for Midband', *BT Technol J*, 1998, **16**(4), pp. 48–57

7 CHEN, W.: 'A proposal for ADSL Issue 2 to include the low complexity ATU-R', ANSI contribution T1E1.4/96–199r1 (July 1996)

8 SCHNEIDER, K.: 'Simulated performance of HDSL transceivers', ANSI T1E1.4/97–444 (December 1997)

9 FOSTER, K. T. and STANDLEY, D. L.: 'A preliminary experimental study of the RF emissions from dropwires carrying pseudo-VDSL signals and the subjective effect on a nearby amateur radio listener', ANSI T1E1.4/96–165 (April 1996)

10 FOSTER, K. T.: 'On the subjective effect of RF emissions from VDSL – a practical comparison between transmissions from two laboratory demonstrators', ANSI T1E1.4/96–191 (July 1996)

11 FOSTER, K. T. and STANDLEY, D. L.: 'Further practical measurements on radiated emissions and path loss', ANSI T1E1.4/96–317 (November 1996)

12 REFI, J. J.: 'Mean Power Sum Far-End Crosstalk of PIC Cables as a function of Average Twist Helix Angle', 29th International Wire and Cable Symposium, *Proceedings*, November 1980, pp. 111–116

13 REFI, J. J.: 'Power Sum Crosstalk of PIC Cables as a function of Cable Design', 31st International Wire and Cable Symposium, *Proceedings*, 1982, pp. 237–244

14 SHANNON, C. E.: 'A mathematical theory of communication', *BSTJ*, 1948, **27**, pp. 379–423 and pp. 623–656

15 SHANNON, C. E.: 'Communications in the presence of noise', *Proc IRE*, 1949, **37**(1), pp. 10–21

16 HUNT, R. H. *et al*: 'The potential for high-rate digital subscriber loops', *IEEE ICC Conf*, 1989, pp. 17.1.1—17.1.6

17 FOSTER, K. T.: 'On the potential for multi-megabit/s transmission in BT's access network', MSc thesis, BT/University of London (1995)

18 WERNER, J. J.: 'The HDSL environment', *IEEE JSAC*, 1991, **9**(6), pp. 785–800

Chapter 4

DSL spectrum management – the UK approach

J. W. Cook, R. H. Kirkby and K. T. Foster

4.1 The issues

DSL systems do not live in isolation from their neighbours. We have seen in Chapter 3 that in fact there is crosstalk between systems operating in the same part of the access network, and that this crosstalk can be categorised into two important types, near end crosstalk (NEXT) and far end crosstalk (FEXT). The distinction is important because these two types of crosstalk limit available transmission capacity in different ways. Duplex transmission suffers from the more severe NEXT but benefits from duplicate use of bandwidth for each direction of transmission while the less severe FEXT allows a higher throughput but only in one direction at a time or frequency.

It was shown that a critical frequency could be found below which it was favourable to transmit duplex against NEXT and above which it was favourable to transmit dual simplex against FEXT.

Crucially, though, this critical frequency was found to be loop-length dependent. The longer the loop the lower the critical frequency becomes. This is important because in most real access networks not all loops are the same length and indeed loops of different lengths often share the same cables and so crosstalk between them may well occur. This fact blurs the concept of the critical frequency to cover a band. The band stretches from the critical frequency for the longest loops in the cable, below which duplex transmission is beneficial for all, to the critical frequency for the shortest practicable loops above which dual simplex transmission is beneficial for all. Between these two limits is a band of uncertainty where optimal transmission for some loops may cause sub-optimal transmission for others in the same cable (see Figure 4.1).

Transmission capacity would be wasted if duplex transmission were used in the band best for dual simplex or vice versa. However, any plan will inevitably be a compromise for the loops' different lengths, and in the band of uncertainty some waste is inevitable, at least for some of the loops.

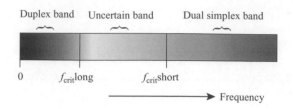

Duplex band   Uncertain band   Dual simplex band

0   f_{crit}long f_{crit}short

→ Frequency

Figure 4.1   Frequency bands and critical frequencies

In the duplex band the rate of transmission is substantially symmetrical as all of the bandwidth is available in both directions all of the time. However, in the dual simplex band a choice has to be made for the direction of transmission for which each part of the band will be used. It would be possible to choose that all the band is used for transmission in just a single direction, making transmission very asymmetric. Alternatively interleaved sub-bands of similar width in each direction could be used so that the net effect is nearly symmetric transmission. But once this choice is made in a cable, all the DSL systems in that cable at least would have to use the same band allocations else where there is conflict NEXT will be created and transmission compromised. The choice defines the symmetry of all transmissions on the cable, and once set cannot be easily changed unless all the services in the cable are changed at once. Clearly the choice is a far reaching one and once made is unlikely to be altered in the future as it would affect so many customers.

A way to look at the complete picture for the cable is that the set of all transmission systems operating in the cable jointly create the dominant common noise environment (due to crosstalk) they each suffer. Introducing a new system type into the mix of deployed systems can change the joint noise environment and affect all the systems in the cable. In fact even a single variant system could have a drastic adverse effect if it causes crosstalk in a frequency range that had previously been comparatively quiet and exploited by already deployed systems for receiving weak signals.

The expected noise environment obviously affects both the viability of existing deployed systems and also the deployability of future systems. Changes to the noise environment could deny service to existing customers where systems had previously been deployed with a different noise environment expectation. Such changes could also adversely affect the deployability of future systems and hence even the commercial viability of a whole class of technology. A careful plan is required to control ongoing deployments to safeguard existing services and preserve the commercial value of deployments still being planned or envisaged.

It is interesting to consider what would happen if such a plan did not exist. Suppose that any kind of transmission at any frequency were permitted at any point in the access network. This would mean that no system operator could realistically expect to operate any part of the spectrum in a dual simplex, FEXT-limited mode, since he would have to plan for the likely event that other pairs in the network would be used to transmit in the opposite direction at the same frequency, causing NEXT. As was seen in Chapter 3, FEXT limited transmission is preferred on long pairs (~5 km) at all frequencies above about 50 kHz and above about 200 kHz there is no NEXT limited capacity at all on these pairs. This would effectively condemn the user with a long wire pair to very low transmission speeds. Meanwhile, on a very short wire pair, NEXT limited transmission is effective up to 1 MHz or so and much greater transmission capacity is available. So the effect of the lack of a plan is detrimental to all users of the cable, but disproportionately detrimental to those that happen to have longer pairs. While there will always be an advantage in receiving service via a shorter pair, a good plan helps to maintain a better compromise between short and long pair transmission possibilities, as well as giving potentially greater throughput for all.

In fact a complete lack of any kind of spectrum plan could have an even more drastic adverse effect in that the greatest transmission capacity would go to those systems that used the loudest transmissions. This is because they gain a signal to noise ratio advantage, through having the largest transmit signal, while still only suffering the common crosstalk noise environment. This advantage is obtained at the expense of all the neighbouring systems since the loud transmitter raises the common noise environment disproportionately, and their weaker transmissions lose signal to noise ratio. Retaliation is possible through the affected systems also raising their transmit signal levels, further raising crosstalk levels. This process is sometimes known as the 'cocktail party effect'. Signal levels all go up but no net gain is obtained, just a lot of chaos as systems that previously functioned cease functioning because they can no longer transmit loud enough.

Breaking this cycle requires a plan. The plan has in some way to constrain what transmission levels and frequencies may be used in what parts of the access network. This will inevitably itself restrict the deployability of some system types in some locations, but the plan will provide some level of confidence in the levels of crosstalk noise that may in the future exist in any part of the access network and so banish the chaos and the cocktail party effect and replace it with a comparatively stable and predictable environment that operators can use to plan their deployments and so that customers can enjoy continued and uninterrupted service.

This plan is spectrum management. Spectrum management must somehow constrain what kinds of signal can be injected into the access network, and hence the systems that inject them, and where they are located. If it does not do this it will not be effective.

It should be noted that once such a plan exists it will rapidly become immutable, or nearly so. This is because the existing plan, as has been described, sets the effective noise environment for all existing systems and the business cases for future systems. Any attempt to change the plan will meet with resistance from nearly all existing players whose deployments or planned deployments would or could be adversely affected. As a result it seems unlikely that the plan could ever be significantly changed once agreed and acted upon, unless there were some overriding outside reason to do so that affected the majority of the players.

A technology that can have a profound effect on spectrum management is that of the repeater. A repeater is an active device that is inserted in a pair to amplify the signal being sent so as to overcome the effects of signal attenuation along the loop. By doing this the adverse effect of crosstalk from other nearby systems can be drastically reduced. The disadvantage of the repeater is that although it offers an advantage to the repeatered pair, systems operating on adjacent pairs without repeaters are disadvantaged by the increased crosstalk caused by the repeater itself. So to deploy repeaters is to need many of them. From a spectrum management perspective repeaters cause a shift in the critical frequencies of the repeatered pair favouring duplex transmission but also introduce a large number of extra points in the access network where signals are being injected. As we have seen that spectrum management is about controlling and planning signals injected into the access network plainly the use of repeaters drastically complicates spectrum management planning.

One thing a spectrum management plan does not do is set out the deployment rules per se that must be used by operators. It may *bound* the deployment options through the limits it sets but as crosstalk noise is statistical in nature, variously in terms of the distribution of systems being used in a given cable, in terms of the individual crosstalk coupling between pairs, and even in terms of the imprecise estimates of loop characteristics on which deployment decisions must be made, the crosstalk expected even in a spectrally managed network is still a statistical quantity. In order to obtain a very low risk that crosstalk noise will adversely affect service deployment it would be necessary to be very conservative with deployment rules, while a more bullish approach might deploy with a higher risk of failure on much longer loops. This choice and the deployment rules derived from it remain the domain of the operators.

4.2   The choices

Having established that to be successful a spectrum management plan must somehow restrict what can be deployed and where, the question remains how should it do this? There are probably as many ways of doing this as there are pairs in the target network to be managed, however there are some broad strategies that can be described:

(a)   limiting the specific types of system (e.g. by references to standards) and where they may be deployed;
(b)   constraining the spectrum of deployed systems with hard power spectral density (PSD) masks for given situations;
(c)   setting out an algorithm by which the interference properties of the deployment of a given DSL at a given location can be evaluated and the relative interference between it and other systems assessed in a consistent way and placing limits on the consequences (e.g. the amount of range reduction permitted by comparison with some reference case for all the systems concerned).

The first approach, (a), has the merit of being extremely simple to describe and lay down. However it is supremely inflexible. It says nothing about how systems other than those that are specifically described in the plan may be used and by implication such systems are excluded even though they may be very similar disturbers to other systems that have been described.

In the same way such a plan is almost certainly doomed to become dated very quickly as new DSL types emerge which were not known when the plan was devised.

But perhaps the most damning problem with such an approach is that it favours the described technologies over all others and so could be considered directly discriminatory against a technology that is not included in the plan but is in every way a more moderate disturber than another system which is. This could make such a plan open to accusations of being an unfair barrier to trade.

At the other extreme the last approach, (c), is ultimately flexible. By using the algorithm the limits that must be placed on a DSL of any given properties should be determinable. New, non-standard, modified or propriety systems can be evaluated without any need to revise the management plan. No system is unfairly excluded.

However, just because of its ultimate flexibility, it is almost inevitably also very complex. The algorithm will be complicated to develop and agree between the parties concerned. In order to be even approximately effective and fair to all concerned the resulting algorithm will be very complicated to evaluate or even interpret. This means that there will be much scope for errors in the algorithm itself or in its interpretation in a given case. And any attempt to police such a plan would be thwarted by the difficulty the policing authority would have in interpreting and applying the rules.

The second approach, (b), is a compromise between the two extremes. It does not explicitly include or exclude any particular technology. It is perfectly possible and in fact quite simple to determine if or where a new, non-standard or proprietary system can be deployed and so is reasonably future proof. And yet the basic method is reasonably simple to specify and understand and does not involve any complex algorithms.

It still has a trace of the disadvantages of method (a) in that it is very rigid; any failure to comply with the mask is a simple plan failure. There is no room for compromise as such a compromise would require some trading algorithm that would quickly lead in the direction of method (c). This inflexibility is actually not as severe a barrier as it may seem since minor transgressions of the plan masks can be addressed by reducing the transmit power of the offending system slightly; most modern DSL systems have the flexibility to do this.

Method (b) also has a trace of the disadvantage of (c) in that compliance with the plan is more difficult than simply quoting a standards reference. The PSD must be measured in a prescribed way and compared with the plan masks. This may sound simple but measuring PSD is a far from simple process especially if accurate calibration is required on which significant commercial risks are at stake.

Within the three broad categories of approach to spectrum management plans set out above there are a number of choices to be made about the scope of application of the plans.

Geographical application can be national, regional or even by exchange area or ultimately even by individual cable. In other words a single plan could be devised that would apply to all regions nationally, or separate plans could apply to different geographical parts of the network. Obviously there is a practical limit to which this can be taken but there might be justification in having regional plans based on the nature of the human geography of the area such as rural, urban, suburban, residential, commercial. For example a plan suitable for use in a commercial area might make use of bandwidth in a more symmetrical way (e.g. for general data exchange applications) while a plan for use in a residential area might have a more asymmetrical flavour (e.g. for broadcast type applications). Having developed just a few such plans a country or other geographic area could be divided up into parts and one of the predetermined plans assigned to each segment based on some measure of its geography.

Application to pairs in a network can also be either uniform or categorised across cables. Crosstalk between pairs in a cable is greatly influenced by their physical separation within the cable build. This means that if a subset of pairs is carefully chosen DSL deployments on those pairs could enjoy a lower resulting noise environment than if deployments were unplanned or allowed on all pairs. Any corresponding spectrum management plan could then take this lower level of crosstalk into account in its design. This mode of deployment though is dependent on network cable pair arrangements being maintained and recorded along the length of the cables, and does inherently restrict the penetration of DSL services. Once the spectrum plan was set any attempt to increase the plan's preset penetration at a later date would lead to difficulties with crosstalk into previously deployed systems.

Another way of categorising pairs would be by length. Different length pairs could then have different parts of the management plan assigned to them. This approach could help with the 'band of uncertainty' issue by allowing different uses of bandwidth on different length pairs.

A final approach that is worth mentioning is that of assigning various spectrum management plans at different *times*, e.g. at different times of day. This could address the commercial/residential issue by for example having a highly symmetric plan for the daytime and an asymmetrically biased plan during the evening. This has been discussed but not adopted anywhere to the authors' knowledge.

4.3   The UK choices and assumptions

The fundamental assumptions on which spectrum management in the UK was built were agreed during debates in the regulatory forum: the DSL Task Group responsible for advising OFTEL through the Network Interoperability Consultative Committee about technical issues associated with local loop unbundling in the UK. The agreements were made during discussions in early 2000 and broadly were as follows:

- That the plan must be as nearly as possible system agnostic in that the method (a) from the previous section should not be used. This was mandated by the lack of flexibility of such an approach.
- That all geographical areas would be treated the same, primarily as no basis for regulation could be found that could justify treating different areas differently.
- That all pairs in a cable must be treated identically, i.e. that pair selection was not permitted. To a large extent this decision was necessitated by the random jointing practices used in the UK network which meant that controlling and recording pair usage in cables would have been insurmountably difficult.
- That repeaters would not be permitted. Originally this was primarily motivated by the need to keep initial approaches to the spectrum management plan as simple

as possible, and in any case there was no pre-existing population of repeaters that forced their consideration. However, at the time of writing it seems that the tendency to immutability of spectrum management plans has already made itself felt and it now seems unlikely that repeaters will ever be permitted.

• Finally it was set as an objective that the UK spectrum management plan should be simple and clear to maximise clarity and simplify enforcement.

4.4   The UK Access Network Frequency Plan

The UK ANFP [1] limits the signals allowed into BT's telephone access network. The document is owned and authorised by the UK regulator, Oftel. It was produced by a cross-industry committee, acting to advise Oftel; where the committee agreed, Oftel accepted its recommendations; where it could not agree it proposed choices from which Oftel chose (a process called determination).

4.4.1   Scope

All connections to BT's telephone access network are subject to the ANFP limits, whether the connection is made by BT or not.

At time of writing the ANFP only permits use of spectrum up to 1.1 MHz, the spectrum of interest to technologies up to and including ADSL. Use of higher frequencies is anticipated in the future (see Section 4.5 Extending the ANFP to VDSL) and that spectrum is reserved for future use by forbidding its use now.

4.4.2   Controlled interfaces

The ANFP identifies the interfaces where connection to the shared cables is permitted, and for each it specifies a single power spectral density (PSD) mask. The ANFP limit is that the spectral power flowing through the interface toward the cables shall not exceed the mask.

These interfaces are the main distribution frame (MDF) at the exchange and the network termination point (NTP) at the customer's premises. The MDF is not a legal demarcation point, so non-BT connections also have a third interface, the HDF (see Figure 4.2). The ANFP does not directly address the HDF, and managing crosstalk

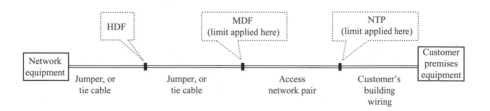

Figure 4.2   Network interfaces to which the ANFP applies

interference in the cabling between the HDF and the MDF is the responsibility of the network operator(s) using that cabling. The ANFP is applied at the MDF.

Of course the lines in BT's access network have two ends, typically one at the exchange and one at the customer's premises.

The ANFP is only concerned with possible impact on other lines in a cable, and does not explicitly consider what may be connected to the interface. The 'spectral power flowing' could equally well be from a single modem connected directly to the interface, from several instruments sharing the line via splitter filters, or from noise picked up via house wiring. The ANFP applies equally to signal intentionally injected into a line with intent to communicate over it and to noise leaked into it.

4.4.3   Masks

Under the ANFP an NTP falls into one of three classes, broadly based on the electrical length of the line (it will be shown later that this classification is the result of BT's historical deployment of HDSL systems). The classes are 'short', 'medium', and 'long' (more recently a fourth class has been added 'extra short'. The general structure of the ANFP remains unchanged however). BT has declared the classification for each NTP in its network. Together with the MDF interface this makes four distinct classes of interface, and the ANFP declares a mask for each of them (see Figure 4.3).

The masks are each defined numerically, in a table of values that specifies corners of a polygon (on axes of log frequency versus log PSD). The masks are shown as a graph in Figure 4.4.

4.4.4   The central approximation

Spectrum management is bound up with being able to predict the noise environment experienced by a system. In general such predictions are probabilistic, partly because crosstalk is only known statistically, and partly because the systems producing the noise are an unknown mix of several species of different DSL systems.

To simplify the 'mix of systems' considerations, the ANFP makes a simplifying approximation: that the level of noise experienced at a given frequency is as if all one's neighbours were injecting power at the level of one's loudest neighbour at that frequency. So the neighbouring systems' spectra are approximated as being the same, being the envelope of the spectra actually present.

Mask name	Defines PSD permitted at
down exch	the MDF of the exchange
up short	the NTP of near customers
up medium	the NTP of mid distance customers
up long	the NTP of far customers

Figure 4.3   The naming of masks

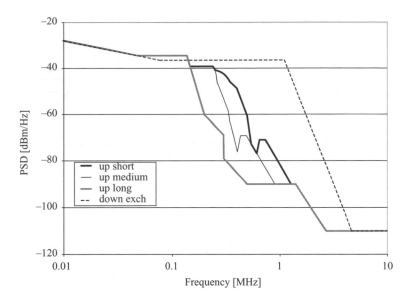

Figure 4.4   The ANFP masks presented graphically

This is quite a good approximation for the purposes of predicting a worst case noise environment – where one would otherwise be searching all the possible combinations of neighbours for the one that hurts one's pet system most. Evidently the approximation is conservative, in that a real mix can never be worse than the approximation. It is efficient in that it only sacrifices about 2 dB of performance (the usual models of FEXT and NEXT in the literature [2,3] feature a power law of $N^{0.6}$ to account for the number of active interferers, so a reduction by half in number of crosstalkers is modelled as a reduction of 3 dB times 0.6 = 1.8 dB). Now consider various mixes of two systems with different spectra, as shown in Figure 4.5.

The graph compares the spectra of two systems '100 : 0' and '0 : 100', in a band of frequencies where one is the louder signal in one part and the other is in another part. Also shown are, for various mix ratios, the spectrum of the effective average interferer. The dotted curve is the approximation for all the mixes, being the maximum of the spectra of the two actual system types. Which of the mixes is most harmful to a given victim system will of course depend on which frequencies that system type is most sensitive to; for a victim equally sensitive to all frequencies in this band the 50 : 50 mix is worst. And the approximation is at most 1.8 dB worse than that. For a differently sensitive victim the difference between worst actual mix and the approximation is less than 1.8 dB.

This approximation obtains simplicity: the approximation is independent of the mix of interfering systems and also independent of the *victim* system.

A valuable side effect of accepting this approximation has been the 'fairness' principle: at a given entry point into the shared cables, if one system is permitted to

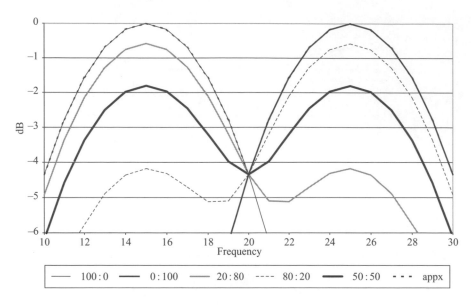

Figure 4.5   Mixes of systems

send a signal at a particular level at a particular frequency then all systems may; there being no benefit to a potential victim from limiting them more restrictively.

4.4.5   Why spectral masks?

Crosstalk statistics only give useful characterisation of the frequency domain magnitude of coupling (for two given pairs phase can also be measured repeatably – and some noise cancellation techniques propose doing exactly this. However, it is not correlated with the corresponding measurement on any other pair of pairs). Hence only this is available to the ANFP – which is concerned only with the aspects of injected power after it has experienced a crosstalk coupling. The loss of phase information would mangle any other transform domain into uselessness (including the untransformed time domain).

In principle one could introduce other limits too, such as a limit on total power sent to line. This was not done in the ANFP principally because the worst-case noise environment would not improve significantly with such limits, and supernumerary limits might exclude future uses of a line for no good reason. Such future uses include sharing a line between two systems via splitter filters (e.g. POTS and ADSL on one line).

4.4.6   Mask construction, overview

The ANFP masks were constructed by a superficially simple procedure:

1. the network interfaces were grouped into clusters of mutual neighbours;
2. the system types to be admitted a priori were identified;

3. for each given system type the deployment rule was identified;
4. taking the deployment rules together the network interface clusters were categorised into 'zones' (the zone indicates which ends of given systems might be installed on at least one line end in the cluster);
5. for each end of each given system type a spectral mask was obtained;
6. for each zone a composite mask was produced (being the envelope of the spectral masks of all system ends which might be found installed at the cluster).

Note this allows each network interface to be assigned a single particular mask.

The simplicity and transparency of this process enabled an informed cross-industry debate in the appropriate regulatory committee known as the DSL Task Group, and eventually regulator buy-in to one of the compromises proposed by that group [4,5].

4.4.7   Mask construction process, detail

4.4.7.1   Network interfaces were grouped into clusters of mutual neighbours

Ideally one would cluster together those neighbours that share common cables and enter them at the same cable end. For BT's network the closest this could be approached was those line ends that share a common piece of equipment, and in practice 'shares an MDF' was used for the interfaces at the exchange and 'shares a distribution point', or DP, for the customer ends.

4.4.7.2   System types to be admitted a priori were identified

The main intention was to identify the system types already deployed in significant numbers, since these cause the existing noise environment. ADSL was included because BT was already committed to mass deployment of it, although the actual deployment when the ANFP was started was small. As an aspiration for the future, the then yet-to-be-standardised technology called SDSL was also included, but only at rates that had minimal impact on the final masks. While the ANFP was being prepared, the masks in the SDSL draft standard were changed. Just this once the ANFP was reconstructed to match – this was the main change at issue 1.1 of the ANFP – but any future increase would disallow SDSL since the ANFP is now effectively fixed.

In addition, during the deliberations of the task group, another system type was identified by some of the members as being similar to already deployed systems but which would be unfairly excluded if only existing UK-deployed systems were admitted a priori. These systems were CAP-based HDSL systems (carrierless amplitude phase is a two-dimensional modulation scheme similar to QAM). Such systems are also operable at a variety of data rates and PSDs and the precise variants that were to be admitted a priori were the subject of some disagreements in the task group. Eventually an impasse had to be broken by a determination by the regulator who decided to sanction the admission of such systems operating at a rate of 1168 kbit/s sufficient to deliver 2 Mbit/s over 2 pairs [5].

This resulted in the following list of systems to be admitted a priori as basis of the ANFP masks:

- POTS
- ISDN basic access
- 2B1Q HDSL systems delivering 2 Mbit/s over 2 pairs
- CAP HDSL systems delivering 2 Mbit/s over 2 pairs
- 2B1Q HDSL systems delivering 2 Mbit/s over 3 pairs
- ADSL over POTS
- SDSL technology, but limited to those rates with minimal impact on the ANFP masks.

4.4.7.3   Deployment rule was identified for each given system type

By design, POTS can be deployed everywhere. ISDN and ADSL have such long reaches that for the ANFP they were taken to be deployable everywhere. HDSL is very much range limited, however, and acknowledging that improves the reach of ADSL. So the deployment rules used here are: POTS, ISDN, and ADSL: deployed everywhere.

For ADSL the 'ATU-C' end is always in an exchange, and the 'ATU-R' end is always at the customer premises. This is just a statement that the wideband ADSL channel goes from exchange to customer and never the other way round; BT does not deploy reverse ADSL. As it happens a reverse ADSL system would not work in a network with normal ADSL, through catastrophic mutual interference. However, for viable normal ADSL it is important that the ANFP also forbids any other signal like a reverse ADSL system. (The other technologies discussed here are spectrally symmetrical, except for low frequency details like power feeding.)

For 2B1Q-HDSL BT's historical deployment rules are couched in terms of the 'electrical length' of a line, the measured insertion loss of the line at a specified frequency. This is a satisfactory measure in BT's network, which does not use bridged taps.

- 2-pair 2B1Q-HDSL: measure insertion loss at 100 kHz, deploy if IL < 26 dB
- 3-pair 2B1Q-HDSL: measure insertion loss at 100 kHz, deploy if IL < 29 dB

For SDSL, no deployment rule given beforehand. Instead SDSL was admitted into the various zones at those rates where it causes negligible extra impact.

Similarly for CAP-HDSL systems the PSD requirement was also admitted into the various zones at those rates where it causes negligible extra impact and as determined by the regulator.

4.4.7.4   Network interface clusters were categorised into 'zones' taking the deployment rules together

We distinguish line ends by the set of system ends that may have been connected to it. By inspection of the deployment rules above there are four distinct classes of end of a line, depending only on where the end is geographically:

Location of end	Set of system ends which might be present
At an exchange	POTS, ISDN, all HDSL, ATU-C end of ADSL, (SDSL at rate to be determined)
At a customer premises line length under 26 dB	POTS, ISDN, all HDSL, ATU-R end of ADSL, (SDSL at rate to be determined)
At a customer premises line length between 26 dB and 29 dB	POTS, ISDN, 3-pair 2B1Q HDSL and 2-pair CAP-HDSL, ATU-R end of ADSL, (SDSL at rate to be determined)
At a customer premises line length over 29 dB	POTS, ISDN, ATU-R end of ADSL, (SDSL at rate to be determined)

A cluster of line interfaces will have line ends from one or more of these classes; so the set of system ends that may be already deployed at the cluster is the union of the sets of the ends. In this case it is the same as the set for the shortest line. This is a coincidence from the development here; it would not be the case if, for example, the real deployment rules for ADSL or ISDN had been included; then there would have been no consistent definition of 'shortest'.

Figure 4.6 illustrates this 'union of sets' idea. Consider a cable section end from which emerge four lines: two have modems A and B connected, one has receiver R connected to it, and we wonder what transmitter may be connected to line X without causing significant extra noise into R. Obviously introducing any new transmitter on an adjacent line will cause some extra noise, but using the central approximation we will accept that this extra noise is insignificant if X's spectrum is less than or equal to the envelope of the spectra from A and B. It follows that if we are given two

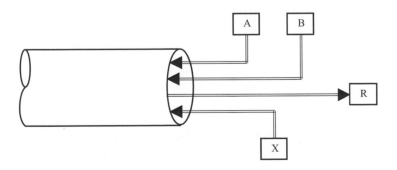

Figure 4.6   Construction of the zones

sets of systems that are already to be admitted at this access point then we should accept any transmitter here which is at or below the envelope of all the systems in both sets.

This leads us to the categorisation of clusters into zones:

Zone	Set of system ends which might be present
'dn exch' : cluster at an exchange	POTS, ISDN, all HDSL, ATU-C end of ADSL, (SDSL at rate to be determined)
'up short' : cluster at a DP with at least one line shorter than 26 dB	POTS, ISDN, all HDSL, ATU-R end of ADSL, (SDSL at rate to be determined)
'up medium' : cluster at a DP with no lines shorter than 26 dB and at least one line shorter than 29 dB	POTS, ISDN, 3-pair HDSL and 2-pair CAP-HDSL, ATU-R end of ADSL, (SDSL at rate to be determined)
'up long' : cluster at a DP with no lines shorter than 29 dB	POTS, ISDN, ATU-R end of ADSL, (SDSL at rate to be determined)

- The UK ANFP zone categorisation has already been done, for the convenience of those companies wishing to use BT's network under local loop unbundling. BT did it using such information as was to hand [6], and every line end now has its zone categorisation available. This categorisation is definitive – the presentation here on electrical line lengths is merely the guiding principle.
- If BT's deployment rules for ISDN and ADSL had been included each would have doubled the number of customer premises line end classes[1], giving 13 zones. No advantage was identified to any system for the extra complication, and the zone categorisation process would have been much more difficult.
- Just because a line is in a particular zone does not mean it has a particular length; it could be longer than its neighbours in its cluster. So zone category is not a safe basis for deployment rules, not even for the HDSL systems which underlie the zone definitions.

4.4.7.5   A spectral mask was obtained for each end of each given system type

The existing standards that limit a POTS signal are complicated, specialised, and do not fit comfortably with the approach used in DSL standards and the ANFP. POTS signals are also accompanied by sub audio signalling and power feed. However,

---

[1] Each of their deployment rules uses an electrical length defined at a different frequency. There is no guarantee that the different electrical lengths correlate, so all combinations would have to be considered. Note that BT uses this kind of measure of electrical length so that a rule can be defined which fits a particular technology's needs even though the network is made up of different gauges of wire with different frequency characteristics.

since these low frequency phenomena[2] do not give rise to significant crosstalk[3] the ANFP has provisionally ignored them: the ANFP masks go down to 100 Hz but the associated definitive lab measurement technique only goes down to 5 kHz. The ANFP could be revised to protect voice-band services if POTS were threatened.

ISDN and the HDSL are spectrally symmetric systems (power feeding excepted), so have the same mask for both ends (see Figure 4.7). Early drafts of the ANFP used masks from the international standards to which these systems are manufactured, but these were found to be overgenerous given the spectra actually produced by healthy systems; the later revisions have used 'figure hugging' masks produced by an international expert group [7].

ADSL has different spectra for the exchange end ('ATU-C') and customer end ('ATU-R'). The spectral masks provided in the international standards are simple polygons and look superficially like the masks in ISDN and HDSL standards. However, ADSL uses a modern modulation method that fills its masks, so the standard mask is used in the ANFP (see Figure 4.8). Note that ADSL is standardised with several 'user options', which have different masks. The masks used in the ANFP are

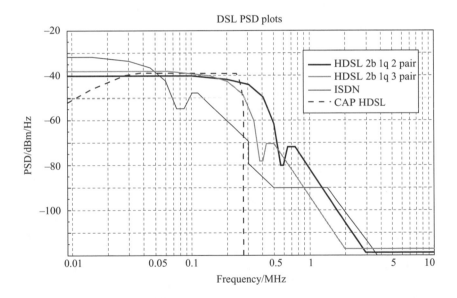

Figure 4.7   Masks for ISDN and all forms of HDSL

---

[2] The sharp edges of dial pulses and ringing cadence do have high frequency content which can induce errors in neighbouring DSL systems; modern DSL systems use error correcting codes to survive this 'impulse noise'.

[3] By design of the cables in the access network: which are so good that as a side effect they are still usable at a thousand times their design frequency!

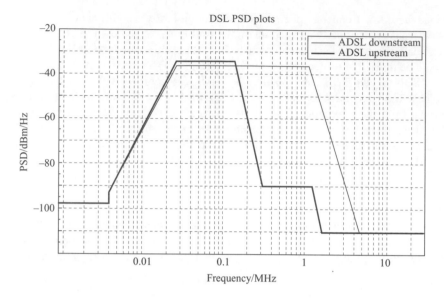

Figure 4.8   Masks for ADSL

ADSL as BT deploys it – so, for example, the standard allows the option of 'ADSL over ISDN' but use of this option is not permitted under the UK ANFP.

SDSL is a spectrally symmetric system, with user options to vary the data rate; the different data rates have different spectra. SDSL is defined in its (draft) international standards with a figure-hugging mask (see Figure 4.9) so that is used in the UK ANFP. Further, the ANFP is in general interested in a range of rates for SDSL, so the masks of interest in constructing the ANFP are the envelopes pertaining to ranges of rates.

SDSL is defined as having the same line power for any line rate[4], and a spectral mask that is the same (on log f/dB scales) except for shifts in frequency and power. SDSL uses a pulse amplitude modulation, and keeps the same number of levels at each bit rate, varying the rate is achieved just by varying the rate of pulses. So given the mask for an SDSL system at one rate, the mask at a different rate can be produced simply. For example, to double the rate the mask is moved up in frequency (by one octave) and down in PSD (by 3 dB). Any number of such masks will have a common tangent, and this has a slope of −10 dB/decade. And the envelope for a contiguous range of rates is the mask of the lowest rate, up to its point of contact with the common tangent, a section of the common tangent, and the mask of the highest rate above its point of contact with the common tangent.

---

[4] Actually for net user rate at or less than 2048 kbit/s, with an exception to this rule at the single rate of 2.3 Mbit/s. However, this rate produces a line signal that is not negligible in any of the zones, so is excluded from the ANFP; and excluded from this exposition for clarity.

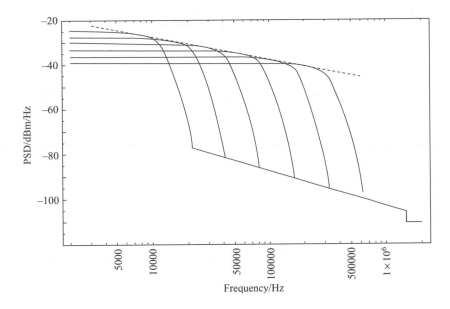

Figure 4.9   Masks for SDSL systems

Unlike the other masks, the mask for SDSL was not imposed on the ANFP as a given. Instead a interim version of the ANFP was produced without SDSL, and then various ranges of SDSL rate were considered as additions; the most liberal ranges that had minimal effect[5] on the trial masks were then included in a second iteration. The rate ranges chosen were:

For zone	Data rate on the line (kbit/s)	Symbol rate (kBaud)	Net user rate (kbit/s)
dn exch	64 . . . 2056	21.33 . . . 685.33	56 . . . 2048
up short	64 . . . 2056	21.33 . . . 685.33	56 . . . 2048
up medium	64 . . . 1505	21.33 . . . 501.66	56 . . . 1497
up long	64 . . . 784	21.33 . . . 261.33	56 . . . 776

4.4.7.6   A composite mask was produced for each zone

Given the above reasoning, this step is automatic. For illustration, consider just one zone's mask, say that for the 'up medium' zone (Figure 4.10).

Figure 4.10 shows the assumed PSD masks of each of the systems that make up this mask and also of the resulting ANFP mask that is their envelope.

---

[5] Done by curve fitting by eye.

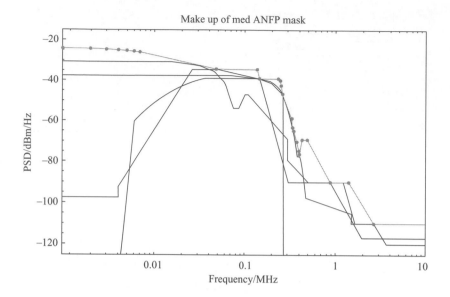

Figure 4.10   Composition of 'up medium' *mask*

4.4.8   Post construction check

The same economic argument that forces one to accept the existing deployment of systems should also, logically, make one protect the existing systems as disabling them is also too expensive. It happens that BT's existing deployment[6] remains viable under the UK ANFP, a fortunate inheritance from the conservative nature of BT's existing deployment rules.

However, this was not an automatic outcome: the local loop unbundling directive is all about getting more systems into use, and if it is successful then the network (and indeed networks all over Europe) will be becoming more heavily populated with DSL systems – so existing systems will get more live neighbours, and a louder noise environment. It is not automatic that an existing deployment can survive this noise increase. This reasoning is independent of how any particular spectrum management rules are written, so finding this problem while attempting to construct the rules does not necessarily invalidate the method.

Early in the development of the ANFP it appeared that BT's deployment of HDSL systems would be compromised: the longest lines of HDSL as BT had deployed it would not survive in a cable full of systems which fully populated the spectral mask as the HDSL standard defines it. This particular problem was found to be in the generosity of the masks in the standard – the masks were deliberately set high so that

---

[6] Those systems properly deployed under BT's deployment rules. Any system deployed outside these rules may well fail in the future.

real systems would pass that test unless a genuine fault occurred; real systems did not fully exploit the mask. The solution was to obtain masks that more closely follow what real systems do, the so called 'figure hugging' masks.

It may seem that basing the UK ANFP on BT's existing deployment is somehow biased in BT's favour. This is not so: the public UK ANFP is aimed at enabling local loop unbundling in the future, and the rules do not favour BT's future deployment over other operators. The inclusion of BT's existing systems was out of consideration of BT's past investment and the existing customers so served. Naturally BT's existing DSL deployment is based on older systems, and a reassignment of their spectra would allow more efficient use of the cables; the replacement of HDSL systems in particular was openly discussed in the cross-industry committee which advised Oftel, and eventually decided against because it would cost too much and take too long.

The astute reader will notice there may generally be a design decision in the step 'identify which systems are deployed in significant numbers', because there is a potential trade-off between the systems admitted and the deployments that will work. For example, in a country where the network has old style PCM systems[7] (perhaps G.703 carrying T1 or E1) there would be a very poor noise environment for more modern DSL systems; to develop an analogue of the UK ANFP for it, one possible way forward is to declare these systems obsolete, and plan to remove them. A common approach is 'grandfathering'; that is live with them for now, but these systems are excluded from the plan, are not to be deployed in the future, and are to be removed as and when they cause problems.

4.4.9   Measurement of PSD

The main ANFP development does not consider the practicalities of measurement, assuming that good engineering judgement can fill in the details later. The ANFP is supposed to be a 'normative' set of rules, however, where an unwelcome measurement result forces remedial action even on an unwilling system operator. So the details cannot be left to casual engineering judgement but have to be specified authoritatively.

The first practical detail is 'resolution bandwidth' (RBW), which effectively specifies how far apart spectral features must be before they are separately distinguishable. If one were considering the needs of a particular victim system one would choose a particular value for RBW to describe that victim's ability to distinguish spectral features: for example, standard ADSL has an intrinsic susceptibility RBW of about 4 kHz.

The UK ANFP chooses its primary RBW value as 10 kHz, mostly for consistency with other spectral management standards. This value is applied in band, where

---

[7] BT has none in its access network. At one time it had a lot, mostly deployed in the 'junction' network which links telephone exchanges. Those few systems deployed from exchange to customer always had their own separate cables, not considered a part of the access network. Nowadays such needs are met by optical fibre systems.

spectral power is permitted to be high and the limit is to control power generated intentionally.

Secondarily it chooses a much wider RBW of 1 MHz for out of band limits. This is because here the limits are to enforce a noise floor, and real systems do produce low levels of tone leakage – clock breakthrough and so forth. In order to permit real systems, a limited tolerance of these spurious leaks is permitted: the wider RBW will average a tone over a hundred times the bandwidth[8], so if it is the only leakage then the system will be tolerated even at a hundred times the power as measured with RBW = 10 kHz.

The second practical detail is where to measure. The ANFP is justified by defining measurement of the signals into the shared cables, so ideally would be measured at the actual interfaces of connection of the actual installation under test. However, only the more gross violations of the masks would be clearly detectable by field instruments that can observe a live system non-intrusively. Amongst the reasons for this are:

- field instruments are of lower quality than laboratory instruments;
- an non-intrusive observation is naturally noisier than one that draws significant power;
- the field environment is itself noisy – a measurement at the interface includes power flowing out of the line too, such as radio ingress, harmonics coupled from power lines, and the legitimate signals flowing from the other end of the line.

For this reason the normative UK ANFP measurement for the resolution of disputes is laboratory based.

4.5   Extending the ANFP to VDSL

The DSL task group that developed the ANFP is working on extending it for VDSL use. The extension is known as the ANFP-S to distinguish it from the existing ANFP.

VDSL is intended as a shorter reach, higher speed DSL using much more bandwidth than any preceding DSL. Existing standards define VDSL spectral usage to typically 12 MHz and even more bandwidth may be used for very short reach systems in the future.

The reach of VDSL is so short that from the exchange only a small proportion of loops are within its capability, for most networks, including that of the UK. To get the high bit-rate advantage of VDSL to the majority of potential customers VDSL must be remotely deployed, perhaps in addition to central deployment. The obvious places to remotely deploy VDSL systems in an access network are at flexibility points such as cabinets in the UK network, where the large cables from the local exchange are

[8] A consequence of defining limits in terms of PSD is that a pure tone appears more prominent as RBW is decreased – the same power divided by a smaller bandwidth. Arbitrarily reducing RBW would eventually make a pure tone exceed any PSD limit, no matter how low the tone power. However, PSD is eminently reasonable for data bearing signals.

flexibly interconnected with the smaller cables that distribute telecommunications services to small groups of customers. In order that this deployment approach is open to all potential operators the European Commission has mandated that independent operators shall have access to loops in these smaller cables from the flexibility point and such loops have become known as 'sub-loops' (the S in ANFP-S is for sub-loop).

Remote deployment of VDSL introduces a 5th signal injection point into the access network, a new downstream point at the remote node, in addition to the four points defined in the ANFP. The signal permitted to be injected at this new injection point must have minimal effect on the noise environment expected due to the ANFP. This must be so for exactly the same reasons that the ANFP was needed in the first place, that is to protect existing deployments and to provide a structured noise environment for the deployment of planned systems. In practice this means that the new downstream mask must constrain most of the transmitted power to frequencies above 1.1 MHz.

The upstream signal injection points for VDSL are of course the same points that already have masks under the ANFP, but for VDSL applications the applicable mask in the ANFP-S needs to extend to much higher frequencies.

The masks in the ANFP will define the spectrum usage in a high frequency environment that is almost wholly FEXT limited. As was identified in the opening section of this chapter, in such an environment the spectral assignment effectively defines the symmetry of the available communication capacity over the managed loops. This fact has made the allocation of that spectrum very contentious in the standards fora, the contention arising out of the uncertainty over what the wideband capacity of the loops could or should be used for. Fortunately most of that debate has been concluded with the result that two main optional spectral allocations have been embodied in standards, commonly known as plans 997 and 998 [8]. Plan 997 is a compromise symmetric/asymmetric plan while plan 998 favours asymmetric services while still supporting symmetric services to some degree (see Figure 4.11). Many operators around the world favour the 998 allocation because one of the potential applications for VDSL services is video delivery to the home, which is clearly a highly asymmetrical service.

It has been agreed in the DSL task group that the favoured ANFP-S will be based on the asymmetric plan 998 band configuration, although not exploiting frequencies below 1.1 MHz, and the current draft reflects that.

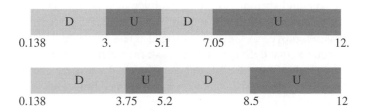

Figure 4.11   Plans 997 and 998

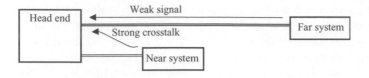

Figure 4.12   Near upstream FEXT versus far upstream signal

There is a crucial aspect of difference between the spectrum management regime of the ANFP and that needed for the ANFP-S. The ANFP only dealt with bandwidth that was either NEXT limited or that was FEXT limited but only in the downstream direction. The ANFP-S upstream is the first time FEXT limited upstream bandwidth has needed to be managed.

This gives rise to a new difficulty because to be fair to the majority the permitted transmissions need to be a function of the customer end distance from the serving remote node. This is so because the level of the FEXT interference suffered by upstream signals arriving at the remote head-end is a function not just of the level at which those signals were sent, but also of the path loss they have suffered since they were sent as shown in Figure 4.12.

So FEXT from a nearby upstream system could completely drown out the received signal from distant upstream systems unless special steps are taken to control it. These special steps are known as upstream Power Back-Off (PBO). The general requirement for such PBO is to restrict the upstream signals of nearby transmitters in some way in order to preserve some usable upstream signal to noise ratio and hence capacity for the distant systems. However, the difficulty in defining a 'fair' way of doing this and the compromises involved in achieving it have proved a rich source of debate in the relevant standards bodies for VDSL. So rich indeed that the ETSI VDSL standards, although agreed and published, remain incomplete and ambiguous in this area, while the T1E1.4 standards have been delayed in their publication not in small measure due to the same issues.

Some reflection will conclude that the objectives of upstream PBO are the same general objectives of spectrum management, that is the control of the common noise environment for the general benefit of all through the restriction of injected signal levels at certain locations in the network. So it follows that the requirements of PBO that have proved challenging to detail in VDSL standards will have to be mandated by any spectrum management standard that strives to make VDSL systems from multiple operators in an unbundled environment also deliver an equitable service.

To achieve this though will clearly require a more elaborate device than the simple fixed PSD masks employed by the ANFP. Instead at least some kind of parameterised PSD mask is required that defines a certain mask for a certain length of sub-loop. At the time the first draft of the ANFP-S was being drafted, however, the standards for VDSL upstream PBO remained incomplete. There seemed little point in the DSL task group attempting to define its own upstream power control for the ANFP-S when it seemed likely to be shortly obsoleted by developments in the VDSL standards. So the first draft of the ANFP-S has only trial status and will probably largely ignore upstream PBO and just define a single upstream PSD mask which will provide a ceiling on upstream transmissions. However, the draft does concede: 'Equipment implemented under this ANFP-S will not be exempt from UPBO requirements when they are specified' [9].

In conclusion the ANFP-S is in draft, is based on the asymmetric 998 band allocation, and will restrict transmissions by means of a new downstream mask that will apply at the remote head-end of the sub loop and a mask extension that will apply at the customer's NTP in addition to the applicable mask of the ANFP.

4.6   Another set of spectrum management choices for comparison

The choice of spectrum management regime depends on the way the network that is to be managed is constructed and on what systems have previously been deployed on that network. As these systems vary from region to region a variety of regimes is coming into existence around the world.

The variety of approaches is partly due to the differences in the underlying networks for which they have been designed. At the time of writing it is difficult to see how the approaches could be combined into a single unified approach across a large geographical region with disparate networks without losing a great deal of the potential of each of the component networks.

This section describes one other approach as a contrast to the UK approach, that developed for North America.

4.6.1   The T1.417 standard

4.6.1.1   Overview

During the last 15 years or so different DSL standards have each addressed the issue of spectral compatibility but it is only relatively recently that ANSI recognised the need for control across the whole DSL family. T1E1.4 began work on an ANSI spectrum management standard in 1997 in order to provide generic technical specifications for DSL systems so that they may co-exist harmoniously in the same multi-pair cable. This draft standard was sent out for Letter Ballot (T1LB 785) in June 1999. This was a complex standard and as usual many comments and 'No' votes were received.

T1E1.4 held extra interim meetings to expedite the comment resolution and the second Letter Ballot was issued in June 2000. The comments have now been satisfactorily resolved and the voluntary ANSI spectrum management standard T1.417 has been published. The standard does not set policy and is purely advisory. However, the FCC's Network Interoperability and Reliability Council (NRIC) may reference part of the standard in future rule making.

It is important to note that the standard does not require or prohibit binder group separation by type of transmission system. A binder group is usually referred to a group of 50 pairs which is normally separated from adjacent binder groups in a multi-pair cable.

The standard does address the unbundling of the local loop but it does not address the issue of line sharing (where one operator provides service in the baseband and the other uses higher frequencies to provide another service). It is anticipated that line sharing will be covered in a future issue of the standard. Other issues requiring attention in a further revision are:

- The effects of non-stationary or short-term stationary crosstalk (e.g. due to rate adaptive xDSL systems).
- The degree of conservatism in the Standard (worst-case versus typical case crosstalk).
- Mid span repeaters and their effects.
- The effects of NEXT on a predominately self-FEXT limited technology (e.g. ADSL).
- The need for new Spectrum Management Classes (see Section 4.6.1.4).
- The need for new Basis Systems (see Section 4.6.1.3).
- Remotely fed loops (e.g. DLCs).
- Power Back Off (e.g. for VDSL).

4.6.1.2   Objectives

The main objectives of the spectrum management standard were to ensure that:

- The addition of broadband service should not harm service to existing customers.
- The ability to provide service should be able to be predicted accurately.
- Once a new system is provisioned, it should remain reliable in the long term.

The standard was also designed bearing in mind the need:

- To enable competition and innovation of transport technologies, services and products.
- To be practical for use by equipment vendors and service providers thereby enabling efficient use of the local loop plant.

4.6.1.3   Basis systems

These systems are also referred to as 'Guarded' systems in the T1.417 standard. The purpose of defining this class was to provide a specified reference for determining spectral compatibility. Basis systems are defined as those that are expected to be used by the largest number of customers in the near term and preferably standards compliant. The current list of basis systems include:

- Voice grade services (e.g. POTs)
- Enhanced business services (e.g. P-phone)
- DDS
- ISDN basic rate access (160 kbit/s, 2B1Q)
- SDSL at 400, 1040 and 1552 kbit/s (older single pair 2B1Q technology)
- HDSL (2-pair at 784 kbit/s/pair for T1 services)
- ADSL (non over lapped spectral mode)
- HDSL2 (2nd generation HDSL using 1 pair for T1 services)

It is expected that this list will be revised as the T1.417 standard matures.

The definition of spectral compatibility warrants further attention. A transmission system is deemed spectrally compatible

a) with a target system if any number of the candidate systems may co-exist in the same multi-pair cable with the target system operating satisfactorily, and
b) with the set of basis systems if it is spectrally compatible with each member of the set.

Signals in both upstream and downstream directions must meet the above criteria.

The definition is somewhat open to interpretation and it remains to be seen whether 'satisfactory operation' requires legal and technical clarification.

4.6.1.4 Methods to demonstrate spectral compatibility

There are two methods currently defined: Method A and Method B. Method A requires that the new system complies with all the criteria defined for any of the defined spectrum management classes. The purpose of this is to allow a quick and easy method to show compatibility with all basis systems. The Classes are intended to be technology agnostic but in fact they are loosely based on the power spectral density of existing transmission technology. There are currently 9 Classes as shown in Table 4.1.

Method B is used for systems that do not fit into an existing spectrum management class. It is an analytical technique described in Annex A of the T1.417 standard. Annex A defines a method to calculate the effect of a candidate system on each of the basis systems in relation to the effect of existing spectrum management class template signals. To represent more typical signals the Power Spectral Density (PSD) template is about 3.5 dB less than the mask. In essence, a threshold is defined for the acceptable level of crosstalk into each of the basis systems.

Table 4.1   Class definition

Class	Example	Deployment guidelines (rules)
1	ISDN, SDSL (up to 272 kbit/s)	All non loaded loops
2	SDSL (up to 528 kbit/s)	11.5 kft*
3	HDSL, SDSL (up to 784 kbit/s)	9 kft
4	HDSL2	10.5 kft
5	ADSL (partial overlap spectrum)	All non loaded loops
6	VDSL (under study)	To be determined
7	SDSL (up to 1.5 Mbit/s)	6.5 kft
8	SDSL (up to 1.168 Mbit/s)	7.5 kft
9	ADSL (non overlapped spectrum)	13.5 kft

*Kilo Foot – a North American length measure. An unusual mixture of metric and imperial units. 26 AWG gauge wire is assumed. The closest metric standard gauge is 0.4 mm. Other equivalent working lengths (EWLs) may be calculated according to the standard T1.417.

4.7   References

1 'Specification of the Access Network Frequency Plan applicable to transmission systems used on the BT Access Network'. Oftel Technical Requirement OTR004:2000 Issue 1.1. Available at URL http://www.oftel.gov.uk/NICC/Public/anfp1_1.pdf

2 CHEN, W. Y. and WARING, D. L.: 'DMT ADSL Performance Simulation for CSA'. ANSI T1E1.4/93-166, August 1993

3 FSAN VDSL working group: 'A new analytical method for NEXT and FEXT noise calculation'. ANSI T1E1.4/98-189, June 1998. (May be downloaded from ftp://ftp.t1.org/pub/t1e1/e1.4/dir98/8e141890.pdf)

4 OFTEL: 'Access to Bandwidth: Delivering Competition for the Information Age' November 1999. (May be downloaded from http://www.oftel.gov.uk/competition/a2b1199.htm)

5 OFTEL: 'Access to Bandwidth: Determination on the Access Network Frequency Plan (ANFP) for BT's Metallic Access Network'. (May be downloaded from http://www.oftel.gov.uk/competition/anfp1000.htm)

6 *Journal of the IBTE*, **2**(1), Jan–March 2001

7 FSAN: 'Realistic ADSL noise models'. ETSI contribution TD37 to Amsterdam meeting of TM6, November 1999

8 FSAN: 'Proposed VDSL band plans'. ETSI contribution TD13 to Montreux meeting of TM6, February 2000

9 Specification of the Trial Access Network Frequency Plan applicable to transmission systems used on Sub-Loops in the BT Access Network. Available at URL http://www.oftel.gov.uk/ind_groups/nicc/public/specs/anfps_1.pdf

10 T1.417-2001 American National Standard for Telecommunications – Spectrum Management for Loop Transmission Systems

Chapter 5

A solution for broadband

A. Kerrison

5.1 Introduction

Broadband has been much talked about in terms of its impact on industry and the home. We know that the government believes broadband is important to the prosperity of the country through the push for a Broadband Britain. But it is far from a simple subject and there is much that has been said either to make a particular point or simply in error. There are many players in the story, and the range of products offered is very broad. This chapter provides a worked example of the introduction of DSL-based broadband services.

It starts by looking at the definition of broadband, correcting some of the myths that have been circulated, and discusses the range of technologies available to deliver broadband. It then focuses in on the DSL story so far with a look at the technology and a short history of the evolution of broadband. It looks at the players in the broadband business model before outlining the broadband products offered by BT and the design of the solution that supports those products. The chapter concludes by commenting on the early experience, drawing comparisons with others and considering some of the possible evolutions of the DSL broadband story.

5.2 A definition of broadband

Picking up on the definition of broadband in Chapter 1, it is worth noting that different people have defined broadband in different ways. Often as fast Internet, or always-on Internet. But whilst these definitions are covered by broadband, they only begin to tell the story.

At its simplest, broadband can be defined as a high capacity communications connection for a user. But a connection to what? True the Internet is one place to which the user may wish to be connected. But it is not the exclusive location and it would be wrong to constrain our thinking in this way. With a high capacity connection it also becomes possible to carry many additional services such as multiple voice channels, or Video on Demand. Such services would not be best served by the Internet and hence require dedicated networks to support such services.

Another part of the definition is that the connection is a bi-directional one. This means that there is ability for the user to interactively control what information is transmitted to that user. This would therefore include services like Video on Demand and Web browsing, yet exclude terrestrial analogue and digital TV as well as Teletext services. Two points to note about the bi-directional nature of the connection. First, the capacity of the two directions does not need to be the same. Second, the route of the two directions does not have to be the same.

So broadband is a high capacity bi-directional connection. This connection can be provided by a number of technologies:

1. DSL or Digital Subscriber Line. This is a modem technology that is used on the telephone line between the user and the local exchange. It differs from the technology used for 28 and 56 k modems in that data is extracted in the local exchange. This means that the technology is not constrained to the voice band-width by the switches in the narrowband telephony network and it can utilise the full spectral potential of the copper line. There are a number of variants within the DSL family that will be explained in the next section.

2. Cable modems. Again this is a modem technology. However, the cable modems use the cable TV access infrastructure rather than the telephony access infra-structure used by DSL systems. The ease of addition of cable modem capability on an existing cable TV system depends on the age of the cable TV system. The more modern systems deployed in the UK generally make this a relatively easy operation and this is one reason why this is quite popular within the UK.

3. Wireless or broadband fixed wireless access (BFWA). The European Telecom-munications Standards Institute (ETSI) are standardising three different access technologies covering both equipment and antennas to provide a 2 Mbit/s capac-ity: frequency division multiple access (FDMA); time division multiple access (TDMA); and code division multiple access (CDMA). In the UK a number of licences have been awarded by the DTI to operate in the 28 GHz band. The technology here is less well developed than cable or DSL solutions, though it does potentially offer a number of advantages: easier deployment of access infrastructure, reach beyond cable deployment, and reach beyond DSL technical limits.

4. Satellite. There are currently a number of satellite services that are planned. The main difference from the above technologies is that the channel from the user is currently limited to being via a different route, which may for example be by an ISDN connection. Satellite offers many of the advantages of wireless, though with a much larger geographic reach from the start. There is also a related technology known as High Altitude Platforms (HAPs), though it is not expected that this technology will be commercially available for a number of years.

5. UMTS (Universal Mobile Telecommunications Service). This is the third generation mobile technology or 3G. Although there has been much written about the sale in various countries of licences, these systems, which started to become available in 2003, do not initially offer the full potential of between 1 and 2 Mbit/s broadband services.

6. Optical fibre. The sheer capacity of fibre probably makes it the ultimate solution for broadband. However, the current demand for this sort of capacity is limited to a small number of business customers, and hence at the moment does not warrant the massive investment to make a fibre to the home solution generally available.

At present DSL and cable are the most dominant technologies. But how will the market develop? What new services will be accepted in the market place? What new applications will demand even greater bandwidth? And which technologies will be capable of evolving to meet the demand? The truth is that the development of broadband opens the door to many new services and capabilities that have yet to be developed. It is impossible to predict the future, but it is possible to look at the way the DSL technology has developed to date and the way it is capable of evolving to support some visions of the future.

5.3 Broadband trials

There are three trials that have significantly influenced BT's thinking on broadband and consequently the evolution of DSL broadband in UK. The trials that took place throughout the 1990s looked at technology effectiveness, service requirements, gathering marketing information, understanding operational issues, and gaining hands-on understanding of the commercial viability of operating a broadband network. Having operated the trials and gained the information BT launched broadband in 2000.

5.3.1 Bishop's Stortford Fibre Trials

The Bishop's Stortford Fibre Trials commenced in the early 1990s with the aim of assessing two alternative approaches to using fibre as the broadband access technology [1]. The two approaches were BIDS (broadband integrated distributed star) and TPON (telephony over a passive optical network). The BIDS system was an active system using a single fibre from the exchange to an active cabinet. From the cabinet each customer was served by a single fibre delivering a choice of 16 TV channels, video on demand, home banking, home shopping, and hi-fi audio. The cabinet provided much of the intelligence of the system, receiving the TV channels from the exchange and providing the required switching based on customer signalling.

The TPON system was a passive system that did not have street electronics. Instead, there was a single fibre from the exchange out into the access network where it was optically split, typically at two locations, using passive optical couplers. The electronics being only at the exchange and customer premises. The advantage of this approach is that it was more easily upgradable to accommodate technological advances without the need to interfere with the access infrastructure – the passive optical network. The passive optical network used one wavelength and carried the narrowband services using the TPON technology at each end. Broadcast TV services were then added by the BPON technology (broadcast over a passive optical network), which operated on another wavelength and would be added through the introduction of BPON equipment at each end of the passive optical network. There were three flavours of TPON trialed; home TPON providing standard POTS, business TPON providing POTS and business products including private services, and street TPON which terminated the PON in the street and provided POTS to the residential users using copper pairs for the final part of the access network. Although street TPON deployed street electronics it was investigated as a way of rolling out a passive optical network part way into the access network at a lower cost yet with the ability to enhance later as the demand for broadband services develops.

The trial achieved a great deal in terms of understanding the business, operational and technical challenges in moving to an optical access network to support broadband, as well as the advantages and disadvantages of the two approaches – particularly the implications of street electronics. It also brought home some of the difficulties in trying to replicate existing services over totally new technology. This experience is still relevant today and is still influencing strategic decisions on the evolution of BT's broadband capabilities. However, it was recognised that the economics of wide-scale deployment of fibre in the access network could not at the time be justified. What was required was a technology that did not demand such huge investments in the access network to support broadband services.

5.3.2   Colchester Interactive TV Trials

At about the same time as the Bishop's Stortford Fibre Trials a feasibility study into the possible use of DSL technology was initiated. Here was a technology that did not require a massive uplift of the existing access network. It simply used the access network supporting telephony to also support broadband through the addition of equipment at each end (see Section 5.5). If the network could be used, then it provided a step that could be taken to meet megabit broadband data rates prior to a market demand for even greater rates and requiring a push towards fibre. The study concluded that at the time a symmetrical approach resulted in an unacceptable level of crosstalk. However, an asymmetric approach, where the downstream data rate is greater than the upstream data rate, reduced this problem to an acceptable level. Much work followed in terms of developing and standardising the ADSL approach until in the summer of 1993 the first ADSL systems became commercially available.

In 1995 ADSL had its first serious test with the Interactive TV trials based at Colchester and Ipswich [2]. Here customers in 2000 homes and 8 schools took part in using a similar set of services to those in Bishop's Stortford including Video on Demand, home banking, and home shopping. When the trials completed in 1996, it was the largest such service in the world. Not only was it a success in terms of size, but it proved the technical feasibility of ADSL to deliver TV based applications and provided a great source of information on the potential market place for these types of service.

5.3.3   Broadband commercial trials

The interactive TV trial proved the technology and the TV-centric applications, however, there was also a rapidly growing interest in the evolution of the Internet and the need to provide PC-centric applications to the home for both work and leisure pursuits. This led to a small technical trial based as Kesgrave near Ipswich before the full commercial trials in 1998/99.

The commercial trial differed from all previous trials in that for the first time the roles of the various players became more clearly established. BT provided the broadband network with the ADSL access from the users and service providers provided TV and PC-based applications to their customers. The commercial trial established 2000 customers in North and West London and proved the success of the PC-based applications at meeting the needs of businesses and residential users. Business users could see the usefulness of the broadband capacity in increasing their efficiency and productivity. The residential market began to take advantage of the always-on nature of these applications which led to a change in their usage patterns and in particular an increase in usage.

With the key issues now much better understood, it was time to focus on the development of effective processes and supporting operational support systems (OSS) to meet the needs of a commercial launch. The developments here together with the incorporation of feedback from the commercial trial took BT into a pre-launch phase that was run with around 30 service providers. This gave the service providers the opportunity to assess the products prior to commercial launch.

On the 29th June 2000 BT launched its ADSL broadband and started to take commercial orders from service providers. This is the foundation from which the broadband products will evolve. Changes will be introduced to reflect moves in the market place; new technologies will be trialed and may subsequently lead to new products. One thing that is certain is that BT's DSL broadband story has only just started.

5.4   Who are the players?

The business model is quite interesting in that it defines a number of roles (see Figure 5.1). The main ones being service providers and network providers. Network providers operate the network and service providers provide service to their customers. There is also a range of equipment suppliers supplying equipment to network providers, service providers and CPE (customer premises equipment) to the customers. The content provider owns the content and strikes an agreement with the service provider that enables the service provider's customers access to the content provider's content. Finally there is Oftel, who ensure that the whole process operates fairly and for the overall benefit of the consumer.

It would be quite simple, apart from the fact that network providers offer a portfolio of products. The reasons for this are twofold: first, by offering a portfolio of products, the potential market place is increased, and second it needs to meet the Oftel requirement for competition. In essence, Oftel's stance is that for consumers to have

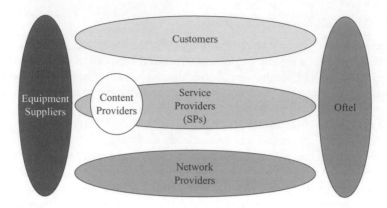

Figure 5.1   Principal broadband players

adequate choice, quality and value for money, there is a need to offer connection at a number of levels. More details of the products offered by BT to meet this requirement are covered in the subsequent section. The implication of this position is that the UK probably has the most open broadband market in the world.

As a network provider, BT offers wholesale products that are aimed either at service providers or at other network providers. Other network providers who see a potential market place may wish to buy some, all, or none of its network capability from BT. They would also be able to buy network capacity from another source. This means that service providers can then buy from a portfolio of products, not just from BT, but from all network providers. It is also worth pointing out that is quite possible for a single company to operate both as a network operator and a service provider. Hence they may chose to buy one product from BT, connect into their own network and provide service to their own customers.

There are around 200 service providers registered in the UK each trying to focus on a particular bit of the market. Some are focused on provision of video-based services while others focus on provision of PC based services. Some target the business market while others target the residential market. BT has its own service provider organisation that can retail broadband products in competition with other service providers. Hence although BT does have a route from network provider to the customer, it is also in competition with other network and service providers. It should, however, be noted that a competitor to BT as a network or service provider might also be a BT customer through the purchase of other network products.

5.5   The DSL technology

The DSL technology works by exploiting the existing copper access network between the customer's telephone socket and the local exchange (see Figure 5.2). Because this infrastructure already exists to support telephony, it only requires the addition of

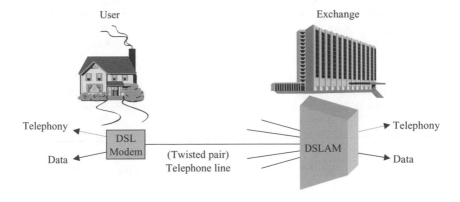

Figure 5.2   DSL access

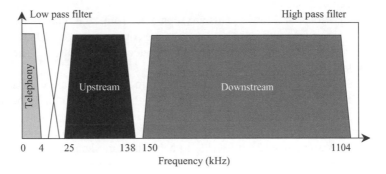

Figure 5.3   ADSL spectral allocation

equipment at both ends of this link to upgrade the telephone line into a high-speed data link.

5.5.1   ADSL

There are a number of flavours of DSL available but the principle flavour now being deployed is ADSL (asymmetric digital subscriber line). It is called asymmetric because of the difference in capacity of the upstream and downstream channels – the upstream channel being of lower capacity than the downstream. The ADSL technology is well described in other articles [3,4].

The ADSL technology that enables this upgrade is similar to the modem technology used on 28 and 56 k dial-up modems. There is, however, one major difference in that the coding is not constrained to fit within the normal voice bandwidth. Indeed, it starts above the voice spectrum (4 kHz) and continues up to 1 MHz. The way in which the bandwidth is allocated between the voice channel, and the upstream and downstream channels is shown in Figure 5.3. The technology and spectrum allocated is described in more detail in Chapter 4.

The fact that the ADSL coding is not constrained to fit within the normal voice channel means that the data channel can be used at the same time as a telephony channel, i.e. simultaneous voice and data communication. However, because the encoding is not restricted to fit within the voice bandwidth, the voice network cannot transport the signal. Hence, whereas for voice band modems, the encoded signal is transported over the voice network to another modem to decode the signal, for ADSL the signal needs to be extracted in the local exchange before it reaches the voice network.

The device that performs this extraction in the local exchange is the DSLAM (DSL Access Multiplexer). The DSLAM performs this separation and returns the voice traffic to the voice network with the ADSL encoding removed. The DSLAM also performs the task of a number of modems coding and decoding the data from the ADSL signals, and aggregates the individual customer data channels onto a single connection suitable for carriage over a data network. More detail of this will be covered in Section 5.8.

The way the voice and data signals are separated from each other is via the use of filters. The filters are required at both ends of the copper pair to avoid interference from the voice signals on the data equipment and vice versa. Figure 5.3 shows how the filters select the correct channel.

The ADSL equipment is capable of carrying quite significant bandwidths, e.g. up to 8 Mbit/s in the downstream direction. However, the limiting factor is the quality and length of the copper pairs. Hence, people living near to the local exchange would be able to achieve bit rates in the order of 8 Mbits/s whereas someone living many kilometres away would not be able to achieve more than say 0.5 Mbit/s (see Figure 5.4).

It is then up to the network provider to determine how to work within this speed versus distance trade-off in order to offer products. Offer a high rate service and only a very limited number of people will be able to take it. At the other extreme if the speed was reduced until everyone could have it, the speed that could be offered would

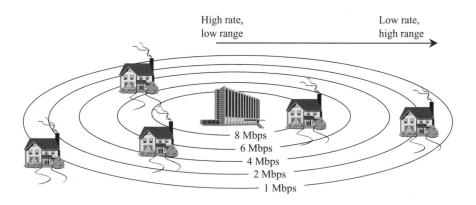

Figure 5.4   ADSL performance trade-off

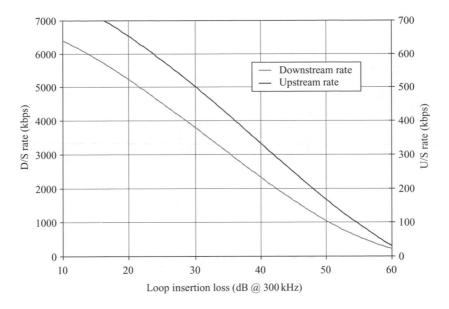

Figure 5.5   ADSL penetration

be so low as to be pointless. An example graph that shows how the data rates tail off as the insertion loss increases is shown in Figure 5.5. Anyone deploying ADSL needs to understand the trade off between speed and potential market place that could be achieved for a particular network.

Hence, although the technology is capable of downstream speeds up to 8 Mbit/s or more, to achieve a reasonable percentage of the potential customer base, the products offered by BT have been based around downstream speeds of between 0.5 and 2 Mbit/s (see Figure 5.6).

5.5.2   Other DSL variants

As was noted earlier, ADSL is but one of the flavours of DSL. The other principal flavours are SDSL (symmetric digital subscriber line), VDSL (very high speed digital subscriber line) and HDSL (high-speed digital subscriber line).

SDSL differs from its ADSL cousin in two ways. First, as the name suggests, it provides symmetrical upstream and downstream data channels. Second, the encoding of the data channel utilises the voice part of the spectrum. Hence services offered over SDSL do not support simultaneous POTS (though it is possible to use part of the data capacity to support a derived voice capability). At present it is expected that SDSL-based services will be targeted towards the business rather than the residential community and that service speeds will be up to 2 Mbit/s in each direction to enable a significant penetration of the customer base.

VDSL operates in the spectrum above narrowband services; hence it is compatible with both POTS and ISDN. It also has options of working either in symmetrical or

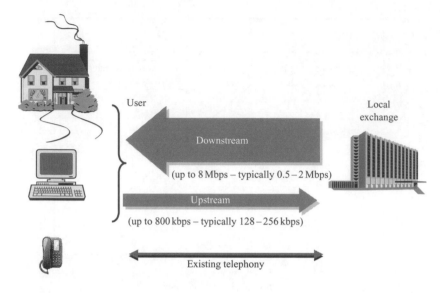

User

Local
exchange

Downstream

(up to 8 Mbps – typically 0.5 – 2 Mbps)

Upstream

(up to 800 kbps – typically 128 – 256 kbps)

Existing telephony

Figure 5.6   ADSL data rates

asymmetrical modes and is capable of achieving speeds of between 6 and 25 Mbit/s. However, as with ADSL (see Figures 5.4 and 5.5), there is a trade off between speed and distance. Hence although great speed is possible, it is achievable over distances much more constrained than for ADSL. The only way of achieving a reasonable penetration therefore requires the deployment of equipment in the access network. This will require a significant investment, and hence will probably be deployed only once there is a high confidence in the demand for such capacity.

There is also HDSL which is an older technology that is generally used to provide 2 MHz private circuit tails. This technology differs in a number of ways from the others. First, the service is generally fixed at 2 Mbit/s in each direction, and over longer distances multiple pairs are used to achieve the 2 Mbit/s capacity. Second, there is generally not a DSLAM to aggregate multiple services into a single pipe. For HDSL there tends to be a matching HDSL modem in the local exchange which decodes the users data channel. These are provided on an individual basis.

5.6   BT broadband services

There are three main wholesale broadband product families that are supported by BT: IPStream, VideoStream, and DataStream. There is also the local loop unbundled product that facilitates other network operators with access to BT's local lines enabling them to provide their own network in competition with the three main products. Other interconnect products are also being discussed, though it is currently unclear if there is sufficient market pull for more than those products already mentioned.

In addition to the wholesale products, service providers, including BT offer retail products through their retail channels. This chapter does not cover the detail of these retail products, but concentrates on the underlying wholesale products that support a huge number of service providers.

5.6.1   IPStream

IPStream is a wholesale product that offers service providers the ability to connect to their customers over an IP network. The product offers the service provider the ability to connect many users through BT's IP network and provide an aggregated delivery to the service provider's premises. The product comes in two parts: the end user access and the aggregate access (see Figure 5.7).

The end user access provides the links from the users into BT's network, and channels the traffic over an ATM network to a remote access server (RAS). This device is IP aware and has the ability to recognise the service provider for which the user is a customer. Traffic is then routed by the RAS to the correct service provider over BT's IP network. The aggregate access part of the product defines the capacity of the link to the service provider.

There is a range of speeds available. For business users downstream rates are 0.5, 1, and 2 Mbit/s. The upstream rate is set to 256 kbit/s. For residential users, the downstream rate is 0.5 Mbit/s and the upstream rate is adaptive between 64 and 256 kbit/s. The actual upstream speed obtained by residential users depends upon the quality of the line and distance between the user and the local exchange. The aggregate access speeds available go from 0.5 to 155 Mbit/s.

All services offered are contended, which means that the network is configured to less than the sum of the individual user access rates. This is perfectly acceptable for fast, but 'bursty' type IP traffic (variable bit rate), but would not be suitable

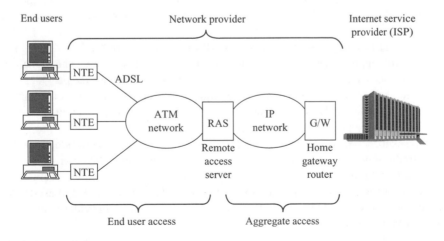

Figure 5.7   IPStream

for applications where data is constant for prolonged periods, e.g. video – though downloading of IP video clips would be considered part of the normal 'bursty' IP traffic.

The business user interface is 10baseT Ethernet, which enables connection to the user's PC LAN. For residential, the interface is USB, which enables a direct connection into most modern PCs. Self-install options are now offered that enable service providers to choose the ADSL equipment that best meets their requirements. The service provider then sends the equipment to the users who will be able to connect this device to their telephone line without the need for an engineering visit. Alternatively, the end user can purchase a service provider package from high street retailers. Already broadband modems are being pre-installed into PCs as ADSL PCI modems.

The business users have a range of addressing options available to suit their particular needs in terms of number of PCs, LANs, and IP services they wish to use. For residential users, the options are focused on the needs of the mass market.

Although IPStream is an IP product, it does not provide a connection to the Internet. The product provides a connection between the users and their service provider. It is then an option for each competing service provider to decide what content and facilities they will offer their users. However, typical service providers might provide e-mail, news groups, and Internet access as well as information, education and entertainment services.

5.6.2   VideoStream

VideoStream covers a portfolio of wholesale products that offers service providers the ability to connect to their customers over an ATM network. The product offers the service provider the ability to connect many users through BT's ATM network and provide an aggregated delivery from the service provider's premises. The VideoStream product is specifically designed and configured to carry real time video from the service provider to the user's TV via a set top box provided by the service provider (see Figure 5.8).

The user has an ADSL connection to BT's network where traffic is aggregated towards the appropriate service provider using the ATM network. The signalling from the user identifies their 'on demand' video request. The service provider then transmits the required content back to the requesting user. It should be noted that the signalling channel is always available, but that the transmission of content to the user is only established over the network 'on demand'.

The user connects to a set top box (STB) provided by the service provider that interfaces to the VideoStream product through an ATM25 interface. The downstream rate is around 2.5 Mbit/s and the upstream rate is set to 256 kbit/s. The downstream rate is specifically designed to enable the flexible real-time delivery of MPEG1 based video content and services. The aggregate connection to the service provider is available over STM-1, 4 or 16 interfaces.

Unlike the IPStream product, VideoStream is designed to support transport of constant bit rate (CBR) services such as a constant video transmission. Where the

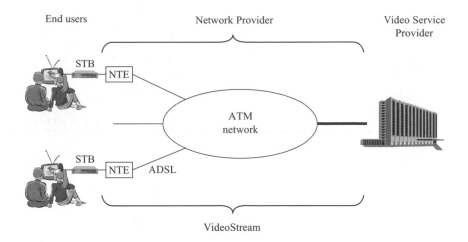

End users   Network Provider   Video Service Provider

Figure 5.8   VideoStream

user is watching a film, for example, this CBR capacity is required for the duration of the film. This means that the use of contention that was used to optimise capacity requirements in the network cannot be used for this type of service. Instead, a technique of overbooking is used. This means that the total number of users exceeds the number of transmission channels from the service provider over the aggregate connection. As not all users will be using the service at the same time, this capacity management is not a problem and use of the transmission channels is managed by the service provider through requests received from the users on the signalling channel.

As for IPStream, it is up to each competing service provider to decide what content, and facilities they will offer their users. However, typical service providers using VideoStream would have the option to provide movies, music videos, time shifted TV, non-broadcast TV programmes on demand, and video games. The interactive nature of the connection also means that familiar video features including stop, fast forward, pause, etc. are also available as part of the service provider's service definition. Finally, e-commerce services such as home shopping can develop to really take advantage of the broadband nature of the connection through the use of video clips and personalised product information.

5.6.3   DataStream

DataStream is a wholesale product that offers service providers the ability to connect to their customers to the service providers network. The product offers the service provider the ability to connect many users through BT's ATM network and provide an aggregated delivery to the service provider's IP network. The DataStream product is specifically designed and configured to permit the carriage of IPStream type services,

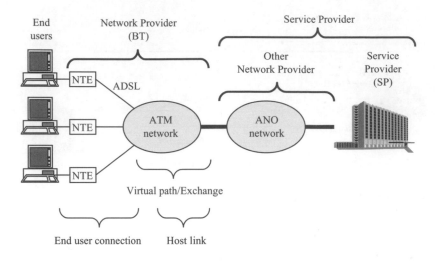

Figure 5.9   DataStream

yet without the compulsion to use BT's IP network. The product comes in three parts, an end user connection, a host link, and a virtual path to each exchange were service is required (see Figure 5.9).

As with the VideoStream product, the users link into BT's network using the ADSL technology and traffic is aggregated towards the appropriate service provider using the ATM network and delivered over a host link connection. The traffic is transported over BT's network through a virtual path that routes the traffic from the DSLAMs in the exchange to the service provider over the host link.

There is a range of speeds available that map to those available on IPStream. For business users downstream rates are 0.5, 1, and 2 Mbit/s. The upstream rate is set to 256 kbit/s. For residential users, the downstream rate is 0.5 Mbit/s and the upstream rate is adaptive between 64 and 256 kbit/s. The actual upstream speed obtained by residential users depends upon the quality of the line and distance between the user and the local exchange. The host links are either STM-1 or 4 interfaces.

As with the IPStream product, the user traffic is contended into the virtual paths from the exchanges. This is again because the product it intended to support service provider services with 'bursty' type IP traffic. The range of DataStream user interfaces offered is also the same as for IPStream.

The commercial model means that even though DataStream is specifically designed to support service providers who wish to offer IPStream type services, it is also possible that a network provider could use the DataStream product, add their own IP network and sell the consequential managed IP product as a wholesale offering to other service providers. Similarly, there is not compulsion to provide an IP network, so it is also possible that service providers could provide alternative services based on ATM.

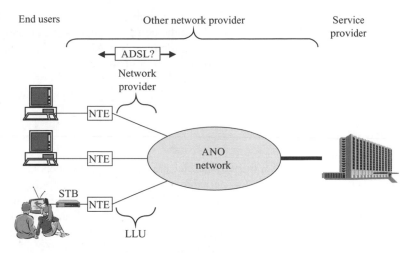

Figure 5.10   Local loop unbundling

5.6.4   Local loop unbundling

As a result of Oftel's desire to promote competition at all levels of interconnection, in August 2000 Oftel mandated full unbundling of BT's local loop. Local loop
unbundling (LLU) enables other operators to rent space in BT's exchanges, lease local
access lines, and upgrade them with DSL technology to provide a range of higher
bandwidth services to the end customer (see Figure 5.10).

With LLU the competing network operator simply uses BT's access line and
can operate both POTS and broadband services for the customer. LLU requires the
operator to provide all equipment in the exchange and indeed the rest of the network
in order to support the services offered to the users; though obviously it is possible
for these operators to buy part of the capacity they need for their network from BT
or other network operator.

There is little restriction in the equipment that can be used other than the requirement for the transmissions from the equipment to comply with the Access Network
Frequency Plan (ANFP). More detail of the ANFP is covered in Chapter 4. The reason
for this requirement is that it prevents unwanted interference that could affect equipment using other copper pairs in the same cable, or indeed interference that could
affect any other equipment in the vicinity. The other purpose of the ANFP is that by
identifying acceptable limits for transmission, it also identifies levels of emission that
must be tolerated by the equipment.

With such a broad range of options, it is impossible to describe how an operator
might use local lines to support broadband services, apart from the fact that they could
simply replicate one or more of the above broadband products.

Related to LLU, is the requirement for shared access. In October 2000, Oftel
published a consultation document entitled 'Access to Bandwidth: shared access in
the local loop'. This document discussed arrangements under which just the higher

bandwidth of local access lines could be made available to a competitor. It differs from the LLU product already available in that the competing operator can lease the spectrum on the copper pair so to provide equipment such as ADSL, without leasing the POTS part of the spectrum. Hence, a competitor could offer ADSL based services whilst BT offer POTS over the same copper pair.

5.7   Network design

In previous sections the ADSL technology and the broadband products have been described. This section looks at the network design that supports the broadband services and describes the way in which the ADSL technology is implemented over the access connections to the users. It should be noted that the full end-to-end design covers the operational aspects of running a network as well as the network aspects. The operational design is covered in the next section.

The key components of the network design are shown in Figure 5.11. They consist of an ADSL user access network, a nationwide ATM network, and an IP network. The different broadband products use these networks in differing ways.

The general principles are quite straightforward. All users are connected to the ADSL network, which is terminated in the local exchange by the DSLAM connected to the ATM network. For ATM services, such as VideoStream and DataStream, traffic is simply carried across this network to the interface with the service provider.

For VideoStream, the purpose of the ATM network is to aggregate the traffic from the DSLAMs within the VideoStream footprints and present the aggregated traffic to a Point of Presence within the footprint. For DataStream, there is a similar purpose of aggregation, but the delivery of the traffic to the service provider's is not restricted to footprints.

For IPStream, the traffic is aggregated over the ATM network to a RAS that routes the IP packets across the IP network towards the service provider. Where the distance

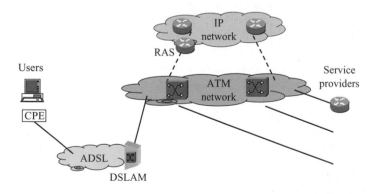

Figure 5.11   Broadband network design

from a convenient router to the service provider is too far, then the ATM network will be used to carry the traffic to the service provider interface.

5.7.1   User domain

For the user, there are two distinct installation options, BT-installed and self-installed (see Figure 5.12). The two options not only identify who does the installation, but also the actual configuration in the user's premises. For BT-Installed, the master socket will (if necessary) be updated to the modern NTE5 that has a replaceable bottom half of the front plate. For broadband users this replaceable plate is replaced with a front plate that includes a splitter. The purpose of the splitter it to split the high 'data' frequencies from the lower 'POTS' frequencies. The user's existing POTS wiring is then connected to the POTS output from the splitter and there is no more modification to the user's POTS connections. The equipment that is attached to the data output from the splitter depends upon the product. However, there must be an ADSL modem connected with one of a number of interface options. In addition to the modem, there may also be the additional functionality of either a router or a set top box, though increasingly functionality is being incorporated into single pieces of equipment, e.g. a modem/router.

For self-installed, the master socket does not need be updated to the modern NTE5, and instead of using a splitter to separate the frequencies this separation is performed by a combination of filters. There is no need to modify the user's existing wiring because low pass filters are simply inserted for each telephone. This requires no more than unplugging each phone, plugging the filter into the master or extension socket, then plugging the telephone into the filter. The data equipment then simply plugs into a

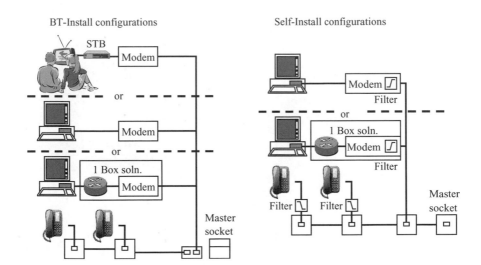

Figure 5.12   User domain

convenient master or extension socket. This equipment must include the functionality of a modem together with a high pass filter. Other equipment or functionality may also be required to provide the user with the connection and service required.

With both configurations, data between the user to the exchange is encoded and transmitted by the ADSL modem utilising the bandwidth above that required for the POTS service. It should be noted that the structure for carrying traffic on ADSL lines is ATM formatting.

5.7.2   Exchange domain

In the exchange the copper pairs from the users are terminated on a main distribution frame (MDF) (see Figure 5.13). This is a flexibility point for copper pairs. From there, the pairs go to the filters that are located with the DSLAM. The filter performs the same function as filters in the user's premises in that is separates the data frequencies from the POTS frequencies. POTS traffic is then returned to the MDF where it is jumpered to a line going into the narrowband remote concentrator unit (RCU) then into the rest of the telephone network.

The data traffic extracted by the splitter is decoded by a modem in the DSLAM, before being aggregated onto the link to the ATM network (each line having its own modem). The connection to the ATM network is an STM-1 connection (155 Mbit/s SDH connection). The traffic is carried within the STM-1 connection using the ATM cell structure that specifies virtual paths (VPs), virtual circuits (VCs) and classes of service, as is traffic from the users. This is achieved as shown in Figures 5.14 and 5.15 for the services.

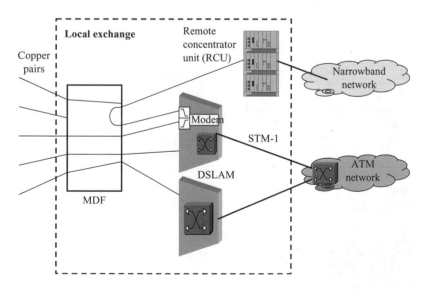

Figure 5.13   Exchange domain

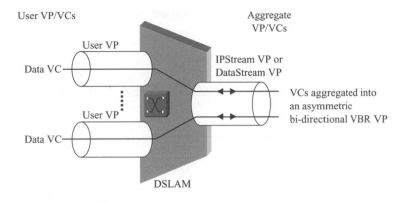

Figure 5.14   IPStream and DataStream VP/VC architecture

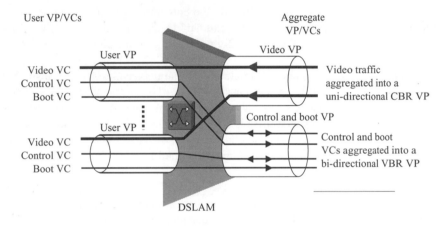

Figure 5.15   VideoStream VP/VC architecture

There will be a number of virtual paths (VPs) established from the DSLAM across the ATM network on the aggregate connection. The actual number depends upon the number and type of services supported on that DSLAM. Each service supported has at least one VP, and the ability to add additional services or additional VPs for an existing service depends upon the remaining capacity within the STM-1 aggregate connection. In all cases, the user has a single VP to and from the DSLAM through which all traffic is carried.

On the aggregate connection there will be at least one IPStream VP established to carry IPStream traffic to the RAS. Depending upon the number of DataStream service providers offering service from that DSLAM, there may be many VPs or none at all. Each DataStream VP would be connected to their service provider over the ATM network. As with DataStream, the number of VideoStream VP established depends upon the number of service providers offering service from that DSLAM.

The only difference is that the VideoStream product requires a pair of VPs to support the service.

With the two sides of the DSLAM defined, it is therefore necessary for the DSLAM to cross connect traffic from the user's VCs within their VP to appropriate VCs within the correct VP on the aggregate connection.

For IPStream and DataStream the user has a single bi-directional VC within the VP. The cross connect is established at the time service is provided by selecting the next available VC within the appropriate aggregate connection VP for the service requested. All traffic for IPStream and DataStream is classed as variable bit rate (VBR).

For VideoStream, the user has a number of VCs for different types of traffic. One downstream for the video, one bi-directional for the passing of control signals, and one upstream 'boot' channel. The control and boot traffic is classed as VBR, but the video is classed as constant bit rate (CBR).

The size of the VP on the aggregate connection is determined by the number of users desired to connect within the VP and the contention ration. The principle of contention ratios applies to both the IPStream and DataStream products. For a 20:1 contention ration the size of the VP is 1/20th of the sum of the end user rates for the number of desired users.

For VideoStream the number of simultaneous channels required to be transmitted from the service provider determines the required size of the VP to the service provider. Where the number of channels capable of simultaneous transmission is less then the total number of users connected there is a need to manage this overbooking. In this situation all cross connections are established, but they need to prevent transmitting on more than the defined maximum. Through signalling with the user the service provider manages this overbooked situation.

The ATM network is used to transport the user traffic to an appropriate exit point. For ATM-based services of VideoStream and DataStream, the data is carried in the VPs as described above to the service provider interface. For IPStream the VP is terminated at a convenient remote access server (RAS).

5.7.3 IP domain

The principles of the IP domain are that an IP router in the user's domain (see Figure 5.16) communicates over the ADSL and ATM networks to the RAS. The RAS then tunnels the traffic over the IP network to a home gateway router on the service provider's premises. The home gateway router then presents an Ethernet interface to which the service provider interconnects. For higher speed aggregate access links, the home gateway router is on BT premises in a home gateway cluster, and the interface to service provider is provided by an NTE router on the service provider's premises.

The session from the user to the service provider is established when the user goes on-line by establishing a PPP session. The PPP packets will first be received at the RAS, which will check with the BT RADIUS that it is valid to forward this and to which service provider this should be sent. Once this is determined, the PPP packets

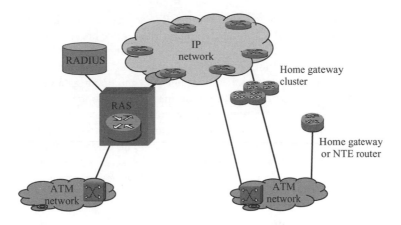

Figure 5.16   IP domain

are tunnelled to the service provider. The service provider needs to provide a RADIUS server such that they can perform additional security check on the user's request such as authentication. Depending upon the option selected by the service provider either the service provider's RADIUS or the home gateway will be responsible for allocating the IP address to the user for that session. Once the request has been validated and the IP address has been conveyed to the user, the PPP session is established with the home gateway router.

There is a limit on the number of simultaneous sessions that can be supported over the aggregate interface to the service provider. This limit ranges from 250 for smaller configurations up to a maximum of 8000.

The network has the ability to support either network address translation (NAT) or no-NAT options. NAT means that as packets pass through the router on the user's premises the address used is translated between that allocated for the session to the home gateway and that used for the PC. No-NAT means that router passes the address without translation.

All IP addresses are allocated for the session with the home gateway using the dynamic address allocation process. However, some service providers may support the ability to provide consistently served IP addresses. This option ensures the 'dynamically allocated' IP address is always the same. The option of consistently served IP addresses is required in combination with no-NAT in order to support some applications including H.323 Voice over IP (VoIP) applications.

There are a number of RASs located around the country, each serving a set of DSLAMs that in turn supports a set of users connected to the DSLAMs. The RAS that serves an exchange tends to be the one that is physically closest, though this does not have to be the case. As there is no control on the service providers that can be connected to and from any user, it is necessary that each RAS has the ability to establish tunnels to each service provider. These tunnels are established dynamically as a result of the first user request from that RAS to that service provider. Hence, when a user request is received and no tunnel exists to the appropriate service provider the PPP packets are held at the RAS until the tunnel to the service provider is established. The tunnel between the RAS and the home gateway router uses the L2TP protocol (layer 2 tunnelling protocol), though the service provider does not normally see this.

The normal termination of the L2TP tunnel is at the home gateway. However, there is also an option known as L2TP passthrough. This is an option where the L2TP tunnel does not terminate on the home gateway router, but is carried further into the service provider's network. This permits the service provider to control the user PPP sessions rather than having them controlled by the home gateway router. With this option there is no home gateway router provided on the service provider's premises. Instead the interface presented is a 155 Mbit/s STM-1 optical interface.

5.8   Operational design

Previous sections have looked at the network elements of broadband. This section looks at the way the network is managed on a day to day basis. The operational design is split into a number of areas reflecting the key business process requirements of broadband: plan and build, service fulfilment, service assurance, billing, etc. These areas are not, however, independent and there are considerable interconnections between the process areas, e.g. the successful addition of a new user needs to be reflected in the billing process such that the service provider can be billed for the user. There are also a number of common themes including the need to interface to the service provider, the need to interface to the network elements, and the need to maintain accurate information on the status of builds, orders, faults, etc.

This section introduces the main process areas and gives an insight to some of the functions covered. It also introduces the supporting OSS architecture (operational support systems).

5.8.1   Plan and build

This area is all about providing an exchange with the necessary equipment and capacity to meet the demand. It has to take into account the supplier lead times in order that the equipment is installed and commissioned on time.

For broadband, provisioning a new exchange requires the planning and introduction of: accommodation in the exchange; the DSLAM; the line cards; the STM-1 connection to the ATM network; and VPs across the ATM network for the services supported from the exchange.

Part of planning and building includes the capacity management so that an exchange does not run out of capacity. Hence there is the need to monitor the capacity and plan the addition of new capacity before the existing capacity is exhausted. The items that require being capacity managed include:

- the DSLAM line card capacity and number of available line card ports;
- the STM-1 transmission capacity to and from the DSLAM;
- the capacity of each VP for the services supported per DSLAM;

- the ATM and IP network capacity;
- the capacity of the RAS;
- the aggregate connection to the service provider.

5.8.2   Service fulfilment

The service fulfilment area covers the processing and fulfilment of the various types of service orders received from the service providers. The principle broadband order types are: new provision; ISDN/Home Highway conversion; broadband service migration; and cease. As part of the process, it is necessary to ensure that the order received is capable of being implemented, i.e. the request is valid, all required information is available, and network capacity is available to fulfil the request by the customer required date.

With orders for broadband services, the first part of the process is the line qualification that covers the checks made to ensure the user is capable of achieving the desired data rates on the ADSL connection. The main step is checking the current status of the line that is held in a database. The database holds the results of a number of checks that have been performed including acceptable line loss, compatible existing services, and acceptable equipment on the line. The response from the database is red, amber or green. Green indicates the line is able to support the required data rates. Red indicates the line is not suitable (e.g. distance is too long). Amber indicates that the line might be able to support the service, but further analysis is required or changes to the network need to be implemented. The amber responses are the most problematic in that they represent potential users. Yet to find out which ambers are capable of supporting service requires additional steps and hence time and cost. One objective is therefore to continuously improve the information held in the database such that the percentage of ambers can be reduced.

If the line is capable of supporting the service, the capacity checks then need to be made. The items that need to be checked are: available line card port; capacity in the appropriate VP; and capacity in the aggregate connection to the service provider. Assuming the required capacity is available, then the appropriate equipment configurations are made, i.e. DSLAM, and RAS which includes the updating of the capacity management databases. Other information that needs to be updated is the association of the user's details with the equipment configured. If the user has requested an engineer installed option, this needs to be scheduled; otherwise it is the service provider's responsibility to ensure the user receives the appropriate equipment. Finally, the billing systems need to be informed so that billing can commence. The processing of service orders represents the largest volume of the operational work for the early years of broadband, so much effort has been focused to ensure that the bulk of this area is automated.

5.8.3   Service assurance

This process ensures that the user has a working service, and if problems are identified, the appropriate steps are put in place to diagnose the source of the problem and then to fix the problem and hence restore service.

Problems can be identified either from the user via the service provider, or from the network itself. The equipment monitors certain aspects of normal operation and if a problem is identified an alarm is generated. It is not possible to monitor all aspects of normal operation and hence this process also has to rely upon the notification of problems from the user. Where the user detects a problem, they report it to their service provider who obtains the appropriate details of their problem. It is possible that the problem identified by the user is not actually a fault, but is a problem of mis-operation. If this is the case, the user is advised and no further action is necessary. If the service provider believes there to be a genuine fault, then it is necessary for the service provider to identify if it is in their domain or the BT domain. If the problem is in the service provider domain, then this needs to be corrected by the service provider, and BT does not need to be advised of the problem. If the service provider believes the problem to be in the BT domain, then it is reported to BT. To help the service provider distinguish between a BT fault and their fault, BT has provided a range of facilities to give status and test information to the service provider.

Problems in the BT domain (either network or user generated) need to be diagnosed to localise the problem. There is not a simple explanation of how this is achieved. There are numerous tests and sources of data that help the diagnosis. Some tests are automated, whilst others need to be manually initiated. This is one of the most important areas to be addressed as the volumes increase, as it is essential that problems can be rapidly and effectively solved. It is therefore very important that effective data is collected on the current types of problem and how easy they are solved in order to achieve continuous improvement in this area.

Jeopardy management oversees the status of the trouble tickets, and tries to ensure that the problems are resolved within the target time. Hence, if the diagnosis takes a while, then there is less available time to solve the problem. There are various ways in which the problem can be solved including a reconfiguration of the existing network, replacement of a faulty piece of equipment, or some other resolution. Once the problem has been resolved, it is necessary to test to ensure the reported problem is cleared. Assuming the problem has been solved, the appropriate databases are updated and the problem is cleared.

5.8.4   Billing

Bills are generated to the service provider based on the components of the services they have purchased, e.g. user connections, aggregate connections and virtual paths. The bills also have both connection and rental elements. This process manages the collation of the data such that billing is accurate, then follows up on the payment of the bills.

The information required includes the total number of user connections of each type together with the start or finish dates, the number of aggregate connections of each type together with the start or finish dates, the number, capacity and distance of the VPs together with the start or finish dates. As well as this base information, it is necessary to also consider the minimum contract periods, discount entitlements, the ability to make amendments, and any violations of the service level guarantees.

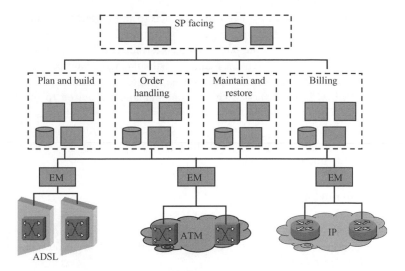

Figure 5.17   Operational system support

On a regular basis, this information is collated and a bill is sent to the service provider. It is then necessary that this process follows up on the payment of bills. It is also important that there is an effective audit kept in case of disputes.

5.8.5   OSS architecture

The OSS architecture is aligned to the above process areas (see Figure 5.17). Hence the main support systems can generally be partitioned into the different process areas. The purpose of the OSS systems is principally to reduce the cost of performing a particular process, though some systems may be introduced at additional overall cost because of a need to improve the quality of the process. It is these simple objectives that drive future development of the OSS systems.

This means that there are systems for billing, systems for service fulfilment, etc. There are also some legacy systems that are not as easy to categorise into a single area. But this is simply because when they were developed, the system was required to perform more than one process function. Within each area there could be a number of systems and databases. The actual number depends upon the design decision about how best to implement the requirements of that area. Such decisions would depend upon what already exists in that area, what capacity the current systems have, and what changes to functionality is required.

Above the systems supporting the main process areas, there are a number of systems dedicated to interfacing with the service providers. These systems support the reception and new orders, the reception of fault reports, as well as providing status information on the network, progress on orders and other management information. The systems support both automated (electronic) interfaces as well as manual interfaces to the service providers.

The supporting systems also need to interface to the underlying equipment in order to carry out certain functionality, e.g. configure a DSLAM, detect an alarm, perform a test, etc. The way the support systems interface to the equipment is through the element managers. Each element manager will manage a number of pieces of network equipment. The equipment supplier usually determines this, e.g. the different suppliers of DSLAMs have their own element managers required to interface to their equipment. Each element manager then presents a common interface to each of the support areas. The common interface to the element managers helps to reduce the complexity in the support systems by developing only a single interface, irrespective of the number of equipment suppliers.

5.9   Early experience

Following the launch of BT's broadband products, it is interesting to look at the early experience and feedback there has been on each of the products and compare with other broadband situations.

5.9.1   IPStream

The IPStream product is the most popular of the broadband product with service providers. There are probably two reasons for this popularity. The product offers the business user not only business IP applications at a faster rate than can be achieved with traditional dial-up interconnection, but also introduces new ways of working, and the service is always-on. For business users, these lead to an increase in efficiency and hence savings which will always be popular. The second reason is with the residential users. With relatively low prices from service providers it is seen as affordable and worth the increment above dial-up charges.

Further improvements in the product will continue to reflect the feedback received. The introduction of self-install had a significant impact on the volumes. Here a service provider is able to choose a modem from any supplier and either send it to the user to connect to the telephone socket on an ADSL enabled line or sell it to the user through high street stores. This has widened the choice of equipment available to the users. The cost of the self-install product is also less than the previous engineer-installed IPStream equivalents which led to price reductions. Hence the combination of reduced price, simplicity and increased choice for both service providers and end users has led to a big increase in the market place.

The other area that will help drive up volumes is the increase in higher bandwidth applications. Hence in a similar way that there is a need to use faster and faster PCs to effectively use the latest PC applications, so the development of IP applications will increasingly demand higher capacity IP connections. As this happens, users will progressively want to migrate from dial-up access up to broadband access.

5.9.2   VideoStream

This was the portfolio most in demand in the early days of the commercial trials and broadband launch. However, there has been limited take up, probably due to the relatively high cost.

One significant part of the service provider's costs is the cost of the VideoStream products. Hence the question: what can BT do to either add value to the product or reduce the price? In terms of reducing the price of the product, this needs to be triggered by some form of cost reduction in running the network and there are many such initiatives going on to try and achieve this. Some of the main ideas to reduce cost are: redefine the products such that it could be implemented with less cost; progressive automation should lead to reduced running costs; and the cost trends for the equipment should reduce over time. However, a degree of savings from automation and equipment cost trends has already been factored into the prices for the products, hence in the short term the best way of reducing the price would be to redefine the products such that it could be implemented at lower cost to BT.

It is reasonably clear that there is a market for Video on Demand services in the UK, but it is becoming apparent that there needs to be more work to optimise the commercial/technical solution. It is therefore likely that the debate around this type of product will continue for some time.

5.9.3   DataStream and local loop unbundling

These are products that enable either service providers or network operators to inter-connect into BT's network at differing points. The idea being that they would provide more of the network and not have to lease it from BT. Although, there have been a number of operators who have shown interest in these products, the early experience is that demand small.

One reason is the economic climate throughout the telecoms sector during the period of early rollout of these products. This has resulted in a lack of investment finance for competitors. Hence plans to compete have been delayed. For example, where service providers may have wanted to buy a DataStream product and invest in their own IP network, they are at present preferring to take the lower risk option of leasing the IPStream product, or completely putting their plans to develop as a service provider on hold.

It will be interesting to see how the market for these products evolves as the sector's economics take an upturn.

5.9.4   Competition in the UK

The main competition to the DSL solution for broadband in the UK is from cable modems. Cable TV operators have a significant coverage of cable TV systems, especially in the major conurbations. The systems deployed are of the modern type and readily able to support cable modems. Hence it is not surprising to see the cable operators are making efforts to exploit their existing assets through the addition of the cable modems to support data service. In the long term, however, it is unlikely that cable systems will be able to achieve the same penetration that could be achieved by DSL solutions simply because it is uneconomical to deploy cable systems outside major residential areas.

Strangely, another competitor to DSL in the UK is the standard dial-up modem. Although this does not offer broadband capability, it does offer Internet access that has attracted users in their millions. Attractive flat rate pricing packages have been introduced which mean users can use as much as they want for a fixed monthly fee. This flat rate is a requirement from Oftel that was introduced to encourage users onto the Internet. It has been very effective, but it is important to understand if this requirement is restricting the migration towards broadband.

5.9.5   Broadband in Germany

There is one highly successful DSL story in Europe that is worth looking at – Deutsche Telecom in Germany.

The launch of broadband in Germany by Deutsche Telecom was in 1999, but it was following a re-launch in August 2000 that the numbers of users really took off. By 2002 there were 80 000 new users being added every week hitting the 1 million, then the 2 million, user figures. There are many technical similarities with the BT solution, but it is the differences that have helped Deutsche Telecom achieve this dramatic take up.

The first difference is Deutsche Telecom's focus simply on a fast Internet product. Although there are other products, all the processes are focused towards the provision of the fast Internet offering and other products are managed as exceptions from the main route. Also, the German regulator has not demanded the level of interconnection to other operators/service provider that has been required from Oftel. One example is the requirement from Oftel for the DataStream interconnection product. Undoubtedly this has led to greater customer choice in the UK, but greater complexity and cost, whereas the product focus from Deutsche Telecom has reduced complexity and enabled greater early take up.

In terms of pricing, for many users in Germany, it is cheaper to use broadband than it is to use dial-up. It is quite clear that this pricing regime has encouraged the shift in the Internet market from dial-up to broadband. It also demonstrates that there is a significant broadband market.

Finally, competition in Germany from cable modems is very low. The main reason for this is that Germany had a significant deployment of the early type cable systems. This fact means that the deployed systems are not readily upgradeable to support the cable modems, and hence little competition.

5.10   Next steps to the market

If one looks back to the early 1990s, trials were looking at the roll-out of a fibre based access network, ADSL was an embryonic technology that was still very much a laboratory experiment, and the Internet had a limited number of business users. The reason for making this observation is that in those days, few people would have predicted this future. There were numerous analyses of the potential broadband market at the time broadband was launched. All showed the anticipated demand for broadband within the UK to be extremely healthy. Moreover they showed that DSL and cable modems will provide the dominant access mechanisms, with DSL slightly above cable modems (see Figure 5.18).

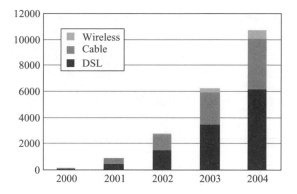

Figure 5.18   Potential UK broadband market

Source: IDC, 2000

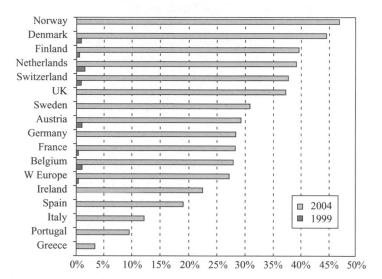

Figure 5.19   Percentage of households with broadband connections

Source: IDC, 2000

The analyses showed that in 2004 the UK will have the greatest broadband penetration of the European G7 countries (see Figure 5.19). This expectation is supported
by the fact that, in December 2001, the UK had the greatest penetration of Internet
amongst the European G7 countries (see Figure 5.20).

With the research at the time pointing to a healthy future for DSL broadband, the
focus has been on taking up that opportunity and kick-starting the market. Pricing
offers, marketing initiatives and the introduction of product developments are helping
to create an upwards trend in the number of orders received. These initiatives are a

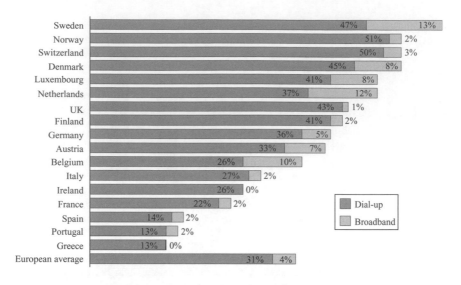

Figure 5.20   European household Internet penetration, 2001

Source: Forester, Dec 2001

combination of BT's activities as well as active marketing by several service providers who are BT's broadband customers. As a result, in June 2003 BT achieved its one millionth ADSL end user.

Although take-up has not kept pace with the early market predictions, the general view is that the potential market is achievable, but in slightly longer timescales. There are, however, wider challenges that need to be tackled if DSL fulfils the current expectations and there is also a need to understand alternative technologies and the role they may play in the broadband future.

5.10.1   Rural provision

One of the challenges is to look at ways in which the technology can be tailored so as to economically serve more of those who do not currently have ADSL availability. The two areas that need to be tackled are:

1. Identifying a solution that is capable of meeting the risks of uncertain demand in the smaller exchanges, including both the configuration in the exchange and in connecting BT's ATM network.
2. Identifying methods to provide service to users connected to serving exchanges, yet whose copper pair is too long to support the data speeds required.

The current rollout has been achieved by targeting the top exchanges in the most populated area. As rollout focuses on the next largest exchanges, so the potential market added with the exchange will become smaller and smaller. Also, there is greater risk in smaller exchanges that there simply will not be sufficient take up of broadband to justify the investment. With the market still in an early state of development, there is little evidence to show what levels of take up might be reasonable to expect. This means in small exchanges there is a greater risk that the required user numbers will not be achieved to provide a return on the investment. Alternatively, if demand is high, it may be necessary to upgrade exchanges earlier than expected which could lead to writing off assets earlier than planned.

One solution BT introduced was Pre-Registration. The idea here being that if demand in an exchange area is uncertain, a way of measuring it would be for potential users to register an interest in broadband. This measure of interest can then be tracked against a trigger level for that exchange to be economically viable. Once the registered interest hits the trigger level BT enables the exchange. This simple process has enabled a large number of communities to receive ADSL broadband that would otherwise have looked unviable. In spite of the increased risk, equipment suppliers are making equipment available in smaller configurations. There is also the ACT NOW project, which is a partnership between BT and key public sector organisations, to assist small and medium sized businesses in Cornwall. The project involves the rollout of ADSL to 12 exchanges and it is hoped will provide a blueprint for the development of broadband services in rural Britain. The Government is also keen to stimulate the roll out of broadband to as much of the population as possible. It is therefore likely that solutions will be found to enable a much greater percentage of the population to be served by broadband exchanges.

Although it is possibly that economic and technical solutions will be identified that will lead to an increase in the number of DSL enabled exchanges, it is unlikely that such solutions will lead to all exchanges being enabled. At present, the government are tending to accept that DSL will not be universally available and rely upon the availability of alternative technology to provide broadband to those who are out of reach of DSL and cable modems such as fixed wireless or satellite.

The second problem area to be tackled was provision to users in DSL-enabled exchange areas whose lines were too long to provide the required service. The introduction of the rate adaptive IPStream and Extended Reach helped by using an ADSL feature that gets the best data rates from the line. This means that typically users who are between 3.5 and 6 km from the exchange can now get ADSL broadband.

For some specific locations (e.g. a business park), it might also be possible to achieve ADSL service through the introduction of out-stationed equipment. But it is clear that out-stationed equipment would only be of limited use in increasing the penetration. Hence it will again be necessary to rely upon alternative technologies. It is possible that either the SDSL or VDSL alternatives may provide a future solution as well as the fixed wireless and satellite.

5.10.2 Additional service drivers

Voice is currently provided to users as POTS in parallel with the ADSL broadband capability. There is, however, a demand both from business and residential users for more voice channels to be provided. There are a number of options on how this could be achieved. Voice over IP (VoIP) and Voice over ATM (VoATM) are two options for carriage of the service, but many other questions remain. Exactly what services are required? What voice quality is acceptable? How many derived voice channels are required for users? Is ADSL capable of supporting the service, or is there a need to use SDSL? At what point is the connection to the narrowband network? What interfaces need to be made available for interconnection to other players? The list can go on and on. The point to note is that there is a possible market here if appropriate products can be defined. It is therefore likely that developments in this area will lead to service offerings, but it is currently too early to suggest how this might be done.

One of the evolutions currently being discussed in various broadband fora is the requirement for triple-play. The three parts being voice, data and video, i.e. all business, information and entertainment services being delivered through the same broadband connection to the user's home. Such a capability is really a development of the existing broadband video and data products with the addition of derived voice all in a single product. Depending upon the capacity requirements of the individual parts of the product, this may well have a significant requirement to increase the capacity of the DSL channel to the user. This could be achieved by providing DSL services at higher speeds to a smaller percentage of the population, or if demand is sufficient, it could drive the introduction of VDSL. The key question to resolve this dilemma is how many video channels need to be delivered simultaneously.

5.10.3   DSL alternatives

BT ran SDSL technical trials during 2001 in order to understand the technical implications of supporting symmetric broadband services. These trials of symmetric services were targeted at business users and it is this group that was targetted when the service was launched in September 2003. There is, however, an expectation that eventually services such as 2-way video and multiple derived voice channels will also lead to a demand for symmetric services from the residential market.

SDSL is also being looked at as a way of providing greater reach from the exchange. Two possible techniques could help achieve this. First, the use of multiple pairs between the exchange and the user, and second the use of line regenerators. Although technically possible, these options are a long way from introduction, but may offer a solution prior to VDSL.

The ADSL standard has itself evolved – ADSL2, ADSL2+ and ADSL-ER (Extended Reach). These standards mean that the basic ADSL technology can use a greater part of the spectrum to provide higher bandwidths over potentially greater distances. However, one of the key considerations is how much of the additional available spectrum can be used without compromising either the ANFP or other technologies (e.g. VDSL) that may use that part of the spectrum.

Proprietary DSL techniques may also offer an alternate solution for specific problems such as providing a service to users over long copper lines in enabled exchange areas. The main consideration here is the size of the market for a proprietary solution and if it is big enough to enable an economic solution. There have been small-scale trials and demonstrations of VDSL including one in Greenwich Millennium Village. From these trials it is clear that the only realistic way of providing VDSL is from equipment located in the access network and fed by fibre. This requires a significant uplift to the access network and hence requires considerably more investment than the ADSL or SDSL technologies that simply make use of the existing copper pair access network. However, as the demand for higher and higher data rate increases, it is likely that VDSL will have a significant role to play.

One other advantage of the VDSL solution is the penetration of fibre towards the user. Some of the techniques for feeding the out-stationed VDSL equipment are based on PON technology. Hence, it is still possible that VDSL could be used as a step towards achieving a fibre based access network to the users much as was originally conceived in the early 1990s (see Section 5.3.1).

5.10.4 Other broadband technologies

Cable modems are already established as a viable broadband access mechanism, and can be expected to provide a significant percentage of the broadband connections within the UK.

UMTS will develop, but the technology lags behind the broadband capability of fixed access solutions of cable and ADSL. Hence it would not be unreasonable to believe that although it is likely to be the mobile broadband technology, it will not replace fixed broadband technologies.

The BFWA technology will also develop but the government does not expect the technology to represent more than about 6% of the UK market. It is expected that the main use will be as an access technology for those who cannot obtain DSL or cable.

At present, two-way satellite communication is under development which will make it more competitive as a technology. However, the government believe it likely to be a very expensive solution compared to DSL or cable and therefore sees the satellite market restricted to competing with wireless in areas of DSL unavailability. Another problem for the satellite technology is that it is significantly behind the other technologies in terms of maturity. Hence, by the time it is sufficiently developed to compete with ADSL based services, it is quite possible that VDSL systems will be available and with a much greater capacity.

Fibre is likely to be a solution for the larger businesses, though it is unlikely to be used for widespread broadband access until there is sufficient user demand to go at least to the VDSL type date rates and probably beyond. The sheer capacity of a fibre-based solution does, however, make it an attractive solution in the longer term.

5.11 References

1   HOPPITT, C. E. and RAWSON, J. W. D.: 'The United Kingdom Trial of Fibre in the Loop', *BT Engineering Journal*, 1991, **10**(1)
2   LIVINGSTONE, A.: 'BT Interactive TV', *BT Engineering Journal*, 1996, **15**(3)
3   FOSTER, K. *et al.*: 'Realising the potential of access networks using DSL', *BT Technology Journal*, 1998, **16**(4)
4   PIRIE, A. and CHRISTOU, S.: 'Asymmetric digital subscriber line technology', *Journal of the IBTE*, 2000, **1**(2)

Chapter 6

VDSL – The story so far

D. Clarke

6.1   Introduction

The introduction of ADSL (Asymmetric Digital Subscriber Line) to drive forward the interactive broadband revolution is now well underway. ADSL is providing consumers with unprecedented high speed access to the Internet, and has sufficient capacity to deliver video and high speed data to the majority of homes. But ADSL is not the end of the story; demand for bandwidth will continue to grow and telecommunications network operators are developing the ability to deliver capacity far beyond ADSL using an exciting new technology called VDSL (very high speed digital subscriber line).

Like ADSL, VDSL operates on existing telephony wire pairs and is therefore a natural step for network operators who own an extensive telephony infrastructure. The key difference between ADSL and VDSL is that VDSL relies on deployment of optical fibre deep into the access network to reduce the distance over which data has to travel on the wire pairs, thereby reducing cable losses and increasing capacity. Figure 6.1 shows the typical make up of the UK telephony access network.

As can be seen in Figure 6.1, the topology of the UK telephony access network includes a number of flexibility points where cables are jointed and customer wire pair connections are managed. These flexibility points include the main distribution frame in the local exchange (MDF), underground joint chambers, primary cross-connect points (PCPs) known as 'cabinets', and distribution points (DPs) – typically the final drop into the customer premises.

The UK network includes both underground and overhead cables and is fairly typical of the networks to be found in most developed countries. This fact is relevant because it is important that VDSL technology is internationally standardised to reduce equipment costs, and it is helpful that most network operators have similar network constraints.

The distance from cabinet to customer is typically less than 1 km and provides the opportunity to access the final drop to connect VDSL transmission equipment.

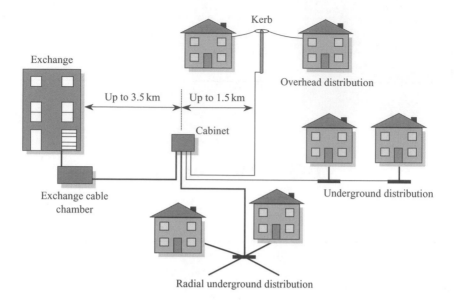

Figure 6.1   The UK telephony access network

Fibre will used to backhaul the broadband services from the cabinet to the nearest broadband access node.

6.2   The first VDSL demonstrator

The VDSL story begins around 1994 when it was realised that broadcast quality video could be delivered to a very high proportion of the UK population using the existing telephone network, without the disruption then being caused by cable digging. However, restrictions in BT's operating licence prevented it from offering broadcast entertainment using its network, and it was decided that an impressive demonstration of broadband capability was required to stimulate public debate.

Unfortunately, VDSL technology was not then commercially available and it was decided in the summer of 1995, to pull together a small highly skilled technical team at Adastral Park (formerly BT Laboratories) tasked with building an end-to-end VDSL demonstrator by the end of that year. An informal collaboration was established with DigiMedia Vision Ltd. (now a division of Tandberg), who loaned state-of-the-art MPEG-2 video coding and multiplexing equipment.

In less than four months a VDSL transceiver was designed and constructed from the ground up with general purpose components and using novel design techniques and training algorithms (now patented). In its very first test the modem operated at 23 Mbit/s on 500 m of cable for over 48 hours without a single error – an impressive achievement.

This modem is shown alongside the latest commercial equivalent in Figure 6.2.

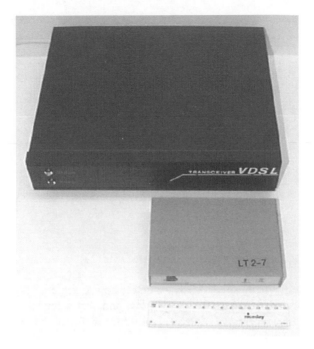

Figure 6.2   The 1995 BT VDSL modem (top) with modern equivalent

This early VDSL demonstrator delivered two simultaneous high quality digital TV programmes, plus high speed Internet over 600 m of typical telephony cable, and was sufficiently impressive for it to be shown to members of a House of Commons Select Committee early in 1996, and many other important influencers in the following months.

The prototype BT VDSL modem was based on the single-carrier modulation technique (SCM). The SCM variant selected was carrierless amplitude and phase (CAP-16) noted for its simplicity. A close collaboration with Nortel Networks' DSL team in Harlow was established who implemented a similar demonstrator using the competing, but more complex, discrete multi-tone modulation technique (DMT). This co-operation with Nortel provided valuable implementation insights for both SCM and DMT technologies, enabling BT to play an informed role in progressing VDSL standards.

This ground breaking work also provided the initial stimulus for the comprehensive programme of technical work carried forward since 1995 which underpins BT's acknowledged leadership in VDSL network technology and standardisation.

6.3   Technical challenges

Not surprisingly, moving DSL transmission technology forward beyond the state-of-the-art requires finding solutions to new problems. The technical challenges

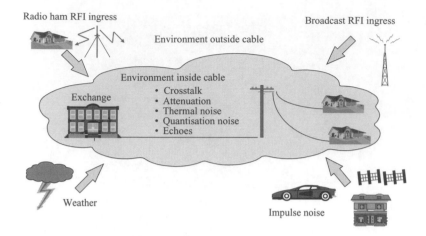

Figure 6.3   Technical challenges for VDSL

for VDSL are similar to those for ADSL – notably how to achieve high capacity transmission in the presence of high frequency noise and very steep losses in the twisted
wire pairs which were originally intended only for low frequency voice telephony.

As highlighted in Figure 6.3, the technical challenges for VDSL fall into two main
categories: (i) external factors: high frequency interference external to the network,
and (ii) cable factors: impairments due to the inherent characteristics of the cable
infrastructure, the most significant of which is crosstalk.

6.3.1   External factors

The external factors that need to be taken into account include radio frequency interference (RFI) – typically originating from high power broadcast transmitters; and
impulse noise interference – typically originating from transport systems, home appliances and of course the weather. There is nothing that can be done to eliminate these
external factors, and VDSL systems must therefore be designed to withstand their
influence to operate reliably. Key enablers being robust error protection techniques,
and ensuring that the systems are deployed with adequate head-room above the noise
– known as the noise margin.

Determining the appropriate noise margin is critical but it is also non-trivial
because it requires the network operator to acquire intimate technical knowledge
of both modem performance and network performance. This technical knowledge
must then be combined with knowledge of the business opportunity to engineer the
noise margin for a given service capability. Excessive margin, and fewer customers
will be reached, undermining the business case. Insufficient margin, and systems will
be unreliable in service due to varying noise and cable conditions.

6.3.2   Cable factors

High capacity transmission systems such as ADSL and VDSL operate at much higher
frequencies than the voice telephony which the cables were designed to support.

But when BT embarked on development of broadband capability in the early 1990s it was discovered that little research had been carried out anywhere in the world to quantify telephony cable performance at frequencies above 1 MHz. And yet to commit to huge broadband investment without being able to accurately predict the transmission performance of the modems was judged to be unacceptable. BT therefore initiated the most comprehensive programme of technical investigations ever carried out for DSL transmission systems by any network operator in the world.

These technical investigations included:

- A comprehensive network survey involving high frequency transmission measurements and recording the physical condition of over 7000 telephone pairs throughout the UK selected to be representative of the network as a whole.
- High frequency crosstalk measurements on 70 live 100-pair cable infrastructures throughout the UK – a task previously thought impossible.
- High frequency impulsive noise measurements and analysis.
- Investigations to assess unwanted emissions from ADSL and VDSL systems

Finally, there has been an extensive and ongoing programme of laboratory and field testing with the assistance of modem vendors, to relate knowledge of real network characteristics to modem performance to predict ADSL and now VDSL service capability.

The BT measurement work to verify DSL service capability has to date not been mirrored anywhere else in the world. For example, it was discovered that far-end crosstalk (FEXT) is actually significantly higher in the UK network than had been predicted by extrapolating previous data. This vital information is built into UK planning rules for DSL systems. Moreover, BT's measurement data underpins the UK's Access Network Frequency Plan (ANFP) – the regulatory framework which all UK operators must obey in their DSL deployments to avoid chaos.

The practical effect if the vital information on network transmission performance and crosstalk had not been discovered, would be that high rate DSL systems deployed in the UK could have become progressively unreliable, and even fail, as more and more systems were deployed, with disastrous consequences for operators and customers alike.

In recognition of the significance for the UK of this work, BT was awarded the IFE/National Physical Laboratory Measurement Prize in 1998.

The BT work has been selectively published to influence key international standards. Not surprisingly, the findings have stimulated urgent measurement work in other countries to verify that their broadband deployments are in fact sustainable in the long term.

6.4 VDSL technology

VDSL and ADSL modems use similar transmission techniques but the frequency of VDSL operation, and hence capacity can be much higher because VDSL has been optimised to operate over much shorter cables. The two key factors influencing the design and performance of VDSL are the duplexing scheme, and the frequency plan.

6.4.1   *VDSL duplexing scheme*

The duplexing scheme is the method employed to achieve two-way transmission on a single pair of wires. Like ADSL, VDSL uses frequency division duplexing (FDD) to achieve two-way transmission on a single cable. FDD was chosen over time division duplexing (TDD or 'ping-pong'). TDD was seriously considered for VDSL because the proportion of capacity allocated to upstream and downstream can be easily varied in software. But to operate correctly, TDD modems sharing the same cable must all be synchronised to avoid severe crosstalk. It transpired that this is a serious disadvantage in the unbundled environment because adopting TDD would have required the industry and regulators to agree a synchronisation plan in each country as well as a frequency plan (which would still be required) – not an attractive prospect.

This was the first-ever DSL standards decision to be influenced by regulatory factors. Now, all key DSL standards decisions must take into account regulatory implications – a new dynamic which came into play again in the choice of VDSL frequency plan.

6.4.2   *VDSL frequency plan*

The most significant hurdle in VDSL standards has been to agree on the frequency plan. The frequency plan is the allocation of frequency bands to the two directions of transmission on the wire pair, and determines how much of the total available capacity is allocated to each direction of transmission – the service capability of the system.

Ideally, there should be only one frequency plan for global application, or only a small number of plans, to minimise modem complexity, facilitate interoperability, and to avoid global VDSL market fragmentation. Unfortunately, reaching agreement on the frequency plan for VDSL was always going to be difficult because the choice of plan impacts on market positioning of competing VDSL vendors. It was therefore highly contentious.

In the autumn of 1999, after months of argument and counter-argument, ETSI accepted an FSAN[1] recommendation to adopt a compromise of 4 bands (2 downstream plus 2 upstream) in a frequency range extending to 12 MHz. ANSI adopted the ETSI position, but neither body was able to progress further to define the actual transition frequencies.

Finally, in February 2000, 15 FSAN network operators met in the USA to decide the VDSL frequency plan to avoid further delay. They found that only two frequency plans were required to meet global service requirements. This included the requirements of competitors interested primarily in business rather than consumer services.

---

[1] FSAN (Full Service Access Networks) is a Telco initiative to accelerate standardisation. It consists of 21 Telcos working with major vendors. See *www.fsanet.net*.

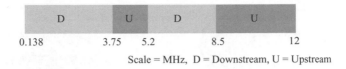

Scale = MHz, D = Downstream, U = Upstream

Figure 6.4   Frequency plan for VDSL – plan 998

The two selected VDSL frequency plans are known as 'Plan 997', which is suitable for mixed services, and 'Plan 998' which is suitable for asymmetric services, but which also accommodates symmetric services. Plan 998 is shown in Figure 6.4.

ETSI has documented both plans in the European VDSL standard. Plan 998 has been adopted by ANSI as the single plan for trial-use in the USA, and is favoured by NTT for use in Japan. Plan 998 is also favoured by BT and most other network operators who wish to deliver video services to the consumer market.

It should be noted that only one VDSL frequency plan can exist in a given cable to avoid mutual interference.

With the VDSL frequency plan agreed, the few remaining standards issues for VDSL can be resolved.

6.5   VDSL equipment considerations

In typical deployments world wide, VDSL equipment will be remotely located. The typical VDSL network architecture is shown in Figure 6.5.

In the UK, the VDSL equipment will be located at the cabinet which is typically less than 1 km from the customer. Over such a short distance VDSL will have much greater capacity than ADSL which has to traverse the much longer distance between the exchange and the customer.

Locating VDSL at the cabinet requires optical fibre to be installed in the access network to connect to the nearest broadband node. Analogue telephony remains unaffected by installing a splitter at the cabinet, enabling the basic telephony service to continue to be supplied by the local exchange.

VDSL will not be widely deployed in exchanges in the UK, mainly because the service capability of VDSL when located in the exchange is not attractive because of more severe noise (mainly from ADSL), and long cable runs to exit the exchange. Exchange based VDSL therefore has limited service capability. Moreover, systems operating at the exchange interfere with similar systems operating in the same cable at the cabinet, due to crosstalk. Generally, a choice has to be made to avoid such crosstalk: for ADSL, exchange-only operation is appropriate, while for VDSL, cabinet-only operation is appropriate.

VDSL can co-exist with ADSL in shared cables because VDSL frequencies will be configured so as not to overlap with ADSL frequencies thereby avoiding crosstalk.

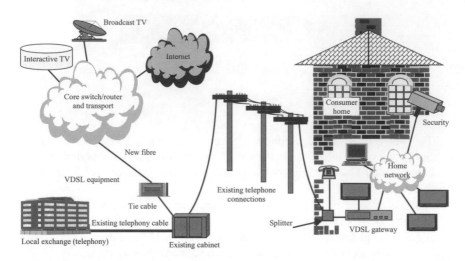

Figure 6.5   Typical VDSL network architecture

6.5.1   Issues for remote electronics

BT has limited experience remotely locating electronics in the access network, so new operational processes are being developed. And VDSL equipment is being specified to be physically compact and rugged to withstand the hostile environment. Minimising heat dissipation is very important and this influences the choice of technology.

Two implementation options are being considered for the remote VDSL equipment:

1. VDSL 'Brick' a modular and ruggedised sealed for life unit that can be located in the existing underground joint chamber near the cabinet (or in any convenient location) (see Figure 6.6).
2. A new electronic cabinet located above ground near the existing cabinet.

In the UK, there are around 88 000 street cabinets which each serve up 600 customers – typically 350. If take-up of VDSL services is (say) in the region of 10–20% on average, this implies cabinets or modular bricks able to support on average 30 to 100 VDSL connections. In practice, BT will plan for all customers to be migrated to fibre/VDSL to eliminate exchanges and copper cable all together for its future, 21st century network platform.

6.5.2   Powering

In the UK, a connection to local mains power is an economic option for VDSL, but power feeding from a central node will offer operational advantages.

6.5.3   Regulatory considerations

The reality of local loop unbundling has to be taken into account in deploying VDSL to provide other network operators with access to wire pairs at the cabinet. This is

Figure 6.6   Typical installation of modular VDSL brick

known as 'sub-loop' unbundling and BT has offered to provide a tie-cable to connect its cabinets to other operators' VDSL facilities. Some technical constraints must be applied to control crosstalk, but multiple operator access at the cabinet is feasible provided that only BT personnel can access the cabinet to avoid compromising security, or the reliability of telephony connections.

One less than attractive implication of sub-loop unbundling is the potential growth of 'cabinet farms' as many operators seek to install their own VDSL capability at the same BT cabinet location.

6.6   Controlling emissions

BT has been diligent to minimise the potential for unwanted emissions from VDSL (and ADSL). The VDSL standards specify a much reduced level of transmitted power at higher frequencies – the signal is up to 20 dB less than ADSL. Provision has also been made to notch the levels by a further 20 dB, for example, in bands used by radio amateurs.

In its efforts to be diligent, BT installed at considerable cost the world's only purpose designed open-area DSL emissions test facility at Adastral Park. This unique facility enables DSL egress and ingress measurements to be carried out on representative configurations of telephony drop wires in a quiet rural environment to verify they pose no interference threat to users of the radio spectrum.

A radio-transparent building was also installed at this facility, with typical internal wiring, to contrast the low level of emissions from DSL systems with the much higher

Figure 6.7   Open-area DSL EMC test site at Adastral Park

levels emitted by commonly deployed LAN systems such as Ethernet. It was also discovered that common low voltage lighting systems generate much higher levels of interference than DSL, and yet are perfectly legal because they meet internationally agreed EMC limits.

The open-area DSL EMC test site at Adastral Park is shown in Figure 6.7.

6.7   International standards

International standards are vital to leverage economies of scale through global deployment. But as mentioned earlier in this chapter, VDSL standards have been difficult to achieve, in part because two vendor groups are competing to win the market for their technology: the VDSL Alliance which supports adoption of multi-carrier modulation (DMT), and the VDSL Coalition, which supports single carrier modulation (SCM).

Network operators are interested in facilitating the lowest cost solution that meets the required service capability and market window and are, in general, agnostic on choice of technology. But network operators are interested in cost and availability, and have therefore been active in standards via FSAN to ensure that both technologies are given due consideration – both to accelerate progress and to ensure fair play.

The ETSI VDSL standard was agreed in November 2000. The ANSI VDSL standard is currently in the final stages of ballot. Both standards are broadly compatible.

A key role of the ITU-T is to harmonise global standards, and the VDSL 'Foundation' recommendation G.993.1, achieved consent in December 2001. This is a very significant milestone which enables vendors and network operators around the world to confidently commit to VDSL equipment development and deployment.

6.7.1 The Full Service-VDSL committee

Even with VDSL standards now in place, network operators are not standing still waiting for vendors to implement interoperable end-to-end equipment. Fourteen global network operators, service providers and over 60 vendors have formed the Full Service-VDSL Committee (see *www.fs-vdsl.net*). This forum was focused on developing a common end-to-end specification for a low cost interoperable DSL network capable of offering consumers a competitive bundle of consumer services including video and high speed data.

Reducing operational complexity to minimise costs of ownership was a key focus and technical working groups collected the operators' common requirements and created specifications for system architecture, VDSL transceivers, operations, and customer premises equipment (CPE) – it may be surprising to learn that up to 40% of the equipment costs for such a network are in the CPE.

The FS-VDSL Committee published its specification in 2002 after 2 years work: an aggressive timescale driven by the perceived window of opportunity for network operators to enter the video services market. In July 2003, the ITU-T adopted the FS-VDSL work as ITU recommendations, the first end-to-end DSL video-centric network to be standardised.

6.8 VDSL service capablity

The primary motivation for VDSL deployment is to provide a competitive bundle of video-centric consumer services. This implies an asymmetric capability, with highest capacity towards the customer – hence network operators' preference for frequency plan 998.

In the UK, it is expected that VDSL will deliver around 14 Mbit/s downstream plus 3 Mbit/s upstream over a distance of around 1 km from the cabinet. Even higher rates are achievable in other countries where crosstalk is less severe. This is a very impressive capability: assuming live video content requires digital coding at ~3.5 to ~4.5 Mbit/s, BT anticipates being able to offer 2–3 simultaneous video streams, plus high speed data, plus high quality video teleconferencing and security services, to nearly 100% of the UK population.

6.9 Deployment prospects

In the USA, Qwest Communications International has already commercially deployed an excellent video service using VDSL which connects over 50 000 homes in the Phoenix and Denver areas. Consumers receive over 150 digital TV channels with 3 independent video streams with integrated telephony features, plus 256 kbit/s and

1 Mbit/s symmetric Internet; all at a price competitive with cable. Qwest have been very focused on the economics to ensure that their service is not only perceived to be superior, but is also cost-competitive with cable. They have been a leading player in the FS-VDSL Committee where their operational experience was invaluable.

Bell Canada have started commercial deployment of their ExpressView™ digital satellite TV service using VDSL in multiple dwelling units. The key advantages being the avoidance of new cabling – the VDSL utilises existing in-building telephony wiring – plus the potential to develop new revenue streams using the interactive two-way service bundling capabilities of VDSL.

Qwest and Bell Canada are almost unique amongst major network operators in being video service providers in their own right, enabling them to enter the video services market using VDSL relatively quickly. For other network operators, including BT, the focus is likely to be on a wholesale model which is significantly more difficult to implement, especially in the current adverse regulatory climate in the UK and Europe. But despite the barriers, a number of VDSL trials are already underway or planned around the world. VDSL deployment is accelerating around the world, with almost one million lines already deployed in Korea.

NTT has started VDSL deployment in Japan, focused initially on 10 Mbit/s symmetric Internet delivery to apartment blocks, reusing the existing internal telephony wiring. This important in-building application for VDSL is already very popular in Asia where deployment is expected to ramp-up very strongly in coming years.

6.10   Acknowledgements

The author would like to acknowledge the contribution of colleagues in the Advanced Copper Technologies Unit at Adastral Park, notably Unit Manager Kevin Foster who led the critical start-up phase of the ETSI VDSL standard, and is a leading authority on DSL EMC; John Cook, architect of BT's VDSL transceiver, and internationally recognised expert on DSL transmission and spectrum management; John MacDonald who has led implementation of BT's VDSL test facilities and trials. Thanks also to John Warren whose team carried out the network surveys in all weathers. And many others too numerous to mention who have contributed to BT's broadband DSL endeavour. An inspired team of which I am very proud to have been a member for the past seven years. The work would not have been possible but for the vision of key managers in BT to provide sustained funding and guidance, notably BT Wholesale's Noel Jackson and Phil Laws. Finally, I wish to record my personal thanks to Dr Tom Rowbotham, BT's Director of Technology (retired), who supported and encouraged me during some very difficult times in the battle to achieve international agreement on VDSL frequency plans.

Chapter 7

Implementing local loop unbundling – an account of the key challenges

A. Cameron, D. Milham, R. Mistry, G. Williamson, K. Cobb and K. O'Neill

7.1 Introduction

Local loop unbundling (LLU) is a regulatory device that allows telecom operators without physical network infrastructure to lease access to the physical assets of other operators. It is called unbundling as it reverses the normal 'bundling' process whereby network owners usually only sell telecom services to customers in bundles which consist of switching, transmission, billing as well as the use of the physical infrastructure.

There has been much discussion of LLU recently as telecom regulators, mainly in the USA and Europe have introduced it in an attempt to boost competition in local access services and broadband access in particular. Much of the discussion has focussed on LLU from the new entrant operator perspective. This chapter describes the implementation of LLU from the incumbent operator's perspective. The authors have all been key members of the team that introduced LLU services to BT. The experience described here although based on BT and the UK regulations will be of general interest and application to other countries and other operators.

7.2 Broadband competition and LLU legislation

Telecommunications regulators have pursued LLU regulation in the belief that it will boost competition of broadband access services. However, broadband competition already existed both in service provision and infrastructure. In mid 2001 the BT deployment of wholesale asymmetric digital subscriber line (ADSL) services had reached in excess of 65 thousand end users via a wide range of service providers.

At the same time infrastructure competition in the form of cable TV networks had created around 5 million non-BT local loops.

The European Union passed a regulation [1] in 2000 imposing an obligation to offer LLU services on all fixed network operators which had been assessed as having significant market power. Prior to this the rules were different in each country. Some such as Germany and Denmark had started the process of unbundling earlier (in 1998), others such as the UK were already implementing LLU as a result of national rulings [2]. Most other European countries had made no significant progress towards LLU.

The LLU regulation was drafted and introduced very quickly. First drafted in early 2000 it was passed on 18 December 2000, came into force on 29 December 2000 and operators were required to comply by the beginning of 2001. In the UK the regulation had wider scope than Oftel's LLU requirement and effectively brought the launch date forward by six months. BT complied with the EU regulation by offering four variants of LLU – classic unbundling, line sharing, sub-loop unbundling and sub-loop unbundling with line sharing. Details of the services are published on the BT Interconnect web site [3]. Most other European incumbent operators also published details of their LLU service offers shortly after the regulation came into force.

7.3   Commercial considerations

Without LLU competitive operators have two basic choices. Building new access networks offers great flexibility but is very capital and labour intensive. Reselling the services of the incumbent operator offers less flexibility but is much less capital and labour intensive as it benefits from the economies of scale inherent in a large network. LLU appears to offer a middle course allowing the competitive operators some of the benefits of each approach while mitigating some of the drawbacks. Some operators have seen LLU as an entry strategy which will allow them to gain customers and revenue before making the investment to build their own networks as well as allowing them to differentiate their services from the incumbent operator. In any case, LLU requires significant investment and commitment from the new entrant operator. It can be difficult to build a business case and even large operators are not guaranteed success. In the USA there were several large operators which built their businesses purely on LLU. During 2001 all bar one of them ceased trading.

The attractiveness of LLU to an operator depends on the a series of complicated and interrelate factors such as the availability of wholesale broadband services, the maturity of the market for dial-up Internet access, and leased line and cable modem services.

The reaction of the incumbent operator to LLU will depend on their particular situation. A large integrated incumbent Telco is likely to see LLU as a threat as it allows other operators to get relatively risk free access to customers. It may be thought that a wholesale provider of access networks would be supportive or neutral towards LLU as it simply offers a different route to the market. However, even for the wholesale access operator LLU can be daunting. Significant investment is required in processes, operational support systems and technology to support new LLU products and new ways of working. The cost of these developments must be recovered from product revenues.

Almost without exception incumbent operators do not volunteer to offer LLU. LLU is inherently a regulatory product and the prices for it are almost always set by the regulator rather than by a free market. It is difficult to estimate the market attractiveness or volume of a new product and this leads to great uncertainty in the market forecasts available to operators and regulators. This further complicates the pricing as the cost of development work must be amortised over the volume of the product. As with all regulated products there is a risk that incorrect pricing will distort the market and send an inappropriate 'build or buy' signal to the market. Incorrect prices will result in networks that are not cost effective and may not allow the incumbent operator to recover their costs. Cost allocation methods have been the subject of many protracted arguments in most countries where LLU has been introduced.

It can be seen that while LLU can offer an attractive opportunity for new entrant operators it requires commitment from all parties including incumbent, new entrant operators and regulators.

7.4   Involving industry

If successful customer service is to be provided using LLU, it is vital that incumbents and new entrant operators work closely together. The best way to achieve this is for the operators to work together when designing the processes and any supporting operational support systems for the new services. At an early stage of the UK implementation of LLU a number of working groups were formed involving BT, Oftel and representatives of the many new entrant operators. Each working group had to negotiate a difficult path to achieve commercial agreements between BT and its competitors, while maintaining inter-operator consensus. Two broad streams of work were initiated:

1. A technical task force was set up under the auspices of the Network Interoperability Consultative Committee (NICC) to develop technical specifications for the new products:
    - a plan to control crosstalk interference in the network;
    - technical specifications for the unbundled loops;
    - a code of practice for maintenance of unbundled loops;
    - a code of practice to deal with alleged cases of crosstalk interference.
2. Definition of the processes and support systems was delegated to the Oftel-sponsored Operator Policy Forum (OPF). This group had to produce a number of complex agreements including:
    - exchange area data to support LLU operator planning processes;
    - processes to supply LLU service to specific customers;
    - operations and maintenance processes;
    - testing processes supporting both provisioning and maintenance;
    - billing processes.

An essential part of this work was to document and agree the business policies to be adopted by all parties to support LLU services.

7.5   Metallic access network

All developed countries have near ubiquitous metallic access networks and they represent one of many Telcos' largest and most valuable assets. Once limited to providing voice access and voice band modems, the metallic access network is now capable of carrying multi-megabit data services thanks to the introduction of digital subscriber line (DSL) technology made possible by recent developments in microelectronics and in digital signal processing. Each customer access line consists of a pair of wires running from the customer premises to the nearest local exchange. The line consists of a number of elements that provide connection, jointing and cross-connection functionality. The key architectural features of BT's metallic access network in the UK are shown in Figure 7.1. In the UK, access network cables terminate on a main distribution frame (MDF) within the local exchange building and this represents the cross-connection point between the external network and transmission equipment. There are flexibility points in the external network at primary cross-connection points (PCPs) usually located in cabinets seen in the street and at distribution points (DPs) located close to end-user premises, typically on telephone poles. The wire between the DP and end-user is known as a customer feed or customer drop. Access network cable sizes range from 10 pairs to 4800 pairs. In general, cables near a local exchange building have larger pair counts and smaller conductor sizes. Conversely, nearer to customers, the network cables have smaller pair count and larger conductor diameter.

7.6   LLU variants

The two major variants of LLU are *full* unbundling and *sub-loop* unbundling. Each of these variants may be offered exclusively or with shared use. Full unbundling

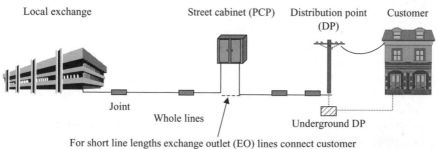

Figure 7.1   *BT's metallic access network*

involves leasing the use of a complete access pair from a local exchange site to the end-user premises. The LLU operator may choose to make exclusive use of the line or may choose to share it with the incumbent operator. In shared access, the incumbent uses the lower frequency part of the line to provide conventional telephony service, while the LLU operator uses ADSL technology to provide a data service.

In sub-loop unbundling the LLU operator is given access to the line at a sub-loop access point which will usually be located at a flexibility point between an exchange building and the end user premises. This may be attractive to operators who wish to use very-high-speed DSL or VDSL technology. If the sub-loop access is shared, the incumbent will use the lower-frequency part of the loop to provide a conventional telephone or basic rate ISDN service.

LLU operators will usually require certain ancillary services to allow them to make use of unbundled loops. A key service is space for the operator to place its network equipment. In the case of full unbundling, the LLU operator may be able to lease some space within the exchange building; this is called *collocation*. Alternatively, the operator may be able to place its equipment in a nearby building or street cabinet. This is called *distant location*.

The LLU operator does not need physical access to the main distribution frame. Tie cables are used to connect a handover distribution frame (HDF) in the LLU operator accommodation to the MDF. If distant location is used, the tie cable may be provided by either incumbent or by the LLU operator in conjunction with the incumbent.

Close co-operation between the LLU operators and network owner is particularly necessary during the ordering and maintenance stages. Much of this interaction is being automated using operational support systems that can be connected to the LLU operator's systems. The approach to systems is described later in the chapter.

LLU operators also need core network transmission links to connect the unbundled loops to the rest of their network. These links may be leased from any suitable operator or may be built by the LLU operator themselves. The LLU product variants are described in more detail below.

7.6.1   Classic unbundling

The basic form of local loop unbundling, sometimes called *full* or *classic* LLU (Figure 7.2), involves leasing the whole capacity of the access pair to another operator. The LLU operator can use the access pair to provide a range of services. The services offered are limited only by a few rules designed to protect the integrity of the network. The network owner remains responsible for the maintenance of the unbundled local loop from the network terminating point in the end-user premises to the connection point to the LLU operator equipment at the HDF or at a joint in a tie cable.

7.6.2   Line sharing

If ADSL technology [4,5] is used it is possible for an unbundled loop to be shared between incumbent and the LLU operator. In shared access an LLU operator provides

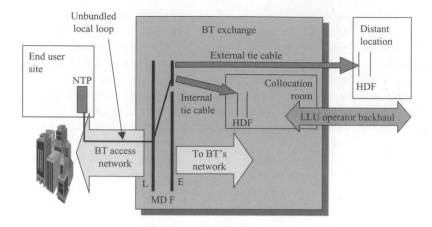

Figure 7.2   Classic unbundling showing key network components

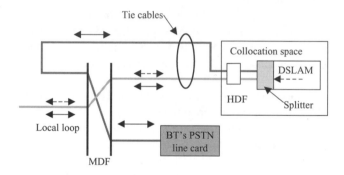

Figure 7.3   Line sharing – application to the exchange end of the local loop

an ADSL-based data service to an end-user while BT provides analogue telephony using the same access line. Line sharing requires extra tie cables and wiring to be provided and these are shown in Figure 7.3.

Line sharing is made possible by the use of 'splitter' filters which are installed at each end of the line (Figures 7.3 and 7.4). Without a splitter neither service would function correctly: the ADSL channel would suffer from data errors and the analogue telephone channel would suffer from echo. In the UK, BT has published specifications for the performance of the splitters and the LLU operator is responsible for providing splitters that comply to the agreed specification.

7.6.3   Sub-loop unbundling

In sub-loop unbundling, the LLU operator connects to the loop at a sub-loop access point located between the MDF and the end-user premises (Figures 7.1 and 7.5).

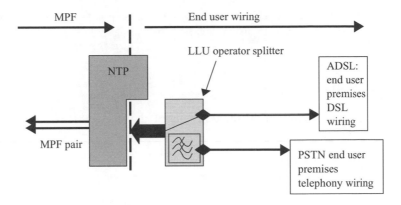

Figure 7.4   Line sharing – end user's wiring and components relative to the network termination point (NTP)

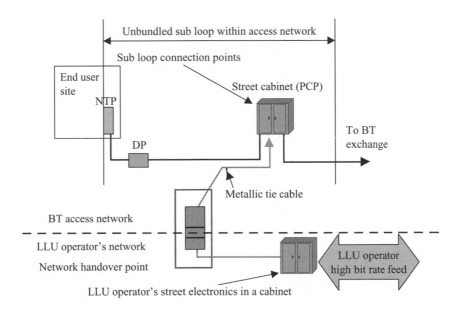

Figure 7.5   Sub-loop unbundling networks and interface

Sub-loop access is suitable for use in conjunction with a high-speed DSL technology such as VDSL [6]. Sub-loop LLU lines may be shared or dedicated. Under shared sub-loop access the incumbent will provide an analogue telephone service to the end-user while the OLO provides a data service.

Collocation is not relevant to sub-loop unbundling as access points are located away from local exchange buildings. To use sub-loop unbundling an LLU operator must provide a street cabinet close to the sub-loop access point and connect it using a tie cable. This is analogous to distant location for classic LLU.

7.7   Technical challenges

This section describes four of the most significant challenges that BT addressed in its LLU implementation:

- the design of a plan to control crosstalk;
- implementation of physical collocation;
- modifications to the BT operational processes;
- the design of computer interfaces between BT and the other operators.

Incumbent operators in other countries have faced similar problems but the detailed solutions may be different.

7.7.1   Access network frequency plan

LLU will often be used in conjunction with DSL technology to provide data services. BT has, for many years, been a leader in research into DSL crosstalk interference and in the formulation of deployment rules to control it [4,5]. Before LLU was introduced to the UK, BT was solely responsible for ensuring that its transmission systems were used in a way that avoided excessive crosstalk on its network. With the introduction of LLU, it was no longer appropriate nor possible for BT alone to control system deployments. A new set of rules was required that all operators could follow. The new rules had to be clear, easy to implement, fair and had to take account of likely future developments in the access network.

The NICC DSL Task Group [7] was the cross-industry technical forum that was tasked by Oftel to design an access network frequency plan to meet these requirements. The task group was chaired by BT and was open to all interested parties. A wide range of telecommunications operators and equipment vendors actively participated with BT and Oftel. The access network frequency plan (ANFP) for the BT network is described in detail in Chapter 4.

The ANFP for the BT network controls crosstalk by defining limits on the power, frequency and location of DSL systems. The ANFP works by placing limits on the worst-case crosstalk noise in the network. By ensuring that new developments in the access network are no more harmful to their neighbours than current systems, the ANFP provides the stability that allows all operators to plan for the future with confidence. The ANFP places the onus on individual operators to ensure that their DSL deployments are compliant and remain so. The DSL task group designed a code of practice to be used when investigating allegations of interference.

When the ANFP was first mooted, line sharing and sub-loop LLU were not required so the first issue of the ANFP made no specific provision for them. Similarly, international standards for VDSL technology were at that time not agreed and the first plan excluded VDSL. At the time of writing the DSL task group is extending the plan to accommodate VDSL and sub loop unbundling.

7.7.2   Maintenance procedures

Cost effective and timely maintenance is an important component of any telecommunications product. This can be difficult to achieve in the LLU situation as responsibility for maintenance is divided between two or more operators. Clear demarcation points between areas of responsibility are crucial to effective transparent processes and to ensure a good customer experience. In the UK the NICC DSL Task Group addressed this concern early in the development of LLU. The unbundled loop is defined between two demarcation points located at the network termination point in the customer premises (usually a telephone socket) and at the handover distribution frame in the LLU operator room or distant location. A clear and unequivocal specification of the performance of the unbundled pair was written and agreed by the DSL task group. It was expressed in terms that could be measured by either BT or the LLU operator and took account of the skills and tools available to the workforce. The specification provided a base from which DSL Task Group developed a code of practice for maintenance activities that allowed each operator to understand their responsibilities under fault conditions.

7.7.2.1   Test equipment

The technical specification of the unbundled line forms a part of the contract between network owner and the LLU operator. It is important that the network owner is able to test the lines before provision and during maintenance activities. BT has invested heavily in test equipment for use on its conventional telephone lines and it was logical that the same equipment should be used to test unbundled lines. The connection between the test equipment and the metallic line is a part of the local exchange concentrator but this is not connected to unbundled lines. The short term solution to this problem was to provide a physical test point connected to the unbundled metallic path near the point where the tie cables were connected to the MDF. When a test was needed on an unbundled line a BT engineer had to visit the MDF and make a manual connection between the line test systems and the metallic path test point. A further visit was required at the end of the test period. This solution was simple to implement but relatively slow and manually intensive to operate.

BT has decided that the longer term solution to the test problem is to connect all the unbundled lines through a specially provided test access switch matrix. This device simply automates the process of connecting the test equipment to the line. This could not be adopted at the initial launch of LLU products owing to the time taken to specify and procure a switch matrix and in particular to integrate it into BT support systems and processes. The test matrix will allow tests to be carried out more quickly than the manual test solution allows.

This test solution is appropriate for both full unbundling and line sharing. It is not appropriate for sub loop unbundled lines which do not connect to the local exchange site. A different approach must be developed if these lines are to be tested without manual intervention.

7.7.3   Collocation

Collocation is considered to be one of the most important services in LLU. LLU operators must place equipment close to the MDF to which they wish to connect. There are three broad types of collocation which are *physical collocation* where the operator rents space inside the local exchange building, *distant location* where the operator's equipment is housed in a nearby building or street cabinet and *cageless collocation* where equipment belonging to the LLU operator is located in the same room as that of the network owner.

In the UK, BT has more than five thousand local exchange buildings and these were never designed or intended for third-party access. When designing collocation products BT studied the similar products offered by other incumbent operators in the USA and continental Europe and adopted the best practices from these operators.

The introduction of collocation requires that standards for equipment and operations be agreed. In the UK the standards for equipment and the inter operator processes for collocation were developed and trialled by BT and representatives of the LLU operators in a sub group of the Operator Policy Forum during 2000.

A key aim of the product design was to minimise the need for equipment specifically for designed or modified for LLU and wherever possible off the shelf hardware was used to minimise supply chain bottlenecks. Statutory requirements and customer-affecting considerations such as safety, fire detection, planning law, security and risk management all had to be satisfied. In addition, BT's operating licence places it under an obligation to secure its network equipment.

The first collocation rooms and distant location facilities were completed in December 2000. At the completion of the collocation trial BT was well placed to design and build substantial volumes of collocation facilities in response to firm orders placed by LLU operators.

7.7.3.1   Physical collocation

Physical collocation allows LLU operators to place equipment in local exchange buildings and allows their authorised people to gain access to relevant parts of the buildings to operate their networks.

Providing collocation facilities means much more than simply making floor space available. Operators will require power, ventilation and related building services. For security reasons some degree of segregation is required between LLU operators' equipment and network owners' equipment. The agreed design criteria for physical collocation includes:

*   LLU operators must be able to source their equipment from the global marketplace.
*   A wide range of rack and cabinet sizes to be offered as well as differing power and environmental operating characteristics.

- Different and innovative deployment models were required as each LLU operator would be competing for business within each exchange area. These models might range from a small 'toe in the water' presence to an installation able to serve several thousand customer lines from the outset.
- No 'one size fits all' constraint.
- Recognition that some LLU operators would wish to have a dedicated room tailored to their stated need, offering a reasonable level of physical security.
- Some operators might wish to collaborate and share a collocation space in an area.

An *LLU Hostel* product was designed which formed the primary means of responding to demand and meeting the outlined criteria (Figure 7.6). In essence, the LLU Hostel enables an LLU operator to simply order one or more equipment *bays* – an area of floorspace suitable for equipment racks. Electricity is provided to each bay via an AC final distribution fuseboard providing a designated maximum power load. Expectations that LLU operator demand would be focused on exchange buildings serving the main population centres led to an initial minimum hostel size of two suites. This lower limit was subsequently dropped and a reduced version of the product was introduced to satisfy demand for hostel bays in smaller exchange buildings or where space constraints lead to the adaptation of rooms not originally designed for housing network equipment. Key design features of the LLU Hostel range of collocation spaces includes:

- The use of forced air cooling will generally be made.
- An HDF located at, or close to, the end of each bay. The HDF is used as the interface for internal tie cables provided by BT to the MDF.

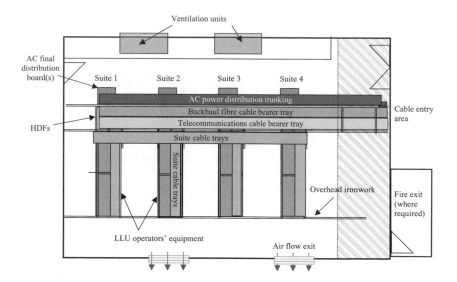

Figure 7.6   Outline design of an LLU Hostel used to collocate LLU operators' equipment in a BT exchange building

- An initial provision of tie pairs to the exchange side of the MDF. Additional ties can be ordered separately.
- Fire/smoke detection.
- Access control. BT has modified its building security systems to allow the other operator's people to access the relevant parts of the buildings without needing to be escorted by BT people.
- An AC power distribution board from which LLU operators take their power feeds.
- A cable support system using 'unistrut' ironwork in a simple configuration. The end-product is a 'ready to go' environment into which an LLU operator can readily deploy and commission equipment.

7.7.3.2   Distant location

Distant location was initially conceived to address the situation where some BT exchanges would offer insufficient accommodation to meet OLO demands for physical collocations. It uses an external tie cable from the BT MDF to premises or street furniture obtained by an LLU operator, near to the BT building. Providing accommodation for distant location is the responsibility of the LLU operator although BT may be sub-contracted to undertake the installation.

Distant location does not require the LLU operator to place equipment in or even to enter the BT building so there are many fewer restrictions on the design of the LLU operator equipment and operations. Distant locations have been built in nearby buildings and in self-contained street cabinets.

7.7.3.3   Cageless collocation

LLU operators in many countries have lobbied to be allowed to place equipment in the same rooms as the network owner. They argue that this is the best way to ensure parity of treatment between operators. This raises significant security concerns as local exchange buildings were almost always designed and built for sole use of a single network operator. Cageless collocation is no simpler to implement than physical collocation as provision must be made to connect power services, metallic pairs and backhaul services to the LLU operators equipment. Design calculations to ensure that sufficient air conditioning capacity are more complex in a cageless environment. Safety and security issues are more significant where the LLU operator shares a room with the network owner. In many cases the LLU operators' equipment and people have to satisfy more stringent testing or vetting before being allowed into the equipment rooms.

7.7.4   Process design

Like any large network operator, BT has a large number of complex processes and systems which govern the operation of its access network. Figure 7.7 shows the main processes that are affected by LLU.

Figure 7.7 does not show the many supporting processes nor does it cover the detailed procedures that are affected. However, it is a starting point from which

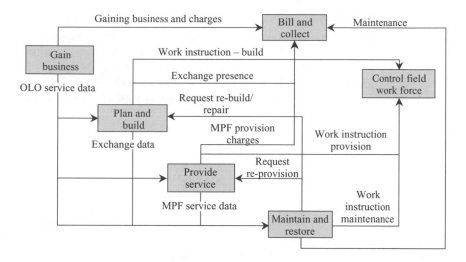

Figure 7.7   BT processes and typical interactions

to determine what would need to be modified. Typical examples of these changes include:

- how appointments are made with the end-user for equipment installation by BT;
- how faults are determined and resolved (that is, whether a fault is in the LLU operator's service area or on the network provided by BT);
- what to do if the end-user contacts BT directly rather than contacting the LLU operator.

These and similar issues were discussed at several OPF meetings which resulted in the interface between BT and the LLU operator being defined. Although this work was completed in April 2000, industry and Oftel requested numerous changes resulting in major reviews. A simplified end-to-end description of the unbundling process is shown below.

- Operator registers for LLU service with BT.
- LLU operator orders points of presence (collocation or distant location).
- BT or contractor builds points of presence.
- LLU operator installs equipment.
- LLU operator requests unbundled loop.
- BT provides unbundled loop.
- LLU operator provides service.
- LLU operator reports faults to BT.
- BT investigates problems.

A number of activities such as billing and field force control are inherent in these steps – threads running through every major process. Add to this the number of scenarios identified for unbundling (over 100) and the scale and complexity of the operational environment starts to be seen.

7.7.4.1 LLU registration

The gain business process covers registration for LLU service with BT. New contracts have been developed which cover the provision of network information required to enable operators to develop their business plans, and the ordering and maintenance of new LLU products from BT. If operators decide to go ahead with their business cases then registration is completed and BT will set up service accounts for billing purposes.

7.7.4.2 Points of presence

Once registration has been completed the LLU operator can order physical colloca- tions or distant locations from the BT exchanges in the geographical areas in which they are interested. At each intended point of presence, a two-stage survey and build- ing process with identified timescales was agreed. This was necessary to establish an audit trail should queries arise.

When LLU was initially launched more than 40 operators registered an interest and there was clearly potential for a large volume of collocation orders. The OPF group agreed that a process was required to manage this bow wave of orders before demand settled down and business as usual processes could be adopted. A demand management process was devised, which was to be operated by the Electoral Reform Society and Oftel, to decide in which order BT would tackle equipping its exchanges. This process was put in place in September 2000. In fact the volume of orders was massively below the industry forecast level and BT was able to handle them without using the bow wave. From December 2000, LLU operators were able to order dis- tant location on a business-as-usual basis. Based on a fuller understanding of what was practicable, in February 2001, BT announced that it could handle demand for collocation space as business-as-usual.

7.7.4.3 Unbundled loop provision

Provision of metallic pairs is carried out in three stages: the pre-order enquiry, deliver customer order and order acceptance. At the pre-order enquiry stage the operator is provided with information about the customer access line. The operator may wish to have a new metallic pair or to take over an existing one. The line information enables the LLU operator to decide if its service will work over the metallic pair to be provided. The operator may decide not to go ahead with the order if its service will not work because of the distance of the end-user from the exchange. If the operator decides to proceed, BT provides the pair as requested and the operator has two days to decide to accept or reject the pair. This period provides sufficient time to enable the operator to test its service.

The provision of the pair is usually quite straightforward. However, a number of issues need to be considered. The end-user is now a customer of the LLU operator and BT has to rely on the operator to make arrangements for access to their premises. Further complications are introduced if an order is combined with other products such as number portability, carrier pre-selection and/or calls and access.

7.7.4.4 Fault handling

Accurate and timely handling of customer's reported faults is fundamental to achieving customer satisfaction. LLU services rely on a combination of network elements including the metallic path provided by the network owner, additional components supplied by the LLU operator and perhaps some components supplied by the end-user. When the end-user experiences degraded service, it may be unclear which component is the cause.

In the UK, the OPF group developed processes to reduce the scope for multiple handing of the fault report and to support efficient and timely resolution of the fault report. In full unbundling the LLU operator is responsible for receiving fault reports from the end-user and should pass them to BT only if the LLU operator has reason to believe that a fault exists on the BT network. The situation is more complex where line sharing is used as both BT and the LLU operators are providing services to the end-user using the same metallic path. The end-user should report service degradation to the appropriate service provider in the first instance. Maintenance and repair activity on one service is likely to disrupt the other service so close co-operation between BT and LLU operator may be required. BT already has systems and processes to deal with problems affecting the narrowband (telephony) service due to network faults. These processes have been adapted to enable repair engineers to return the unbundled loop to specification based upon electrical performance. Once a repair is complete the unbundled loop is returned to the LLU operator with a report of the relevant electrical characteristics as proof of repair. Furthermore the LLU operator can purchase further test products to enable it to obtain additional information about a metallic pair. The LLU operator is, of course, solely responsible for dealing with faults caused by that operator's equipment or in the operation/configuration of the broadband service by the end-user.

7.7.4.5 Line sharing and sub-loop unbundling process issues

Processes were developed for line sharing and sub-loop LLU. Line sharing adds complexity to many LLU processes as two service providers use a single access line. New scenarios are introduced making the whole operational environment significantly more complex. Operators must now deal with requests for new shared lines, conversion of a telephony line to line sharing and reversion back to their original state, or a complete cease of all services. When a fault occurs there is potential for confusion. For example, it may be unclear whether the fault is on the network owner or LLU operator part of the line; the end-user may not know which operator to report the fault to and each operator may consider that the fault lies on the other's network. Processes and procedures are required to resolve any issue quickly.

7.7.5 Inter-operator OSS interfaces

The quality of service experienced by the end-user is highly dependent on the ability of all parties to provide, maintain and bill for LLU services in a cost-effective manner at high volume levels. Close co-operation between operators is required both at initial service provision and during any maintenance activity that may be required. It was recognised that a high degree of automation would be required to support large volumes of loops. This required that LLU operators be able to connect their operational support systems to BT's computer systems which were not designed to support LLU. Over 20 major computer systems were impacted and the integration task was one of the most complex undertaken since BT first installed customer care and billing systems in the 1980s.

7.7.5.1   Technical approach to OSS

The definition of industry processes for LLU was an essential precursor to the definition of the technical interfaces. The LLU Automation Forum (LLUAF), a working group established under the OPF, was set up to address this specific need. An innovative approach was adopted for the specification of these interfaces.

Rapid progress had been made in the development of business-to-business e-commerce technologies and supply chain integration. It was recognised that OSS integration might be simplified if they were employed. The ordering interfaces were based on the Commerce One Common Business Library CBL. This is a recognised e-commerce library that allows the specification of purchase orders and order status using the XML (extensible mark-up language). The library comprises a large number of definitions of general use in constructing ordering processes and messages. For example, CBL defines a structure of a purchase order and semantics for data types such as customer name, address, order reference codes and the semantics of delivery and order dates [8]. The Commerce One library needed some adaptation to make it suitable for use in telecommunications services. These extensions are based upon conclusions reached in a research study in an EURESCOM project P908 on business-to-business OSS gateways [9].

Publication of the LLU message sets for the agreed industry processes was completed in January 2001 and industry feedback and comment was solicited through the LLU Automation Forum and the UK Telco API Forum web sites [10]. Promotion of the work through the ITU-T, and the TeleManagement Forum lead to world-wide acclaim of the leading-edge innovative approach being taken by BT and the UK industry.

7.7.5.2   Integration to backend OSS

The development of automated interfaces between operators is only part of the OSS challenge. Many changes have to be made to the backend (back office) systems of both operators. It is well understood in the IT industry that changing legacy systems is complex, expensive and time-consuming.

The LLU systems development has had to support many new products types. Examples from the UK design include:

- *exchange area data* that relates exchange servicing areas to postcodes and makes the information available securely on a web server;
- *collocation facilities* – this requires the linking of BT forecasting and planning processes to LLU operators' plans so as to provide and reserve frame, tie cable and collocation capacity;

- *appointment reservations systems* for LLU service provisioning and repair including access to exchange buildings;
- *provisioning rules for job management systems.*

In some cases the customer may wish to transfer (or port) their telephone number. This required some integration of the processes and systems for number portability products with those for LLU products. The allocation of a line to a LLU operator requires the suppression of the automated billing which would otherwise be sent to the retail customer for an 'in service' access network path.

Considerable changes have been needed to the repair processes to cover demarcation of responsibilities between operators, co-ordination of testing, and escalation processes with the consequential impacts on the BT maintenance and repair systems. Substantial modification to the process and systems for testing of lines has been necessary.

New management information systems are needed to track statistics for LLU provisioning and repair. These are needed to monitor and track actual service performance and support the provision of information to the regulator. These requirements have resulted in significant new developments and hundreds of changes to existing legacy systems in an unprecedented short interval of time.

7.8 Conclusion

This chapter has outlined the technical, systems and process work involved in implementing LLU in an established access network. Several examples have been given derived mainly from the authors' experience in BT. These services, together with wholesale DSL products, allow competitive operators to use existing metallic access networks belonging to other operators without investing in building their own networks. Implementing LLU involves complex developments of technology, support systems and processes. The work requires agreement between network owners, regulators and representatives from other operators. BT was among the first operators to fully meet the challenge of the EU LLU regulation. The first collocation areas in the BT network were handed over to operators in late 2000, and in January 2001 BT released its first unbundled loop. The next few years will see a massive expansion in the availability of broadband access services and there may be volume demand for LLU products when the telecom sector recovers from the downturn which started during 2000. However, a wide range of broadband access technologies is available including DSL, broadband radio, satellite and cable TV. Only time will tell what role local loop unbundling will play in this expansion. The eventual extent and rate of unbundling depends to a very large extent on the commercial interests and firm orders received from new entrant operators.

7.9 References

1 Regulation of the European Parliament and of the Council on unbundled access to the local loop. Regulation (EC) No 2887/2000

2   Oftel web site. http://www.oftel.org.uk
3   BT Interconnect web site. http://www.btinterconnect.com
4   FOSTER, K. T. et al.: 'Realising the potential of access networks using DSL', *BT Technol. J.*, 1998, **16**(4), pp. 34–47
5   YOUNG, G., FOSTER, K. T. and COOK, J. W.: 'Broadband multimedia delivery over copper', *BT Technol. J.*, 1995, **13**(4), pp. 78–96
6   CLARKE, D.: 'VDSL – the story so far. *Journal of the Institution of British Telecommunications Engineers*, 2001, **2**(1)
7   CLARKSON, D.: 'Access to bandwidth: paving the way to Broadband Britain', *Journal of the Institution of British Telecommunications Engineers*, 2000, **1**(3), p. 5
8   http://www.xcbl.org
9   http://www.eurescom.de
10 http://www.telcoapi.org.uk

Chapter 8

Fibre access networks

D. W. Faulkner

8.1   Introduction

Demand for bandwidth is the key driver for fibre in the access network. Fibre has served the large bandwidth needs of long-haul networks for 20 years. Businesses or buildings, which can concentrate many users' traffic onto a fibre, have also been directly connected. The fibre access systems used have much in common with core transmission techniques such as PDH or SDH and are configured as point-to-point or ring networks. Corporate access customers have traffic concentrations, security and reliability requirements similar to core networks.

The big challenge is to solve the problem of the mass-market with the ultimate goal of fibre to the home (FTTH). Much of the technology needed has been around since the mid-1980s, but until now FTTH has been achieved only in technology trials or small-scale deployments. For FTTH to become ubiquitous, costs need to fall to be comparable with alternatives such as twisted pair or hybrid fibre-coax. Three important costs are the technology, installation and ownership. For FTTH to succeed these costs must be lower than the expected income from services over the life of the system.

Green field (new-build) is a key problem area, which gives a focus to the problem of FTTH. The cost of providing any fixed access network is very high when the civil works are included. New homes will incur these costs anyway so the issue is whether to deploy twisted pair, coaxial cable or fibre, or a mixture of all three.

Fibre promises capacity far in excess of the alternatives and offers future-proofing as demand for broadband services grows. What is needed is an entry-level system, which is cost-effective to justify the investment in a fibre infrastructure, rather than an alternative. To do this it must be comparable in cost and carry a richer set of services than the alternatives. Once installed the return on investment of the fibre network over its whole life could be higher than the alternatives, which would not compete so well as the demand for bandwidth increases.

This chapter sets the scene for a mass-market product by focusing on fibre access technology choices that influence cost and/or service capability.

8.2   Service requirements

Bearer service requirements include telephony, data (both guaranteed and best-effort) and broadcast TV. For business users high availability is a key requirement whereas a broadcast TV service may be of secondary importance. For the residential community a rich set of switched and broadcast video services will be attractive but lower availability will be acceptable if it saves cost.

8.3   Fibre access topologies

The choice of topology is critical because it affects the cost of the network, the services, and the reliability of the access network. Once installed it will be very difficult to change the physical configuration, so topology needs careful consideration. Here we examine the key candidate topologies: point-to-point, passive star, and ring.

8.3.1   Active star/point-to-point

Point-to-point systems use one or two fibres to connect a pair of nodes. The full bandwidth of the fibre is available for future upgrades and a high power budget is available for long-distance transmission. A drawback is that the network is not well-suited to broadcast services.

As the number of nodes in a network grows it is desirable to concentrate traffic to avoid the huge proliferation of fibre, which would arise with an unlimited mesh size. The choice of concentration point will affect the cost of the network. For example, if the concentration is nearer the customer the amount of fibre required can be reduced. However, point-to-point systems require a concentration point with active electronics. Consideration of environmental issues such as powering and temperature control and water ingress is then needed. These issues can be solved more easily if there is an existing building such as a central office offering environmental control. In the green field or overbuild situation these buildings may not exist and an external plant solution such as a pedestal or cabinet will be required at some cost.

The benign environment of an existing central office can perform this concentration function for point-to-point systems. The location of this building has been chosen to suit the needs of twisted-pair transmission. With fibre access using point-point technology, it is almost inconceivable to replace the wire pairs with fibre on a one-for-one basis because of disruption to existing services. Alternatively the installation of a new cables as an overlay containing massive numbers of fibres is very costly

For the mass market, operators are considering alternative architectures than traditional point-to-point for the local loop – the passive star that avoids the cost of active concentrators and can reduce opto-electronics costs by resource sharing.

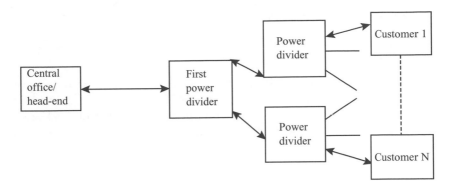

Figure 8.1   PON (tree and branch) topology

8.3.2   Passive star (tree and branch)

Passive star architectures can be implemented with either passive power dividers or with wavelength division multiplexing (WDM) components. Both offer a one-to-many optical concentration function. The passive optical network (PON) approach, shown in Figure 8.1, makes use of passive power dividers in the network and promises lower cost and more flexibility for early deployment. It may use any available multiplexing schemes in the equipment at either end of the network. WDM on the other hand promises a lower power budget and is therefore capable of giving more bandwidth per user on a dedicated basis, rather like a point-to-point fibre. The PON or tree architecture allows many signals to be combined onto a common bearer into the central node by suitable choice of multiplexing rather like conventional cable TV over coaxial cable. Traffic may be concentrated at any point in the network. In general, costs are minimised if the concentration point is near the customers. In the downstream direction, the multiplex is broadcast to all users. In the upstream direction broadcasting does not occur but a multiplex may be formed optically 'on-the-fly' using multiplexing such as WDM or TDM in the terminal equipment. The multiplex must be designed to avoid mutual interference, which would arise if channels are chosen at random.

Capital cost benefits of PONs are as follows:

- can carry both broadcast and interactive services;
- concentration is 'on-the-fly' rather than in an active node;
- no active devices are needed in the network, enhancing reliability;
- matches existing conduit/duct topology;
- increased reach compared with metallic alternatives (20 km is specified in the ITU);
- resource sharing saves cost of fibre, splices and head-end transceivers;
- low fibre count near exchange reduces the number of splices and allows faster repairs.

Operational cost benefits of PONs are:

- flexible to changes in demand for capacity;
- upgrading possible, e.g. via WDM.

Drawbacks to PONs are that the terminals become more complicated electrically and the whole fibre bandwidth is not available for future upgrades. However, fibre bandwidth is very large (Tbit/s) so, for the time being, this architecture can be used to offer customers capacity into the Gbit/s range.

8.3.3   Rings

Rings are used extensively in core and metropolitan networks because they offer a means of interconnecting a number of nodes (cities, towns) in a way that enables large capacities to be transported with resilience. The opto-electronic technology is the same as that needed for point-to-point unless WDM is involved. If a route fails another route can be found by sending traffic in the opposite direction round the ring as shown in Figure 8.2.

The amount of fibre needed is considerably less than point-to-point networks, however there may be more civil works involved in serving a community as two fibre entry points are needed if strict route diversity is required. Furthermore if the ring is to be used to provide resilience, two optical paths and two sets of transceivers are needed to offer loop-back protection which increases the cost. Without loop back protection a unidirectional ring is vulnerable, as a single failure causes all of the traffic to be lost.

8.3.4   Resilient networks

Alternative routing is required by some customers to allow service to continue in the event of a single or multiple failures. Resilience is also required when the traffic concentration is high.

Topologies that offer route protection include point-point or PON systems with two separate fibre paths, dual rings with loop-back protection and mesh networks.

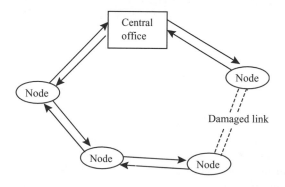

Figure 8.2   Self-healing ring

All of these networks are used but are not being considered for mass-market fibre to the home systems (FTTH) because of cost.

8.4 Deployment strategies

The choice of deployment strategy will determine how much and at what stages capital is spent, the coverage and how much infrastructure is provided. The choice of strategy therefore represents a key economic consideration for fibre access. Three approaches are considered here: green field, overbuilding and upgrading an existing network.

8.4.1 Green field

As mentioned in Section 8.1, green field (new-build) is a key problem area, which gives a focus to the problem of FTTH. New homes will incur the costs of civil works (digging) and outside plant anyway so the issue is which technology to install: twisted pair, coaxial cable or fibre, or a mixture of all three.

Green field fibre construction has been neglected because of the advance of broadband over DSL, cable modems and UMTS. Nevertheless, on a worldwide basis, there are significant volumes of new homes being built and even more homes requiring basic telecommunications services in developing nations.

The problem for system designers is to provide multiple-service capability: telephony, video and data. A rich service set allows a single fibre network to compete more effectively in economic terms with twisted pair and coaxial cable as the traditional means of delivering each of these services. None of these competes with fibre in terms of capacity and fibre offers more resistance to water ingress, lightning damage and electrical or RF interference. The opportunity for fibre systems vendors and operators is to make a single fibre more profitable than a bundle of wire pairs and coaxial cable for the delivery of multiple services. The full services access networks (FSAN) group was set up with this as the final goal. The activities of this and other standards bodies will be described later.

Providing traditional telephony services is troublesome over fibre because of the absence of wire for power feeding. Fibre is dependent upon the local electricity supply and batteries for back-up. Nevertheless, since this problem has been solved for cellular telephony, similar solutions can be used for fibre without incurring the additional cost of a copper access network.

There are numerous ways of providing video services. Both broadcast and switched are possible via PONs. The fibre infrastructure may carry broadcast services in all legacy modulation formats but time division multiplexing (TDM) offers opto-electonic simplicity and high power budgets. Pure TDM unfortunately is not a format used for video services in the mass market yet, although digitally encoded channels in TDM 'bouquets' are found in FDM broadcast systems. Such systems may come a close second best for transmission of broadcast video over fibre. Base-band switched video services direct to a PC are emerging via the Internet.

In the UK, although the franchises for cable operators expired in 2001, most green field build has remained a combination of twisted pair and cable. Twisted pair is provided to almost all locations, while cable is typically provided in the less rural areas. Fibre to the home has remained a niche application because of the continuing success of ADSL and the higher cost of fibre as explained earlier. Only in a few select trial areas has fibre to the home been deployed.

For BT, green field build has been a relatively low volume offering a limited opportunity for new infrastructure. However, there is expected to be a need for around 1 M new homes in the UK over the next decade with large developments in SE England.

A green field solution used by BT is TPON (telecommunications over a passive optical network) in remote cabinets or buildings. It has been narrowband only because of restrictions on cable TV services and an uncertain market for interactive broadband. In most cases the feeder to customers is copper pair rather than fibre.

So, despite the competitive environment that has been opened up since 2001, fibre to the home systems have remained difficult to justify, even in green field where the infrastructure costs are always encountered. This may change in the future as homes require data rates in excess of ADSL capabilities and more broadband services become available.

In the USA, the situation is similar; Telcos are beginning to compete with cable for multiple service provision. This competition is making incumbent operators think about radical restructuring of the access network to offer FTTH particularly in new-build situations and there are a number of trial areas where FTTH has been used.

In mainland Europe and Japan cable operators and Telcos have followed separate evolution paths with separate licenses for cable TV and telecommunications and little competition between the two.

8.4.2　*Overbuilding*

Overbuilding refers to the situation where a new network is being provided in addition to established infrastructures such as twisted pair for telephony and coaxial cable for cable TV services. The incumbent networks may be old and the ability or will to deliver broadband data services is weak. The opportunity for the over-builder is to offer broadband data as an alternative or in addition to the other services. New-entrant operators are introducing point-to-point Ethernet systems to satisfy the demand for data in both residential and small to medium sized business sectors.

The standard PON is based upon ITU G983 series and is capable of delivering 622 Mbit/s symmetrical services by circuit emulation, or with ATM or Ethernet interfaces. These PONs are currently focused on the business market but work is in progress to widen the service set to include telephony and cable TV services for residential green field applications. The cost of fibre access is high, so it is a challenge for any fibre access solution to be profitable in the residential overbuilding scenario. A data-only solution is likely to be simpler and cost less than a full service access network but has fewer revenue streams available to pay for its introduction.

A data-centric approach is possible using point-to-point fibre and LAN technology to connect buildings and apartment blocks to service providers. An active node such as a hub, switch or router at the fibre entry point to the building is used to connect individual offices or residences via category 5 cable (one twisted pair for each direction of transmission). This type of network is gaining popularity but will struggle to deliver high quality telephony and real time video for residential applications because of the absence of circuits. Contention for bandwidth in such networks is difficult to control if the access protocol remains IP.

The G983 series PONs are able to deliver data-only services and multiple line digital telephony by circuit emulation. These systems can be considered as an alternative to SDH and PDH point-to-point rings for overbuilding in a business community where demand for telephony in blocks of up to 24 (T1) or 30 (E1) channels are commonly found.

8.4.3   Upgrading existing network

Two types of network commonly exist as possible starting points, telephony networks and cable TV networks. For the incumbent telecommunications operator there are three approaches to consider: piecemeal, partial overlay and change-out.

8.4.3.1   Piecemeal

In this case point-to-point fibre is installed to one customer here and another customer there as the capacity of the twisted pair is exhausted. Similarities with the introduction of telephony could be expected where, for example, there might be waiting lists for customers who would only be given broadband service when the necessary civil works are completed. This approach does not achieve the lowest initial cost and leads to high frustration. Alternatives to this are a partial overlay where a fibre distribution point is introduced near customers or complete change-out where every telephone line is removed in favour of a fibre alternative. Depending upon the design, this distribution point can serve customers via coaxial cable, twisted pair or fibre. Active electronics and power feeding are required in all but the PON approach. The case where existing twisted pairs are used to carry broadband services is known as fibre to the kerb (FTTK).

8.4.3.2   Partial overlay

The partial overlay suffers from the disadvantage that both copper and fibre networks have to be maintained and many individual house calls have to be made when the broadband service is required. For a very slow introduction of broadband as penetration increases, this approach might be preferable but not the lowest cost overall. Duct capacity will be taken beyond its limit so new ducts are required for fibre. Existing poles may exceed their wind loading so new poles would be required.

Figure 8.3 shows relative costs of upgrading the access network assuming that twisted pairs are already in place. The take-up or service penetration is a key parameter. In the partial overlay situation, where penetration is lower than 20%, FTTH is cheaper than fibre to the kerb (FTTK, active node) or coaxial cable technology. This analysis was carried out based upon price projections and estimated installation costs rather than on actual quotations.

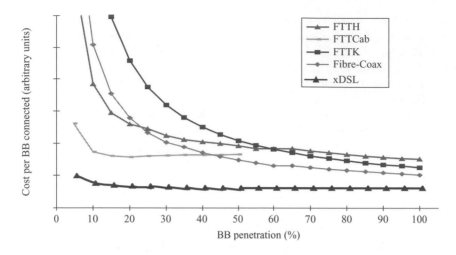

Figure 8.3   Comparison of FTTK, FTTCab and FTTH, HFC access capital costs

8.4.3.3   Change-out

Figure 8.3 shows that in the change-out situation, where the penetration would be 100%, costs are at a minimum, however FTTK and hybrid fibre-coax offer cheaper solutions at lower capacity. FTTK may not be a contender as it will not deliver broadcast video and EMC issues could rule it out. Change-out requires a very high level of commitment from both operator and customers. To gain best economy of scale whole neighbourhoods need connection to the new infrastructure in the shortest possible time. The copper network is recovered to make best use of existing ducts and pole capacity.

8.4.3.4   Upgrading a cable TV network

Many cable networks in the UK include both two twisted pairs and coaxial cable in the final connection to customers. Voice video and data services are possible from the outset. Broadband data may be delivered using cable modems on the coaxial network or DSL on the twisted pair network. Which technology to choose to deliver broadband depends upon costs of the equipment and traffic contention ratios expected.

Most other cable networks only have coaxial cable and are therefore dependent upon cable modems for both telephony and data services. The network has to be transformed from a broadcast-only structure to one also suitable for two-way switched services. Return communication channels therefore have to be provisioned and the network has to be partitioned into multiplex blocks to allow sufficient capacity for additional broadband interactive communications. The cost of doing this is significant and depends upon how much return channel transmission was pre-provisioned.

The role of fibre in making the transformation is normally limited to that required in the primary and secondary distribution networks to feed broadband distribution points from which coaxial or twisted pairs continue to offer the final customer connection. Change-out plans to fibre in the final drop have not been widely reported yet by cable operators as costs are high and existing coaxial cable has broadband capacity to be exploited.

8.5 Developing the mass-market for fibre access

Fibre is more expensive than copper for services such as telephony and cable TV. As yet there is no new application to drive demand for bandwidth that fibre alone can offer. However, familiarity with the office, university or school LAN could be the driver as users who seek to work from home are denied equivalent capacity and struggle to perform simple tasks quickly with a narrowband dial-up link. The advent of DSL technologies and cable modems during the 1990s put 'on hold' the development of mass-market fibre access systems. Both offer broadband data and always-on services over the existing access infrastructure at lower cost than fibre. These networks are not purpose-built for broadband data and suffer from problems of limited capacity coverage, reach and quality of service.

Now the demand for faster modem speeds for applications such as switched video is pushing DSL technology to its limit on existing copper loops with speeds in the range 300 kbit/s to 6 Mbit/s depending upon the actual line transmission parameters and degree of transmission symmetry.

A shorter loop is needed to push copper lines and DSL modems to yet higher speeds. New fibre-fed street cabinets containing VDSL (very high speed digital subscriber loop) line terminations could offer rates up to 17 Mbit/s on UK lines depending upon the degree of symmetry. However, the new infrastructure will be expensive, the bandwidth increase is not large and EMC remains a likely problem.

The shortcomings of DSL mean that mass-market fibre deployment is now more likely. Fibre has for a long time been recognised as the solution to the access capacity problem. It also offers the benefits of longer reach and no electromagnetic compatibility or crosstalk problems.

8.6 Delivering the services over fibre

8.6.1 Telephony

Analogue transmission is the traditional way of delivering individual telephony channels over twisted pair. Conversion to digital takes place at the local exchange. Fibre access systems avoid the complexity of analogue transmission by bringing digital telephony along with other services directly into customer premises.

TDM is today's method of delivering telephony services to sites requiring two or more telephony lines. This would typically be over ISDN, HDSL or point-to-point fibre. ATM bearers can also deliver TDM by circuit emulation and is becoming a key driver for fibre-to-the-business in the USA for T1 repeater and wire-line replacement. In the future, ATM and IP bearers should be capable of replacing TDM.

In the FTTH situation, users are familiar with the benefits of traditional telephony services. The change from a wire pair to fibre delivery would need to be almost imperceptible and a locally powered ONT with battery back-up is needed. At the central office, there is the problem of separating the narrowband and broadband traffic so that it is possible to signal over the existing public switched telecommunications or mobile networks.

8.6.2   Broadcast or cable TV

In the green field FTTH and MTU situations, cable TV is a key requirement. Delivery over fibre rather than coax will require the service to be presented as normal to the set top box. The basic service would therefore need to be carried as a broadcast multiplex. Today's set-top boxes will accept a mixture of analogue and digital signals in the 50–550 MHz band. Analogue signals are not 'fibre-friendly' and are not be easily transported over a PON. An alternative would be to transport a multiplex, which would mimic either a terrestrial or a satellite digital multiplex in a 7 or 40 MHz band respectively. A number of these multiplexes would be required to bring the channel total towards an attractive 250 channel capability.

The fibre distribution system for broadcast cable TV needs to take advantage of fibre amplifiers and PON technology. Delivery requires an additional wavelength in the 1550 nm window. Once the fabrication mass market WDM demultiplexers is solved the stage would be set for massive deployment of TV channels.

For the VDSL systems broadcast TV would be provided using conventional coax, satellite or terrestrial systems. Interactive TV would be available via the interactive data bearer.

8.6.3   Interactive video

Interactive video services are gaining in popularity and are taking on a number of forms including video-on demand, video conferencing, video telephony, video gaming, surveillance and photographic e-mail attachments. These services are very restricted with the existing narrowband Internet but can be expected to improve vastly as the bandwidth available on the Internet increases.

Picture quality and cost depends upon the data rate and whether instant playback is required. Demand for video services is likely to exceed the capacity available. Guaranteed bandwidth and charging schemes will be needed to give higher quality services than can be expected with the bandwidth bottle-necks of the Internet.

Applications such as surveillance and interactive games can operate at narrowband data rates and are therefore inexpensive. Applications requiring a dedicated connection direct to the source such as video-conferencing or video telephony are likely to remain relatively expensive. Multicast services from local distribution points will save cost and offer wider choice than is available via conventional broadcast services. E-mailing photographs and video clips will become easier as the capacity available to service providers increases. Fibre access would give superior performance for all types of interactive video services and will allow multiple sessions with plentiful bandwidth available.

8.6.4 Data

The choice of presentation of the data service depends on the business LAN and home network market. Ethernet and USB are commonly used interfaces.

The requirement for connection to multiple ISPs can best be met using the ATM delivery system. How to control the capacity for each ISP over the same bearer then becomes a partitioning problem for the user. For larger bandwidths WDM offers the prospect of reduced contention and 'guaranteed' access to multiple ISPs.

The need for interactive data services for both small business and residential users is best met with B-PON technology.

8.7 Progress on standards

The FSAN initiative was launched in 1995 as a pioneering co-operative effort involving some of the world's leading Telcos. The initiative first saw the light of day after an informal meeting of representatives from various Telcos. Discussions over the issue of broadband access led to the suggestion that a collaborative project should be set up to ensure a commonalty of approach world-wide.

The founder members were BT, France Telecom, NTT, Deutsche Telekom, Telecom Italia, Telephonica, and KPN, the Dutch Telco. Following this first tranche FSAN has subsequently been joined by Bell South, GTE (Verizon) of the USA and Telstra of Australia. The focus of FSAN is on how to introduce economically broadband services, both broadcast and interactive, whilst maintaining the traditional narrowband service portfolio including POTS and ISDN. Each Telco has views of the nature of what comprises a full service set but accepts that the broadband market is unpredictable and needs therefore a very flexible service capability. Two service sets are envisaged for FTTCab/B and FTTH solutions respectively.

FSAN is a group of 20 international network operators working to develop broadband passive optical networks. FSAN members include: Bell Canada, BellSouth, BT, Chunghwa, Deutsche Telecom, Dutch PTT, France Telecom, Verizon, Korea Telecom, NTT, SBC, Swisscom, Telefonica, Telstra and Telecom Italia. FSAN documentation has been contributed to relevant standardisation bodies, in particular, the ITU-T Recommendation G.983 series.

The Ethernet in the First Mile (EFM) group of the IEEE has been looking at whether new standards are required for the access network to take advantage of low-cost techniques developed for LANs. At this stage it is too early to say whether this approach will succeed in producing the mass-market FTTH solution. However, it is suitable for distribution to and around buildings. The use of gigabit Ethernet technology, active or passive nodes and new gigabit PON technology are all being discussed.

8.8   Conclusion

Green field construction offers a good opportunity for fibre access system deployment to provide broadband services to the mass market. Deployment to existing buildings depends upon the exhaustion of the capacity in the existing copper access network or the end of its economic life. Fibre access will then become attractive with preference going to areas with extensive overhead cables or reusable conduits.

Although the technologies for fibre to the home are reasonably well understood, a key step for incumbent operators is to make legacy telephony and broadcast TV possible over a single infrastructure in addition to new data and interactive TV services. PONs represent an attractive solution because of the resource sharing and their ready ability to carry both broadcast and interactive services.

New entrant operators or over-builders may avoid the need for consideration of legacy services by focusing on data networks and Internet access. The technology to do this is already available for both PON and point-to-point solutions for businesses.

The ability to support multiple sessions and real-time services gives ATM systems an advantage over IP-only solutions, which are well suited to best-effort, single-stream, services.

Chapter 9

Developments in optical access networks

K. James and S. Fisher

9.1 Introduction

Optical fibre access network technology has become more complex over the last few years with bit rates steadily increasing. PON systems are available with bit rates in the range 155 Mbit/s to, currently, 622 Mbit/s. Newer systems, both PON and point-to-point, will allow this trend to continue with 2.5 Gbit/s systems becoming practicable in the near future. With this increase in technology has come an increase in the opportunities for fibre system to cost-reduce more of the network, including the metro space – a fact that increases the appeal of the technology. Some of the newer PON systems are being designed using Ethernet as the underlying technology with a promise of reducing the price of the technology still further. How this will occur in practice, however, is still unclear. What is certain is that optical access technologies are expanding and are sure to find an opening in real networks in the foreseeable future.

Developing equipment for the access network has always been a risk for the equipment manufacturers. The size of the network means that equipment volumes can be huge but the equipment is extremely cost sensitive. Furthermore, the sheer cost of equipment rollout means that network operators do not attempt such a major upgrade of their networks lightly. Indeed, to date there have been no ubiquitous upgrades of the UK access network, rather there have been many small, ad hoc changes. Things are now beginning to change with the introduction of ADSL, although this still utilises the most costly and fault-prone parts of the copper access network.

Whilst some new services are being rolled out to businesses that can afford them, these roll-outs are focused on those customers wanting to use the bandwidths immediately with little effort on overall network upgrades. In the UK, services such as the short haul data service can provide up to 2 Gbit/s to customers separated by no more than 40 km, whilst conventional data networks can supply in the order of an STM1's worth (155 Mbit/s) to high bandwidth customers, again mainly business related.

In terms of residential customers, progress in upgrading the network has been slow, mainly because of the perceived lack of value of bandwidth upgrades. The residential market has been getting ever more cost sensitive, with the restrictions in permitted price increases and a glut of bandwidth in the core now hurting many CLEC companies quite significantly. Against this backdrop comes ADSL, enabling users to get 500 kbit/s downstream and 256 kbit/s upstream at an extremely competitive price.

Back in 1989 people predicted the introduction of Passive Optical Network (PON) technology into the network [1]; however, this roll-out did not occur for several reasons, the main one being that there was no perceived market for the types of services that optical fibre technology could bring. This is now beginning to change.

Current bandwidth upgrades will not satisfy customers for ever. There is a question as to when residential customers will require more bandwidth than that offered by ADSL but it seems to be required sooner rather than later. During the last 10 years, there has been a step change in the computing power available to the average residential consumer. Processing power has increased from 8 Mbit/s systems to the current 2 Gbit/s systems (with 3 Gbit/s around the corner) and there is little sign of the pace of technological advancement slowing. This is already leading to customers being able to manipulate video images in real time. What they cannot do is transport this data, in real time, across the telecommunications network.

Following a similar line, point-to-point Ethernet technologies are also receiving attention with a variety of systems becoming available from different manufacturers.

If developments continue at the same pace, then there will be a need for affordable interconnections at speeds currently unattainable by ADSL. Whilst VDSL could cope in the medium term, questions are already being asked about its suitability, given the high operational costs of the required street located electronics and questions relating to radio interference from overhead copper cables.

Against this backdrop the future development of fibre systems for the access network seems to be important. Fibre access network products are already becoming available but most of these are expensive, difficult to implement and are not full service compatible.

Whilst fibre technology has yet to mature enough to make cheap access systems, the major problem is the lack of volume in the sector to date. It is against this backdrop that there is a need to develop affordable, flexible systems. A variety of companies and people, however, are rising to the challenge. It has become noticeable that there are more start-up companies designing and building equipment for this market sector than ever before and that existing major players are now moving into this field with mainstream products.

Whilst standardisation of systems has been shown to lead to a steep reduction in their price, along with increased availability, standardisation too early can lead to poor products. Hence pre-standards discussions in an open environment can be a useful first step. This can be brought about at specialist fora at which the associated problems and challenges can be discussed and indeed such fora have been created and are making good progress in setting initial requirements and solving technical problems.

The best known standards forum is FSAN. Until recently the FSAN group has been the mainstay of optical fibre access system development. Seven operators, including BT, first met in the spring of 1995 in order to begin the FSAN initiative. FSAN stands for Full Service access network and had an initial focus on passive optical network system (PONs). BT had previously been experimenting with ATM-based PON systems [2] but these systems remained largely in-house.

The network operators invited equipment providers to join the group in order to progress activities in a manner suited to the production and specification of such equipment. Using this mechanism the number of contributors to the FSAN forum could be controlled, preventing the organisation from becoming unwieldy. It is still the situation today that equipment vendors only join FSAN by invitation, although they are free to solicit such invitations.

The IEEE are also now active in this area and have their own forum, the Ethernet in the First Mile (EFM) group, trying to create access network standards around existing Ethernet standards.

These groups will now be discussed in some detail.

9.2 The FSAN group

The FSAN group has been around for 17 years now and is split into optical systems and copper systems (see Figure 9.1). This chapter will deal only with the optical systems. The basic FSAN recommendations were input into the ITU-T some time ago with the baseline documents being G.983.1 and G.983.2. Recent work has resulted in new standards which enhance the usefulness of the basic PON systems in certain circumstances. These will be discussed further in this chapter.

What is of interest is that work in the group is not coming to an end as one might expect. Rather the group are moving on to higher specification PON systems. Work is progressing on higher bit rate PONs although the technology underlying the new architectures (i.e. ATM or Ethernet) has yet to be agreed. Limited further information is included later within this chapter as the key agreements have yet to be reached.

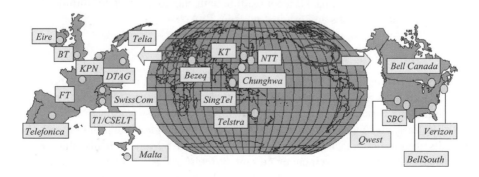

Figure 9.1   FSAN group diagram

9.3   The IEEE Ethernet in the First Mile group (EFM)

The IEEE hope to capitalise on the success of their existing Ethernet products. There is no doubt that these existing products were successful as they are in widespread use and form the basis for most computer LAN installations. Because of their high rate of adoption, these systems and the components from which these systems are built, are extremely cheap and well suited to the task for which they were designed.

Initially, design effort is being expended on creating a 16 way split PON with a reach of 10 km; however, how this will be used in a real access network is unclear as FSAN believe 32 way splitting, with a 20 km reach is optimal for such networks. The IEEE PON is designed to run at 1.25 Gbit/s, double that of the highest speed existing APON.

The idea is that by using these high volume components in different ways, a new Ethernet PON system could be designed. This is not quite the reality of the situation, however, as the functionality of Ethernet systems is not the same as that required by PON systems, Ethernet based or otherwise and significant changes will be required. For example, PON systems usually require a ranging protocol which enables remote multiplexors to access a common fibre upstream bearer channel with correct timing so as not to corrupt each others signals.

Furthermore, the optical power constraints are more severe as the split ratio and the bit rates of the PONs go up which mandates the use of more expensive optical components.

9.4   Recent trends in PON system development

Recent trends in PON specifications have been of a rather technical nature, looking at the many facets of PON architectures, and attempting to make the systems more efficient or easier to use in the real world where deployment issues can entrap the unwary. This section highlights recent trends in PON system specification but the reader should note that none of these new specifications are mandatory. Indeed they may or may not be used individually on a particular system implementation. This can lead to a myriad of systems, all incompatible with each other, but conforming to the ITU-T recommendations and originating from FSAN.

9.4.1   Bandwidth increases

The most obvious change that can be made to a PON system is to increase its overall bandwidth. This would allow more bandwidth per customer on average which should mean that the cost per unit of bandwidth will decrease.

The early APON standards specified 155 Mbit/s to a maximum of 32 users. This results in a bandwidth per customer in the order of 5 Mbit/s. Whilst this can be considered a great improvement on current systems, and is symmetrical in nature, in practice this bandwidth is not a great deal and would not be capable of adequately supporting the 'Triple Play' of voice, data and video services. Furthermore, VDSL

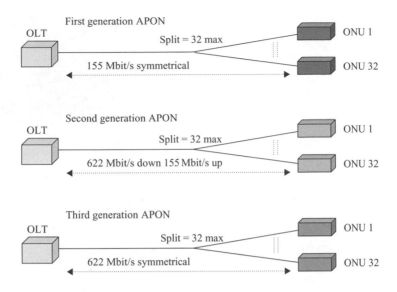

Figure 9.2   Existing APON bit-rate options

can be configured to give bandwidths in this range and therefore the base specification APON is probably not of tremendous interest.

More recently the APON standard has been extended to allow for 622 Mbit/s downstream and 155 Mbit/s upstream asymmetric bit rates as well as a 622 Mbit/s symmetrical option (see Figure 9.2). These are seen as being of greater value, since with the same 32 way split specification these can deliver in the order of 20 Mbit/s to the user. This is closer to the sorts of bandwidth required for a wider mix of services.

Currently FSAN is considering a 1.25 Gbit/s and even a 2.5 Gbit/s PON option, using new architectures not based on their baseline which was G.983.1. These new PONs may have a use in the telecommunications arena probably only as a feeder/backhaul network for VDSL or other access network systems. The IEEE is also becoming active in this area and will soon specify a 1.25 Gbit/s PON, although details are sketchy at the time of writing this chapter.

Point-to-point Ethernet systems are also becoming more prevalent although the lack of standardisation is hindering the creation of consistent product offerings amongst different vendors. Systems comprising of a head-end and street electronic concentrators are being installed, i.e. in Sweden, where they are providing high speed Internet connections to multiple dwelling units (Figure 9.3).

9.4.2   Wavelength allocation

Current fibre systems allow for the introduction of differing colours of light (or wavelengths) onto a fibre or fibre system in a manner whereby they will not interfere with each other.

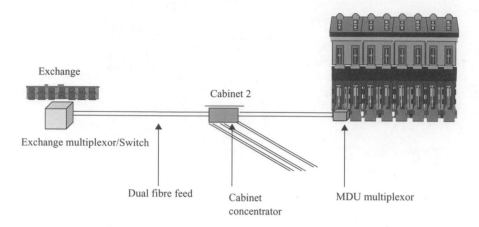

Figure 9.3   Generalised point-to-point architecture

Table 9.1   Original APON wavelength assignments

PON type	Downstream λ	Upstream λ
Single fibre	1310 nm	1550 nm
Dual fibre	1310 nm	1310 nm

Table 9.2   WDN-compatible APON wavelength assignments

PON type	Downstream λ	Upstream λ	WDM part
Single fibre	1310 nm	1490 nm	1500–1550 nm
Dual fibre	1310 nm	1310 nm	1500–1550 nm

The standard APON was defined to run over one or two fibres as shown in Table 9.1.

Bearing in mind that the ITU-T has assigned multiple wavelengths in the band 1500–1550 nm, then it can be seen that there is a lot of spare capacity on a PON. For this reason the APON specifications now recommend the wavelengths shown in Table 9.2.

Table 9.2 shows that the upstream wavelength has been moved from 1550 to 1490 for single fibre systems.

9.4.3   Standards relating to WDM

The FSAN standard relating to WDM is G.983.6 ('A broadband optical access system with increased service capability by wavelength allocation'). This specifies the ITU-T course band of light wavelengths as the basis for the standard and therefore utilises common components from the existing WDM product ranges. Whilst interesting, the standard is not intrinsically difficult to interpret, relying as it does on existing WDM technology, however, its use on a PON network can create some cost-effective solutions to everyday problems. The official wavelength bands are given in Table 9.3. Note that some WDM wavelengths have not been allocated.

The most common example of WDM quoted is that of broadcast television which can be introduced to a PON over a separate WDM wavelength. In Table 9.3 this is referred to as video distribution services. This allows each customer end of the PON to see the broadcast stream. Of course, they can only demultiplex the stream with the correct equipment and even then can only see program groups to which they have

Table 9.3   Official WDM wavelength assignments

Items	Notation	Unit	Nominal value	Application examples
1.3 λm wavelength band				
Lower limit	$\sim$	nm	1260	For use in ATM-PON
Upper limit	$\sim$	nm	1360	upstream
Intermediate wavelength band				
Lower limit	$\sim$	nm	1360	For future use
Upper limit	$\lambda 1$	nm	1480	Reserved band including guard bands for allocation by ITU-T
Basic band				
Lower limit	$\lambda 1$	nm	1480	For use in ATM-PON
Upper limit	$\lambda 2$	nm	1500	downstream
Enhancement band (option 1)				
Lower limit	$\lambda 3$	nm	1539	For additional digital
Upper limit	$\lambda 4$	nm	1565	service use
Enhancement band (option 2)				
Lower limit	$\lambda 3$	nm	1550	For video distribution service
Upper limit	$\lambda 4$	nm	1560	
Future L band				
Lower limit	$\lambda 5$	nm	For further	For future use
Upper limit	$\lambda 6$	nm	study	Reserved band for allocation by ITU-T

The central frequencies in the enhancement band for DWDM application shall be based on the frequency grid given in Recommendations G.959.1 and G.692.

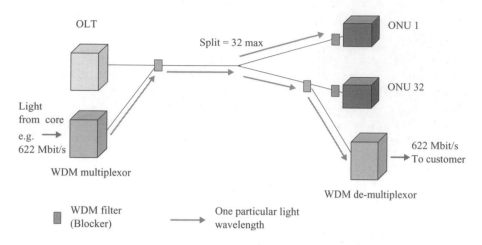

Figure 9.4   WDM example relating to a PON architecture

subscribed. In this case, the wavelength allows common access to all downstream channels. But WDM can be used in another manner.

WDM can be used to provide high speed point-to-point links over PON architectures (see Figure 9.4). If two wavelengths are provisioned for example, then one particular end can receive a high speed data stream from the central office. It can also transmit back to the central office using a second data stream. Given that each of these data streams can be up to 10 Gbit/s or greater, then the PON infrastructure is now effectively also supporting high bandwidth point-to-point connections. Systems providing 16 or more such point-to-point high speed links are possible. This is shown in Table 9.3 as additional digital services. Whilst point-to-point systems could provide WDM services, currently there are none doing so.

9.5   Dynamic bandwidth assignment

Most access network systems, with the exception of point-to-point systems, use fixed, or static, bandwidth assignment. Most point-to-point systems allow static multiplexing and therefore the following section does not directly apply to them. With PON access network Systems, the situation is somewhat different.

A typical APON system would run at a bit rate of 155 Mbit/s, allowing a total customer bandwidth of around 130 Mbit/s. If this were statically multiplexed, the bandwidth per customer on a 32 way PON would be in the order of 5 Mbit/s. Now 5 Mbit/s is not a lot and could be provided by alternative technologies such as DSL.

Dynamic bandwidth assignment relies on the fact that it is uncommon for each user on a PON to require their full bandwidth at any one time. In fact each user can have potential access to bandwidths in excess of their mean bandwidths for extended periods of time and this is what dynamic bandwidth assignments would allow.

Instead of bandwidth being assigned on a per user basis, bandwidth is provided on demand, with the user unaware during normal operation that their share of the overall bandwidth 'pot' has increased.

On an APON system, it is not too difficult to arrange this as the speed of transmission over the PON is quite fast and therefore latency issues are kept to a minimum. The user is unaware of the precise mechanism by which the ATM packets are queued and then transmitted across the PON.

The principle recommendation from the ITU-T that deals with dynamic bandwidth assignment is G.983.4 ('A broadband optical access system with increased service capability using dynamic bandwidth assignment'). Dynamic bandwidth assignment is also known as DBA.

In the DBA recommendation, G.983.1 is modified to allow for extra messages required in order to support the DBA mechanism. The DBA standard is designed to include:

- Performance objectives (for example, bandwidth assignment delay, maximum waiting time).
- Application functionality (for example, dynamic bandwidth assignment for bursty traffic and for ONU/ONT aggregated traffic composed of different traffic classes).
- Fairness criteria and protocols (for example, dynamic bandwidth assignment based on ONU/ONT status reporting, dynamic bandwidth assignment based on OLT monitoring, dynamic bandwidth assignment based on a combination of reporting and monitoring).
- Backwards compatibility.

As a background to understanding the dynamic bandwidth principles, it is essential to understand the way an APON works, at the cell level.

The basic APON standard is the ITU-T Recommendation G.983.1 which specifies a flexible access platform to provide broadband services through passive optical networks. In this Recommendation, the upstream traffic from the ONUs/ONTs to the OLT is transferred in a frame of 53 time-slots (cell slots). Each time-slot consists of three bytes of PON layer overhead and an ATM cell or a PLOAM cell. The upstream bandwidth is shared among the associated ONUs/ONTs. The OLT controls each upstream transmission from the ONUs/ONTs on a time-slot-by-time-slot basis. This is accomplished by sending data grants in downstream PLOAM cells.

PLOAM data grant cells are sent in the downstream direction to all ONUs/ONTs. The data grants are addressed to specific ONUs/ONTs and contain parameters that include the number of upstream data grants and the time-slots for the grants that are assigned to the individual ONU/ONT. Originally the grants were assigned in a static manner and only updated when a new connection is provisioned on a PON or an existing connection is removed from a PON. Once the bandwidth is provisioned, the OLT would continuously send the assigned grants to the associated ONUs/ONTs; subsequently, the OLT would receive corresponding user cells in the upstream. This current granting mechanism was efficient for real-time traffic but lacked any dynamic assignment capability.

However, the above mechanism is not efficient for non-real-time traffic, or where bandwidths are increased beyond the norm. For non-real-time traffic types, the ability to assign bandwidth dynamically is expected to provide higher efficiency than the original static granting mechanisms.

In ITU-T G.983.1, the OLT grant generation and distribution was updated when a new connection is provisioned on a PON or an existing connection was removed from a PON. Once the bandwidth is provisioned, the OLT would continuously send the assigned grants to the associated ONUs subsequently, the OLT would receive corresponding user cells in the upstream.

However, ITU-T G.983.1 was intended to enable a wide range of broadband services, including those that do not have a constant bit rate. For example, the Internet connects to many bursty traffic sources, which are best accommodated by ATM SBR Class 2 or GFR, which have less rigid requirements on cell transfer delay and cell delay variation. Mapping these non-real-time services into a fixed bandwidth channel prevents the ONUs on a PON from dynamically sharing the upstream PON bandwidth. For these non-real-time traffic types, the ability to assign bandwidth dynamically is expected to provide higher efficiency than the current static granting mechanisms.

The DBA protocol consists of three strategies, referred to as non status reporting, status reporting and hybrid types. The non status reporting type strategy is invoked by monitoring traffic in the OLT where lengthening buffer queues can be viewed as a requirement for extra bandwidth. The status reporting type strategy is invoked by status reports sent from ONUs to the OLT, where requests for more bandwidth are explicitly given. The hybrid strategy is invoked by both monitoring traffic in the OLT and processing in the OLT the status reports from ONUs. The recommendation does not specify detailed mechanisms or algorithms for these strategies, but rather message requirements and the required interfaces at specified reference points.

The reader is recommended to read G.983.4 for further details on how to provide DBA over an APON system as the issues are complex (Figure 9.5).

9.6   Protection

It has been identified that PON systems might not give the level of reliability required for certain applications. Whilst the PON electro-optics are generally reliable, it has been found that a major cause of failure is the physical media, i.e. the optical fibres in the access network themselves, and in many cases faults are caused by external influences. Point-to-point systems can allow for protection using back up or standby fibre pairs.

It is not the aim of protection scenarios to restrict the architectures possible in a PON architecture but to support and enhance whatever architecture is chosen for a particular application. For this reason, the protection scenarios offered are not exhaustive, but are representative of those it is considered will be of interest. These must be carefully tailored to the applications for which they are required.

Generally it is only FSAN, and through them, the ITU-T that have considered protection of PON systems in some detail and an ITU-T recommendation is due

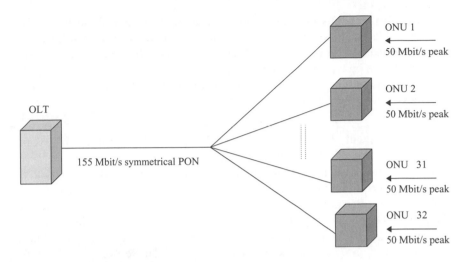

Figure 9.5 Dynamic bandwidth assignment example

shortly. This will be known as 'G.983.5: A broadband optical access system with enhanced survivability'.

The main requirements of a protection system shall be that:

- It should be possible to have several C protection configurations on the same OLT.
- It should be possible to duplicate the OLT-PON-interface and the fibres between the OLT and splitter and to duplicate the entire fibre path between the OLT and ONU for a set of ONUs.
- It should be possible to have a mixture of protected and unprotected ONUs on one B-PON interface.
- The addition or removal of a protected ONU on a PON should not affect other ONUs on the same PON.
- It should be possible to have automatic switching, which would be triggered by a fault detection such as loss of signal, loss of cell delineation, signal degrade (e.g., BER becomes worse than the pre-determined threshold), etc.
- It should be possible to have forced switching, which would be activated by administrative events such as fibre re-routing, fibre replacement, etc.
- It is necessary to avoid unnecessary switching. Because unstable switching affects service quality, unnecessary protection switching and unnecessary revertive protection switching should not occur.
- It should be possible to realise switching without connection loss of the ATM connections.
- It should be possible for the operator to choose between a revertive and a non-revertive switching mode.
- The service halt time should be less than 50 ms if the extra traffic option is not used.
- The events or conditions that trigger automatic switching should be chosen among the G.983.1 OAM parameters.

- The chosen protocols and mechanisms must apply to the B-PON section layer.
- Extra traffic should be carried over the protection entities while the working entity is active and would not be protected. This capability will provide effective usage of bandwidth on the protection entities.
- It must be possible for an operator not to activate the extra traffic option (e.g., to achieve a lower service interruption time).

Extra messages are introduced into the base set of PLOAM messages from G.983.1 to cope with the additional functionality required for protection switching but generally it is the OLT that initiates the switch-over.

There are two basic protection schemes that can be used and the authors do not attempt to judge failure mechanisms within this chapter and make no recommendations as to the better protection mechanism to use. The first is a partial protection scheme and is designated type B protection. It is useful when it is the feeder fibre(s) that are considered at most risk of failure. The more fully protected scheme is known as type C protection. These will be individually described in Sections 9.6.1 and 9.6.2.

9.6.1 Type B protection (OLT protection only)

In type B protection, the protected components are the OLT and the fibre feed to the first split only (see Figure 9.6).

9.6.2 Type C protection (fully protected 1 : 1 and 1 + 1 protection schemes)

Type C allows for protection of the OLT and the fibre infrastructure (see Figure 9.7). It also allows for 1 : 1 protection, where every working system has a standby system, and

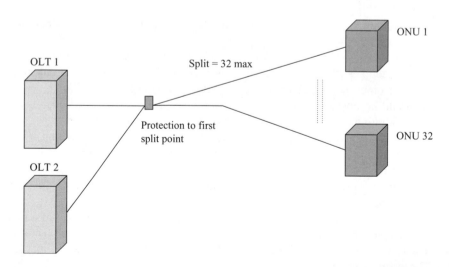

Figure 9.6   Type B protection schematic

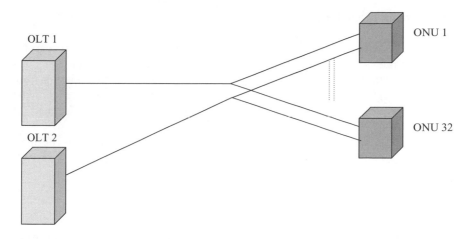

Figure 9.7   Type C protection schematic

$1 + 1$ protection, where only a limited number of protection channels are provided. Standby channels may be used for alternative data services until failures occur. As previously mentioned there can be a mix of the two protection schemes within type C protected systems.

9.6.3   OAM specification

In order to create a complete system specification for telecommunications equipment, it is necessary to detail the managed entities and their scope as well as degree of interaction. For the APON specifications, OAM (operations, administration and maintenance) functionality is specified in two areas. First, we have Recommendation G.983.2 ('ONT management and control interface specification for ATM PON'), and second, we have G.983.7 ('ONT management and control interface specification for dynamic bandwidth assignment (DBA) B-PON system'). These are complementary specifications.

The ATM-PON element management system will only manage ONTs as part of the ATM-PON system through the OLT using the ONT management and control interface.

The OAM requirements present requirements for the ONT management and control interface. First, it specifies managed entities of a protocol-independent management information base (MIB) that models the exchange of information between the OLT and ONT. Second, it covers the ONT management and control channel, protocol, DBA protocol, and detailed messages. Refer to the ITU-T specifications for more detailed information. For point-to-point systems, the situation is not so far advanced and no common specifications exist for OAM applications.

9.7   PON applications

PON systems have never be deployed in quantity in Europe for several reasons. Whilst it is true that these systems need to be multi-service, it is also fair to say that most systems to date have not proven to be so with PSTN services missing from most designs. This situation is now changing and several manufacturers are showing PSTN on their PON product roadmaps, with delivery of such systems scheduled for mid 2003.

Despite the costs and lack of flexibility associated with current equipment designs, there are inherent advantages of PON technology and the next sections look at types of deployments for which PON systems are most suited.

9.7.1   As access devices

As has been said many times, optical fibre in the access network would create the ultimate access network, although controlling the costs of implementing such architectures is key. PON technology has been shown to reduce the costs of access fibre network rollout for two reasons:

- It allows a reduction in the number of fibres back to the exchange giving better duct utilisation and shared head-end (OLT) costs.
- It inherently concentrates the traffic onto fewer data streams hence reducing the cost of metro equipment.

Given the increase in speed of computing devices, and the proliferation of the Internet, it is to be expected that the market for high-speed connections will continue to grow and will become ever more resource hungry, both in terms of bandwidth connectivity and service support. In this market it is to be expected that optical technology will outstrip other access network technologies to become the technology of choice.

9.7.2   As core feeder technologies

Most operators are rolling out ADSL systems and are considering the next step ahead. VDSL is considered a realistic future step for the access network (see Figure 9.8). However, all DSL technologies have one major weakness, they require a metro-type transmission network to feed them. Some in the industry are experimenting with new feeder architectures based on PON designs. Current 155 or even 622 Mbit/s systems are not really powerful enough for these core feeder applications but future PONs based on 1.25 Gbit/s and higher seem to be well suited to these applications.

One could visualise an architecture where the access and metro/core networks could be rolled into one and served by powerful PON systems. Studies are already underway to examine the cost implications of such architectures and if successful, these could pave the way for the introduction of the technology in a much shorter timescale than would otherwise be possible. These systems would need greater split ratios and much longer reach than current PON technology would allow.

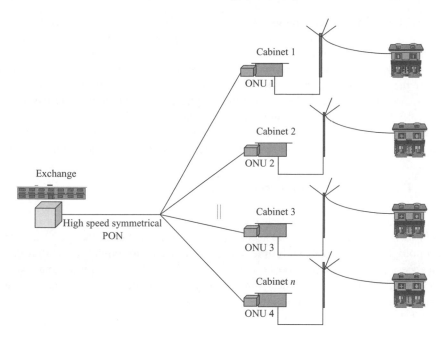

Figure 9.8   VDSL feeder schematic

9.7.3   Point-to-point applications

Point-to-point systems generally utilise multiplexors at or near the street cabinet. This is not ideal for most ILEC applications although CLECs may find the technology useful in the longer term as this fits with their current architectures.

It is fair to say that point-to-point architectures usually compare favourably to PON architectures where there is a strong demand and a central site at which concentration can be achieved, and hence they are often used for campus environments.

Point-to-point technology has been used successfully where there is a need for fast Internet service, however, the systems tend to be poor at providing voice services. For that reason they seem best suited as an overlay technology, leaving existing services on copper. For the network operator, however, the costs of maintaining two parallel networks can prove highly undesirable.

9.8   Future PON technologies

Current PON technology is limited to 622 Mbit/s, however, this is not a practical limit for this technology, merely the highest agreed bit rate supported by the ITU-T. The IEEE and FSAN are already considering systems with 1.25 Gbit/s PON speeds, and higher bit rates are feasible.

Whilst the IEEE Ethernet PONs are rumoured to be a lot cheaper that APON technology, so far no systems have been produced to test this hypotheses. Time will tell whether this technology will actually prove to be cheaper although many components will be common between the two architectures leading to some doubt as to the reduced cost claims.

Already we have seen experimental Super-PON architectures from Alcatel [3,4] where the split ratios can be increased from the more usual 32 to over 100, with thousands or more being possible. Furthermore, optical amplifier technology is making possible PONs with a transmission reach of thousands of kilometres.

WDM technology will allow common infrastructure to serve both residential and business end users, even if those businesses require very high point-to-point data rates.

9.9   Future point-to-point technologies

A constant theme through this chapter is that technology is improving all the time and nowhere more so than in the access network space. It is clear that point-to-point technologies will improve with time, however, they tend to require concentrators in the access domain or long runs of optical fibre, both of which are non-optimal except for green field sites. Despite these limitations, many manufacturers and start-ups are active in this space and future upgrades to this type of equipment are certain.

9.10   Conclusions

The pace of change in the electronics world has never been greater and this is leading to changes in PON technology which will impact significantly on the price/performance characteristics of new fibre access systems. Furthermore, there are many new companies trying to create systems that will become successful in the access network area.

Access network bit rates are increasing from the original 155 Mbit/s (20 Mbit/s for non-standard systems) to 1.25 Gbit/s and seem set to continue to rise. Whilst these bandwidth increases might not be useful in all applications, there is potential for these systems to stray into areas previously the domain of core technologies such as SDH or SONET. It might be that the introduction of new PON technologies results in a reduction in price of the older fibre access systems – making them more cost-effective in the access network space.

Whilst telecommunications operators are presently rolling out ADSL technologies and will be doing so for some time, there is already clearly a case for higher speed access network technologies being introduced into the networks of the world in the near future. Whilst VDSL may have a place in this market, the optical fibre technology must provide the ultimate solution given that the PON was specifically designed for this market segment.

Whilst point-to-point systems will have a part to play, their use as a ubiquitous access network architecture seems unlikely, given the advancements being made in PON technology.

Of course there are questions as to when the cost-reductions and price breaks that always follow new technology will occur, but this cost reduction will happen.

It seems that it is not 'if', but merely 'when' fibre technologies are adopted in the access network as a matter of course, with early adopters perhaps getting the largest customer base. It must not be overlooked that laying optical fibre into an access area is not cheap, and that the economics for a second operator are so much poorer and so a second optical fibre access network in an area will probably never happen.

9.11 Definitions

1 : 1 protection: A system of 1 : 1 protection configuration carries traffic in the working equipment while the protection equipment stands by without serving any traffic. The 1 : 1 protection configuration is a special case of the more general 1 : N protection where the protection equipment is shared among N sets of working equipment.

1 + 1 protection: A system of 1 + 1 protection configuration carries identical traffic in both working and protection equipment.

9.12 References

1   REEVE, M. H. *et al.*: 'Design of passive optical networks', *BTTJ*, 1989, **7**(2), pp. 89–99
2   GALLAGHER, I., BALANCE, J. and ADAMS, J.: 'The application of ATM techniques to the local network', *BTTJ*, 1989, **7**(2) pp. 151–60
3   MARTIN C. *et al.*: 'Realisation of a SuperPON demonstrator', NOC 97, Antwerp, June 1997, pp. 188–193
4   VETTER, P. *et al.*: 'Economic Feasibility of SuperPON access networks', Proceedings of ISSLS 1998, Venice, March 1998, pp. 383–389

Chapter 10

Fibre to the home infrastructure deployment issues

A. Mayhew, S. Page, A. Walker and S. Fisher

10.1 Introduction

To date the use of optical fibre in the access network has typically only proved cost effective for supplying the high bandwidth demands of large corporate customers. For new and established network operators, however, the increasing demand for bandwidth to deliver bearer, interactive and bundled services to business and residential customers is requiring them to seriously consider the high volume rollout of optical fibre based systems. Network operators therefore face some major decisions, not only in terms of the type of fibre transmission systems to deploy, but also how to install a cost effective network of fibre cables, ducting and joints to connect to the customers they wish to serve. Installing fibre in the ground represents a major commitment and a long term investment. Network operators can typically expect the fibre infrastructure to equate to at least 60% of the cost of the overall access transmission system. Planning and building such networks is a major upfront investment, and in today's highly competitive markets operators are faced with the added complications of uncertain take up of services by customers and the likelihood of high customer churn.

This chapter focuses on fibre to the home (FTTH) and the deployment options and challenges for the physical fibre infrastructure. The key difference associated with connecting optical fibre to residential properties instead of business properties is not technical but commercial. There is typically a much smaller potential revenue from a residential property than from a business property. This leads to a need for cost optimisation of both the transmission system and the fibre infrastructure. The chapter also examines the fibre infrastructure issues that need to be considered for a FTTH deployment and describes the fibre access network modelling activities carried out by BTExact's Broadband Network Engineering Unit. The aim of this modelling is to automatically plan and cost optimise the deployment of access networks based on real geographical and demographic data.

BT's telecommunications network in the UK consists of two main segments. First, a core network of approximately 5500 exchanges (switching sites) connected by high speed, mostly optical fibre, links. Second, an access network connecting customers to the switching sites. For large, and increasingly for medium sized customers, this access network is also based on optical fibre. The fibre access network is currently carrying traffic at anything from 2 Mbit/s to hundreds of Mbit/s. For residential customers the access network consists of twisted copper pairs contained within a tree and branch cable network. This network design has not changed appreciably for 50 years or more. Indeed some cables have been in service for that long.

10.2   The access network

The tree and branch copper access network consists of two types of flexibility points (PCP cabinets and DPs) where twisted copper pairs can be patched. There are also a number of cable joints where cable segments can be joined to allow further branching (see Figure 10.1).

The UK copper access network was originally designed and built to support telephony service. Originally, customers wishing to transmit data were restricted to using modems which operated in the 4 kHz voice-band. The first voice-band modems were introduced in the 1950s and were capable of operating at 300 bit/s, but these rapidly developed to 28.8 and 33 kbit/s. Transmission equipment can now exploit the far greater bandwidth capabilities of the network infrastructure between the exchange and the customers' premises. This is accomplished by transmitting at much higher

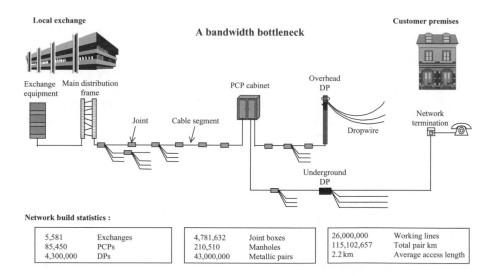

Figure 10.1   The BT UK copper access network

frequencies. For example, asymmetrical digital subscriber line (ADSL) technology uses frequencies up to 1.1 MHz and is capable of delivering up to 8 Mbit/s downstream to the customer, and 640 kbit/s upstream to the exchange over a single copper pair.

A key consideration in the use of ADSL, however, is the trade-off between bit rate and transmission reach. Achievable transmission distance reduces significantly in relation to increasing bit rate. In order to achieve an acceptable coverage of the UK customer population, the highest bit rate offered by commercial ADSL data products (DataStream and IPStream) is 2 Mbit/s downstream and 250 kbit/s upstream. These products are aimed at addressing the considerable pent-up demand within the existing Internet population for fast 'always-on' Internet access. However, faster access to video-rich Web sites, integrated conferencing and multi-player gaming will create demand for even higher transmission capacity. To capture the mass market, service providers need to offer compelling content and applications. This content will typically be visual and readily accessible, and will include video conferencing, video-on-demand, film and TV programming, entertainment services and live net events.

A European Union funded project (GIANT) is aiming to determine residential and small to medium enterprise (SME) bandwidth requirements for the period from 2004 to 2010. This project is in its early stages, but has identified services that require high bandwidths for short periods of time (for instance software delivery and storage area networks). High bandwidths will be required for more extended periods for multiple channels of broadcast quality TV or for high definition TV (HDTV). The bandwidth requirements for the identified bundles of services will be beyond the capabilities of ADSL.

There are a number of technology options for increasing the bandwidth available per customer. These options include very high speed digital subscriber line (VDSL), radio, satellite, free space optics and optical fibre. Multimode fibre would allow the use of low cost transmission equipment, but with limited range and bandwidth. More ideally, single mode fibre has the ability to transmit huge amounts of traffic over long ranges.

Ultimately, only optical fibre offers the capability of cost effective wide scale provision of the full range of broadband services that GIANT has identified. International standards recommendations for fibre to the home (FTTH) have been largely defined by the Full Service Access Network (FSAN) forum. The costs of transmission equipment designed for FTTH should reduce once these standards are agreed by the International Telecommunications Union (ITU) and the transmission equipment is produced in high volumes.

10.3   Advantages and disadvantages of optical fibre

10.3.1   Advantages

The main advantage that fibre has over copper is its potentially enormous bandwidth. By using wavelength division multiplexing (WDM) to transmit data at multiple wavelengths capacities of several Tbit/s have been demonstrated. Even using a single wavelength data rates of several Gbit/s are widely used in core networks. Furthermore, fibre is able to transmit these volumes of data over long distances. Compared to the limited bandwidth and reach of ADSL and VDSL technologies, fibre transmission systems enable far higher bandwidths to be supplied to a greater proportion of customers and therefore offers considerable network future proofing.

Fibre is also smaller and lighter than copper. A sheathed copper pair consists of two wires typically 1 mm in diameter. The alternative is either a single fibre or pair of fibres with a diameter of 0.25 mm. This enables fibre cables to be designed with smaller diameters than copper cables. Size is important because the cables are typically installed into underground ducts which have limited space. Installing new ducts is very expensive since this usually involves digging up and reinstating pavements and roads. Fibre being lightweight also offers handling and transport advantages.

One major problem with metallic transmission paths is corrosion caused by moisture. This can cause interference on lines and premature failure. In a well maintained network, corrosion can be minimized but it is unrealistic to believe it can be eliminated. Fibre does not corrode and has been shown to be able to withstand long periods of water immersion without failure [1].

Furthermore, unlike DSL systems over copper, optical fibre transmission is immune to interference from signals carried on other fibres or metallic transmission paths within the same or adjacent cables. Radio frequency transmissions have no effect on optical signals, and fibre transmission does not generate radio frequency emissions.

10.3.2   Disadvantages

The main disadvantage of fibre compared to copper is the increased difficulty of joining together and terminating fibre at transmission equipment. This requires both a more skilled workforce and more expensive tools. Fibres are generally joined using an electric arc fusion splicer. As fibre usage around the world has increased, the cost of these machines has reduced considerably, although they still remain an expensive investment.

Financially, the cost per cabled fibre is generally greater than the cost per cabled copper pair. The premium depends upon the gauge of the copper pair and the number of copper pairs/fibres within the cable, but may typically be approximately 40%. Due to the capacity of fibre, however, it is feasible to share a single fibre between several customers all requiring broadband services. This would not be feasible over copper.

These disadvantages in themselves do not present a major barrier to the widespread use of fibre into access networks. The biggest barrier is perhaps the existence of usable copper pairs in the ground and the established workforce, operational practices and tooling associated with copper networks.

10.4   Existing fibre usage

The earliest use of optical fibre dates back to the early 1980s. This was for increased capacity in the core network (i.e. the cables that transmit traffic between exchanges).

The first fibres installed were multimode, but these were soon superseded by higher performance single mode fibres. Originally cables generally contained only a few fibres, typically four to eight.

Twenty years on, core networks consist mostly of optical fibre, with cables containing up to 96 fibres or even more in a few cases. The transmission systems operating over these fibres have undergone a continual increase in operating bandwidth to reach the multi-Gbit/s bandwidths currently being deployed. It should be noted that many fibres in the core network are still transmitting traffic from legacy systems at a fraction of the current system bandwidths.

In the UK, access network fibre has been deployed to major business customers for over ten years. These links generally consist of a traditional fibre cable from the serving exchange to a fibre node (external network splicing position) which is located to serve a number of customer sites. Blown fibre is then used to make the final connection from the node for the final few hundred metres to each customer's premises. Alternatively, if a particular customer's service requirements are sufficiently high, they maybe fed with conventional fibre cable direct from a serving exchange.

10.5 Fibre to residential customers

10.5.1 Network deployment options

There are two main deployment options for migrating residential customers from their current copper feed to an all fibre feed. The first is to gradually extend fibre from exchanges into the access network with each fibre (or pair of fibres) feeding progressively fewer customers over time. For example, using industry terminology this would be fibre to the cabinet (FTTC), followed by fibre to the kerb (FTTK) and then fibre to the home (FTTH). In the UK access network this equates to Fibre to the PCP (primary cross connect point), DP (distribution point) and customer premises. Figure 10.2 illustrates this evolution with indicative customer groupings. At the fibre to copper interface active electronics in the form of multiplexors would be used to consolidate, groom and convert signals between the fibre and copper medium. This gradual evolution approach allows customer bandwidth to be increased in three steps in line with increases in customer bandwidth demand.

The second option is to move directly from copper to fibre for selected customers. This requires a greater initial investment but may be more cost effective in the long term. This is because the network, transmission systems and processes would only need to be changed once.

Looking at the bigger picture there are also two main routes for migrating from an all copper to all fibre residential access network. The most likely of these is progressively overlaying the copper network with fibre to serve the increasing number of customers requesting service. The purpose of gradually overlaying the existing copper access network with fibre would be to provide an incremental migration of residential customers from copper to fibre. Operationally, this is initially simpler and would also require less capital investment per exchange than a complete change out. This approach would, however, create a considerable ongoing financial burden

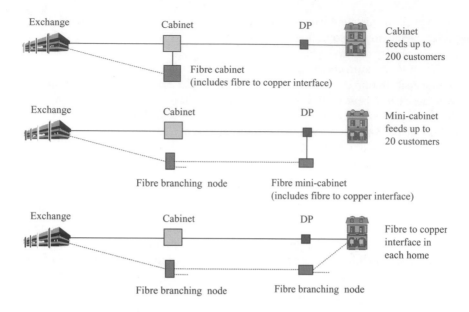

Figure 10.2   Step by step evolution to FTTH

associated with operating two transmission technologies in parallel. The aim for an overlay approach should be to make best use of the existing infrastructure assets (e.g. ducts, footway boxes and poles) where spare capacity allows. Rebuild costs due to the congestion of duct, boxes and exchanges (by the existing copper cables and plant) may be high, especially for high FTTH customer take up. Opportunities for eliminating copper from areas of the network (for new housing developments or during network uplift) should be considered in the future.

The alternative to an overlay approach is to change out whole DP, PCP or exchange customer catchment areas. This approach would be much more efficient in terms of labour and ultimately result in a lower cost per customer connected. It would, however, only be cost effective if a high percentage of customers on the DP/PCP/exchange require high bandwidth services. Changing out large sections of the network in this way also presents major practical and operational challenges, not least how to maintain customer services whilst large sections of the network are cut over from one technology to another. For a PCP change out customers would need to be moved off their existing copper bearers whilst the old PCP cabinet was removed together with the associated copper cables. Service would need to be maintained by some agreed method (e.g. mobile phones or temporary radio links). The new fibre cables could then be installed in the freed duct space. This sounds attractive but is complicated because most E-side cables feed more than one PCP. An additional complexity is that cables from exchanges to PCPs (E-side) and from PCPs to DPs (D-side) often coexist in the same duct segments. These factors mean that several PCPs (or a whole exchange) may need to be simultaneously changed out. This would be a considerable logistical challenge.

For green field sites, installation of FTTH as a day-1 alternative to copper could be considered in the near future. This may be achievable for a relatively small premium once suitable transmission equipment is produced in high volumes.

An important decision to be made is the split between proactive and reactive network provision. To enable customers to be connected in an acceptable timescale some access network build should be installed proactively before customers request service. This network can only be dimensioned based upon estimated customer take-up. Generally the higher the expected take up, the higher the proportion of proactive build that is cost effective. Proactive build is usually less expensive since larger tasks can use labour more effectively than a number of separate reactive task visits.

10.5.2   Network design options

There are a number of different options for designing a residential fibre access network. Architectures can be tree and branch or ring based, dedicated fibre (point-to-point) or shared fibre (PON). Sharing can be achieved using passive optical splitters (passive optical network) or active components (e.g. multiplexors). Fibre will generally be installed in underground ducts, although aerial options using poles or electricity pylons and installation within gas or water pipes are commercially available.

For cost effective wide scale deployment of FTTH, reuse of the existing tree and branch duct network is seen as crucial. Many network operators do not have an extensive installed duct network, allowing them more freedom in network design. However, overall costs could be expected to be higher for these operators than the more established Telcos.

As shown in Figure 10.3, FTTC and FTTH networks at the single system fibre level appear very simple. But at the physical network level of cables and joints, the need to consolidate and manage large volumes of fibre in a flexible manner, creates a far more complex picture.

A schematic of the physical infrastructure elements required to build a FTTH access network are shown in Figure 10.4. This highlights the various types of infrastructure items and how they could be connected together to produce a branching network from the exchange building to the customers' premises.

10.5.3   Infrastructure requirements

Regardless of the network topology chosen, a FTTH network can be considered to consist of flexibility points connected together by cables. A cost effective network is likely to consist of more than one type of flexibility point and more than one type of cable.

10.5.3.1   Flexibility points

The requirements for a flexibility point depend upon a number of factors including:

- The number of input and output fibres.
- Whether fibres are spliced in the flexibility point or routed through by patching tubes.

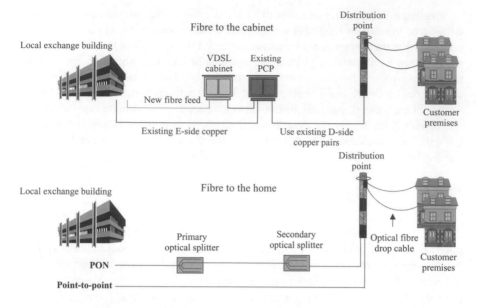

Figure 10.3   Fibre level schematics of FTTC and FTTH

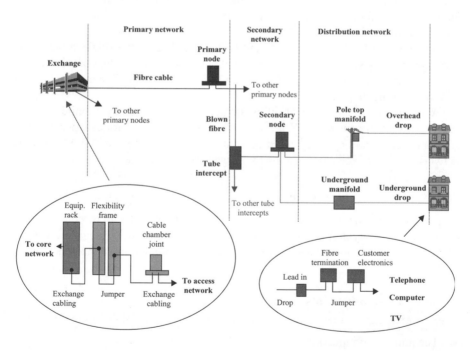

Figure 10.4   FTTH infrastructure connectivity

- The network architecture chosen (PON, dedicated fibre or active split).
- The degree of flexibility needed.
- The location of the flexibility point.
- The existing network infrastructure (underground chambers and poles).

These requirements influence the complexity and size of the flexibility points and hence their cost. Cost becomes more important as the network gets nearer to customers, since the cost of the item is shared between fewer and fewer customers. For all flexibility points ease of use is a prime consideration. Each flexibility point is purchased and installed only once, but it may be re-accessed (to reconfigure or to add customers) many times during its life. The whole life labour costs associated with a flexibility point may consequently far outweigh the purchase cost.

The flexibility points shown in Figure 10.4 can be split into two main types:

- Fibre flexibility points: primary node and secondary node. Fibre splicing or fibre splitting (for PON networks).
- tube flexibility points: tube intercept, pole top manifold and underground manifold. For branching blown fibre tubes to allow fibre unit to be blown from a secondary node to a primary node and from a secondary node to a customer.

Additional fibre and tube branching points may be required in complex branching networks. For instance, spur joints to split a fibre cable to feed multiple primary nodes.

10.5.3.2 Cables

The requirements for cables are largely dependent upon the number of fibres that are needed. The quantity of branching will also influence the choice of cabling technology. Three main choices exist as shown in Table 10.1. A wide variety of cables are commercially available for each of these cable types. Additional factors to consider include:

- fibre density – to make best use of existing duct space;
- speed of installation and jointing;
- labour and installation equipment required.

Table 10.1   Cable options

Cable type	Fibre counts	Suggested use
Blown cable	12 to 144 fibres (in elements of 12 fibres)	Primary, secondary for high fibre counts
Blown fibre	Up to 19 tubes each of 1, 2, 4, 8 or 12 fibres	Secondary and distribution for low fibre counts
Ribbon cable	Up to 2000 fibres	Primary (to spur joint) for high fibre counts

10.5.4   Network deployment issues

10.5.4.1   Primary network

The main issues associated with the primary network are:

- Quantities of fibres required along a route. This influences the choice of cable.
- The number of Primary Nodes that can be optimally fed via a primary cable. This dictates the level of branching required.
- The amount of fibre flexibility required. Complexity of fibre configuration is offset against the fibre utilisation in a branching network.

10.5.4.2   Secondary network

The main issues associated with the secondary network are:

- The amount of fibre flexibility required. Complexity of fibre configuration is offset against the fibre utilisation in a branching network.
- Quantities of fibres required along a route which influence the choice of cable.
- The number of secondary nodes that can be optimally fed via a secondary cable. This dictates the level of branching required.

10.5.4.3   Distribution network

For the distribution network the greatest issue is how to cost-effectively install an optical fibre over the final 20–100 m to the customer premises. The problems encountered differ depending on the customer's existing 'final drop' solution. The scenarios can typically be split into three categories.

- Those fed by copper cables in underground ducts – simple to over-install a blown fibre drop into the same duct.
- Those fed by overhead copper drop cables from telephone poles – the feasibility of an overhead fibre drop as a replacement to the copper drop is being investigated by BTExact (this is a challenging product in terms of survival in extreme weather conditions).
- Those fed by copper cables directly buried in the ground – the use of novel cable installation techniques such as slot cutting may be required to make connection of these customers commercially viable.

The products and practices associated with connecting individual customers back to a secondary node will have a significant impact on the total cost per customer and therefore needs to be optimised.

10.5.5   Other deployment issues

10.5.5.1   Resources

Workforce

- Deploying FTTH to a large number of customers will require a significant increase in the number of personnel trained for fibre installation and jointing. Increased reliance on contractors is likely to be required. Some re-training of copper field personnel can be envisaged, but it should be borne in mind that there will be a continued requirement to maintain the copper access network, since the majority of customers will remain on this for a considerable number of years.

- Even with an increased fibre installation workforce, the scale of infrastructure build that will be required for wide scale FTTH deployment will mean that the timescales for completion with be at best 5 years and more realistically 10 years.

Tools and vehicles

- There will be an increased demand for fibre installation and jointing equipment. The scale of this will depend upon the rate of fibre build and customer connection, plus the geographical spread of the roll-out.
- It will be important to optimise fibre installation equipment for FTTH to reduce the time and resource required to connect customers to the secondary network.

10.5.5.2   Competition and regulation

It is unlikely that a residential customer will ever require more than one fibre connection (the capacity of fibre has been demonstrated to be several terabytes). Competition for FTTH could take a number of forms:

- Local competition in physical infrastructure – more than one company builds a physical fibre network and targets the same customers.
- Dedicated areas – through agreement with government, franchises are awarded for areas resulting in all customers in the area who require broadband services being connected to a single fibre access network. The customer in reality does not care whose network his services are delivered over, only that he has a full choice of service providers.
- No competition – one company is given the opportunity (and obligation) to provide universal fibre access (or acceptable true broadband alternatives for rural areas) with government support and industry/commercial partnerships. This network provider is then required to allow all service providers access to the fibres on a level basis.

Deploying a wide scale FTTH network would involve a very large investment. The incorporation of commercial safeguards may make investors more interested. These safeguards could include government agreement not to regulate the price of access to service providers below a profitable level.

10.6   Geographical network modelling

Given the vast number of optical access systems it is essential to have automated tools in order to be able to contrast and compare competing technologies and fairly assess and optimise the strategic and tactical options for their deployment. The Broadband Network Engineering team have developed a geographical network modelling tool called NetMod. This will automatically determine the cost of deploying access fibre systems, plus the associated cost and physical deployment sensitivities. These activities are commonly referred to as Network Modelling and use geographic and demographic information, together with existing network data to accurately plan network infrastructure build based on existing assets, targeted customer types and expected take up of services.

Network modelling can be performed at three levels. Each of these have their advantages and disadvantages (see Table 10.2). FTTH network infrastructure has been modelled at all three levels by BTExact and this is discussed in the following sections.

10.6.1   Network data

In the UK approximately 22 million customers are connected to BT's 5600 exchanges via copper cables. There are two main types of flexibility point between most customers and their serving exchange. The quantities of these are shown in Figure 10.5.

Table 10.2   Network modelling

Modelling level	Method	Advantages	Disadvantages
Basic	Use average lengths and customer densities.	Rapid results.	Large error bars.
Detailed	Design network using real lengths and customer densities and existing duct routes and roads.	Highly accurate for modelled area.	Time consuming. Difficult to translate to non-similar areas.
Automated	Design network using geographic positions of exchanges and flexibility points.	Narrow error bars across all geographic areas.	Highly complex. Reliant on the availability of accurate data. Requires manipulation of large amounts of data.

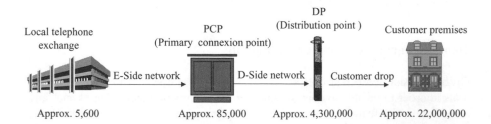

Figure 10.5   BT network data

Table 10.3   Network data

From CSS (copper data)	Exchange	PCP	DP
District identifier	×	×	×
Unique code	×	×	×
Name	×		
Address	×	×	×
Type		×	×
Capacity (working circuits)	×	×	×
Business/residential split		×	×
Broadband status	×		
Location (X–Y co-ordinates)	×	×	×
From INS (fibre data)	Exchange	Node	Cable
District identifier	×	×	×
Unique code (A1141)	×	×	×
Node details		×	
Capacity		×	×
Lengths			×
Location (X–Y co-ordinates)	×	×	×

In order to design a fibre overlay network using the same ducts as used by the copper network, details of these flexibility points are required. Table 10.3 lists the data held by the network modelling team.

10.6.2   Network modelling tool

The network modelling tool (NetMod) was built using VB.NET linking to MapInfo Professional. It aims to model an existing copper access network, dimension it for fibre for the number of customers to be served, and then cost it. A geographical model of the access network is required, since existing network data is commonly not held on computer in any usable form. Network record systems typically hold information relating to the location of various nodes within the network, but no detailed information on how they are physically connected. NetMod uses the node information together with expert algorithms to produce an estimate of the existing network topology. NetMod consists of a number of modules and these are now examined.

10.6.2.1   Demographics module

Demographic analysis can be incorporated into the tool to enable customer targeting based upon established socio-economic groupings. The demographics module calculates the numbers of potential customers likely to take up service at each PCP and DP network node. This is calculated dependent on the total number of customers (served and unserved) within each PCP/DP area, and by the percentage of expected take up entered for each socio-economic group. This module is used for targeting services (and therefore, technology) on specific groups of customers. It identifies the market size for the technology. This is different from the penetration used later, which dictates the operator's market share.

10.6.2.2   E-side geography builder

The E-side geography builder estimates the duct route between the exchange building and the PCPs within that exchange serving area. The estimated route will not be the actual route topology, but a good estimate as to how it might exist. Rather than use an unrepresentative star network connecting all PCPs directly to the exchange, spines and spurs are used to make these connections indirectly via other PCPs. This produces a simple tree and branch network.

10.6.2.3   D-side geography builder

The D-side geography builder is similar to the E-side version but estimates the duct route between each PCP and the DPs it serves.

10.6.2.4   Dimensioning module

This module takes the output from the geography builders and dimensions the required network at the fibre and cable level. It can dimension networks for FTTC and FTTH using both PON and point-to-point systems. It can take account of the strategic use of blown cable and blown fibre technologies, and calculate the number and size of cables and equipment required for each type of network. A key aspect of the dimensioning module is its ability to use its estimation of the tree and branch nature of the duct topology to intelligently plan how the required fibre capacity for a particular system is consolidated into 'sensible' cable sizes. This approach more closely represents how a real network would be deployed. It therefore generates far more accurate costings than many other simpler cost models which assume star network topologies at the fibre level and rarely take account of cable capacity planning.

10.6.2.5   Costing module

The costing module calculates the cost of building the access network for each of the options output from the dimensioning module. This is the simplest part of the process converting quantities to costs using a file of item costs and labour rates.

10.6.2.6   Results output

Results from NetMod can be either numerical or graphical. Figure 10.6 shows an example of a graphical summary output from the demographics module. It highlights the number of potential customers in East Anglia for a particular system service set. With no demographics applied (100% of population) customer take-up numbers are shown within predefined bands, whereas when demographics are applied only targeted customers are identified.

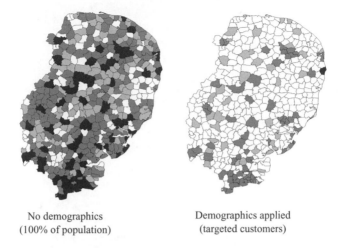

No demographics
(100% of population)

Demographics applied
(targeted customers)

Figure 10.6   Demographics module output

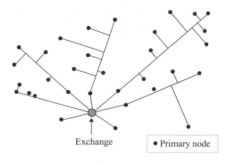

Figure 10.7   Generated E-side network duct topology (exchange to primary node)

These maps are useful for visualising the effect of varying estimated customer take-up of a particular service set/technology. They can also be used to develop strategies for cost effective network rollout based on targeting specific customer types that promise high potential revenue.

Simple networks designed by the E-side and D-side geography builders respectively are shown in Figures 10.7 and 10.8. Figure 10.7 shows a spine and spur network linking a number of primary nodes (positioned at PCPs) to their serving exchange. Figure 10.8 shows a network connecting secondary nodes (positioned at DPs) to their serving primary node (positioned at a PCP). By repeatedly running the geography builders for all the E-side and D-side networks within an exchange area the entire network within the exchange area can be designed. This can then be dimensioned and costed. This process can be repeated for all exchanges within the UK or for a representative sample of exchanges.

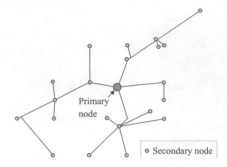

Figure 10.8   Generated D-side network duct topology (primary node to secondary node)

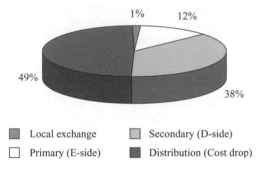

Figure 10.9   PON cost breakdown

10.7   FTTH infrastructure cost breakdown

Detailed cost modelling of FTTH infrastructure has been performed for the town of Ipswich. This area includes five exchanges which serve a total of approximately 70 000 customers. The average cost per customer connected has been estimated by dimensioning a fibre network for an expected level of customer take up. The results of the modelling activity highlight the following cost trends.

Figures 10.9 and 10.10 show simple summary charts indicating the network infrastructure cost breakdowns for PON and point-to-point networks to serve an expected customer penetration of 20% (with no demographics applied). In both cases it can be seen that the distribution side of the network contributes the greatest cost per customer. This is because this network segment is shared between the least number of customers. Optimisation of the distribution network is therefore crucial in order to reduce overall costs. For both PON and dedicated fibre networks, labour costs contribute approximately 50% of the total infrastructure bill. A key part of this process is using NetMod to carry out cost sensitivity analysis and hence start to develop planning rules that ensure cost effective network deployment.

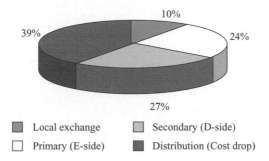

Figure 10.10   Dedicated fibre cost breakdown

Detailed cost breakdowns have also been produced to identify those items of the network build that contribute most to the cost per customer connected. These individual items are now being targeted for potential cost reduction. For a PON network the highest cost items are distribution blown fibre and optical splitters. For dedicated fibre the highest cost items are primary cables and distribution blown fibre.

10.8   Infrastructure cost reduction

It would be technically possible to use current commercially available fibre infrastructure for FTTH. However, much of this infrastructure has been designed for dedicated fibre feeds to business customers. It has been realised that this plant is likely to be over specified for wide scale FTTH provision. Similarly operational practices have been developed for providing fibre feeds to individual or small groups of customers and are not likely to be optimum for FTTH. It can be concluded that cost optimisation is required for both infrastructure stores and labour before FTTH could be seriously considered. It should be noted that a significant reduction in transmission system costs is also required. This is discussed elsewhere [2].

The largest single factor that affects the cost per customer of installed FTTH infrastructure is customer take up. High customer take up in a geographical area enables greater sharing of network segments and consequently greater sharing of the costs associated with those parts of the network. The effect for the town of Ipswich is demonstrated in Figure 10.11 for both a 32 way split PON and a dedicated fibre network. It can be seen that the cost reduction tails off above 50%. For a PON network the infrastructure cost per customer is halved by increasing the customer take up from 10 to 50%.

10.8.1   Stores

Reduction of stores costs could be achieved by one or more of the following:

• eliminating non-essential functionality;
• size reduction;

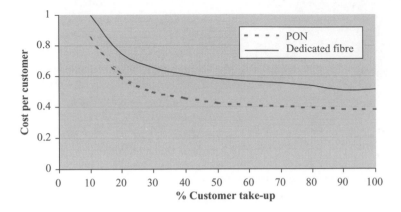

Figure 10.11   Infrastructure cost reduction with increasing customer take-up

- making use of lower cost materials;
- due to reduced production costs resulting from higher volume production.

For cables, the main method for cost reduction is seen as miniaturisation. This has knock-on potential cost savings in reduced underground duct build, plus easier transportation and handling. Several manufacturers are developing miniature blown cables with up to 48 fibres that can be installed into bores of typically 10 mm diameter. BT is currently investigating the possibility of considerably reducing the current 3.5 mm bore size of blown fibre. This would be particularly advantageous for the distribution network.

For flexibility points all the above methods may be significant. The reduction in cable size may enable flexibility point size reduction. Further size reductions may be achieved by optimising fibre management within the flexibility points. For instance it may be decided that single circuit management, currently used by many operators for business customers, is not essential for residential networks. The robustness of the infrastructure items may also be lessened by reducing the thickness of material or by using a lower grade material. The implications of these choices should be fully investigated. It would be false economy to install infrastructure that has a significantly higher failure rate just to save a few percent on initial product costs. At least one manufacturer has launched a fibre flexibility node designed specifically for FTTH networks. It is envisaged that other companies will follow suit as FTTH trials expand and evolve into initial rollouts.

10.8.2   Labour

Labour costs, whether for direct employees or external contracts, form a significant part of the total deployment costs for any network. For FTTH a large amount of external network installation will be involved. The bulk of costs are attributable to this installation work, including duct build, cable installation and jointing/configuring cables. It is, however, important to consider the wider labour

Pre-installation	Research/Feasibility studies Legal/Regulatory issues Strategic planning Network surveys Removal of obsolete plant to free up duct space Product approval Product procurement Marketing/advertising Personnel training
Installation	Detailed planning Liaison with councils (notices etc) Customer order handling Personnel resourcing and scheduling Management and control of installers Installation of exchange plant Installation of external plant Installation of customer premise plant Premiums for out of hours work
Post installation	Acceptance testing Auditing Reinstatement of paving Maintenance Network upgrades/replacements

Figure 10.12   FTTH network life cycle

requirements associated with the whole lifetime of the FTTH network as shown in Figure 10.12. Not only does this add to the labour costs, it demonstrates the number of different functions required; these functions are usually carried out by different personnel.

Reduction of labour costs could be achieved by one or more of the following:

- making the best use of the existing infrastructure (particularly duct network);
- optimising network design;
- more efficient practices;
- more efficient processes;
- better utilisation of personnel;
- deskilling practices to enable lower paid personnel to be used.

10.9   Development of a portfolio of services

The biggest barrier associated with deploying FTTH may not be due to technical or cost constraints but due to customer demand. However much the infrastructure and transmission equipment is developed and cost optimised, unless sufficient customers want the services that FTTH can deliver it will not be commercially viable. Most customers are interested only in the services that they have access to and not the network which provides them. As discussed earlier, services such as software delivery and High Definition TV require bandwidths that would greatly stretch the capabilities of most, if not all, technology options except FTTH. To drive FTTH deployment, one or more high bandwidth killer applications need to be developed which will increase the revenue per customer. Until such applications are found FTTH deployment is likely to continue at a more sedate, though no less stoppable, pace.

10.10   Conclusions

- The deployment of FTTH is largely constrained by commercial and not technology issues. What is needed is both a reduction in the cost of deployment and an increase in the perceived revenue that may be achieved. Once an acceptable profit margin can be seen the speed of deployment will accelerate. This may fuel further cost reductions associated with volume production which may sustain the acceleration.
- The development of a killer application, for which many customers would be prepared to pay a premium price, could generate a compelling commercial argument for deploying FTTH more quickly and as widespread as operationally possible.
- Even with a concerted commitment, universal deployment of FTTH across the whole of the UK is unlikely to be achieved in less than five to ten years even if it were started tomorrow.
- Current effort of the Broadband Network Engineering Network Modelling team is concentrated on understanding the costs involved and seeking ways to reduce these. Optimisation of the distribution network is seen as crucial for cost per customer reduction. Blown fibre is seen as a key enabler for FTTH.
- Infrastructure cost per customer is highly dependent upon customer take up. This tails off at 50%, so ideally active marketing should be undertaken to achieve take up of at least 50% in most geographical areas.

10.11   References

1   STOCKTON, D. J., MAYHEW, A. J. and WILSON, P. D.: 'The reliability of optical fibre in simulated wet field conditions over extended periods', Proceedings of the *EuroCable* Conference 1997, pp. 80–86 (ISBN 90 5199 344 7)
2   JAMES, K. and FISHER, S. I.: 'Developments in Optical Access Networks', *BT Technology Journal*, **20**(4), October 2002, pp. 81–90

Chapter 11

Fixed wireless access

B. Willis

11.1 Introduction

Mankind has been communicating via radio signals for a very long time, between neighbouring buildings, across oceans and even from one planet to another. Theoretically such technology can be used for the last mile of a telecommunications network to connect its customers. In practice, however, it is not well suited to such use and it is only over the last decade or two that tailored solutions have emerged, allowing large scale wireless access to become economically viable.

Wireless access splits into two classes – fixed and mobile. Fixed systems are known by many names including fixed wireless access (FWA), wireless local loop (WLL), multimedia wireless systems (MWS), broadband wireless access (BWA) and local multipoint distribution systems (LMDS). While mobile systems allow users to roam around the network, in fixed systems the radio units are permanently mounted in the same way as copper and optical fibre.

Using radio in the access network presents a unique set of engineering challenges, especially if service quality is to rival wired delivery. Securing suitable spectrum is a prerequisite to establishing a network. The transmission medium presents many challenges which are outside the direct control of the network operator, and those that are of particular importance for fixed applications have to be considered when selecting radio equipment. Careful planning of the network is required to ensure it meets initial and longer term market requirements and the initial technology and architecture choices are key to achieving success. When all these aspects have been considered, a network operator can assess the likely costs and revenue of a wireless system and can answer a more fundamental question – is wireless the best choice of access technology for this application? This chapter will explore each of these issues in more depth.

11.2   Spectrum

All fixed wireless access systems require a portion of the radio spectrum in which to operate. In almost all cases, a licence to use this spectrum must be purchased from the appropriate government agency before the system can be operated or even planned. The size of the radio spectrum is in theory infinite, extending from zero hertz through progressively high frequencies to light, x-rays and beyond. In practice only a very limited part of this spectrum is usable for commercial access systems. Demand for these frequencies is huge and growing all the time with competition to fixed terrestrial access systems coming from uses such as satellite communications, radio astronomy, industrial applications, emergency services, television broadcasting, telemetry, mobile communications, military and medical applications.

The large and ever changing appetite for such a scarce resource leads to a requirement for careful management on national, continental and global fronts, which in turn limits the amount of spectrum available for fixed wireless access use at any point in time. Suitable parts of the radio spectrum, or bands, are usually made available by a national government when agreement is reached with all other previous or potential users of the spectrum and there is judged to be significant interest from operators.

Would-be fixed wireless access operators therefore face the difficult situation where their ability to deploy a network is dependent on being able to obtain a licence for suitable spectrum from the local government, who may or may not make such licences available at some point in future. This makes spectrum an absolutely essential ingredient in any planning or operation of wireless networks.

11.2.1   Licence allocation

Commercially, there are several ways in which spectrum licences can be allocated and these are now considered.

11.2.1.1   Free of charge

Some spectrum can be used without payment to the government that owns it. Some such spectrum still requires the user to obtain a licence prior to use and some is set aside for use without any permissions at all, provided certain technical constraints are met. The main example of the latter are the internationally agreed ISM (industrial, scientific and medical) bands, this spectrum often being known as 'licence-exempt'. Whilst the technical parameters vary by country, the aim is the same – to make spectrum available that may be used on an ad hoc basis without need for prior permission. Example applications include microwave ovens and industrial heaters, wireless LANs and remote control units. Another emerging use of these bands in many countries is for commercial fixed wireless access networks, often focused on delivering data services and based on existing wireless LAN technology.

Within the UK there is an unusual constraint on these bands which limits their licence-exempt use to 'private self-provided communications' in effect precluding their use for commercial access networks without a licence. Thus far the government has only issued one commercial use licence per region within the UK.

11.2.1.2 'Per link' fees

This type of spectrum licence charging tends to be applied to point-to-point systems, where dedicated spectrum is required for each customer connection. The operator must apply for the spectrum required for each radio link and the licence obtained is only valid for that frequency in that particular location. Generally, if the link is removed, the spectrum licence must be handed back to the issuing government. Charging may have temporal, capacity, usage and geographic elements.

11.2.1.3 Auction

Block spectrum allocations are often awarded using a competitive auction. Holders of such licences can use the spectrum wherever and as often as they like, within a given country or region. Such awards usually last for a set number of years, with the issuing authority then claiming back the spectrum for reissue or an alternative use.

There have been many successful auctions for FWA spectrum around the world with considerable income being generated, for instance the 1998 US LMDS auctions (spectrum in the 28–31 GHz range) raised \$600 million. However, the UK 28 GHz auction held in 2000 illustrates the risk to governments of using this process – 26 of the 42 licences offered failed to attract any bids, leaving large parts of the UK without any broadband FWA operators to compete with existing access providers.

Other risks occur depending on the government's original motivation in offering the spectrum for FWA usage. In many cases the objective is to increase competition in the access market by using the potentially low cost of entry offered by wireless networks compared to wired networks. One possibility is that the licence winner will 'cherry pick', only offering service in the highest profit areas. This can mean that only the most densely populated business districts will obtain the benefit of increased access competition. Alternatively, the spectrum might be used for infrastructure connections, such as point-to-point backhaul connections within the operator's network, rather than for its intended application in access networks. Outcomes such as these can be guarded against by introducing licence conditions such as minimum coverage areas that must be achieved within a given time and strict definitions on the permitted use of the spectrum.

11.2.1.4 Competitive tender or 'beauty contest'

An alternative method of awarding block licences is commonly known as a 'beauty contest'. In most cases, an annual fee is charged for the licences, but this is an administrative charge, much lower than the likely auction price. The award is based not on which operator is willing to pay the most, but rather the operator who undertakes to make best use of the spectrum.

In a similar way to the minimum permitted performance licence conditions often imposed on auction winning operators, the beauty contest attempts to ensure that the government's original objectives in issuing the spectrum are met. While an auction process limits itself to specifying a minimum set of desirable parameters that the resultant network must exhibit, the beauty contest award decision is given to the operator whose proposed network offers the most desirable features. Exactly how these desirability criteria are judged will again depend on the original objectives.

One obvious flaw with this process is that an operator could over-promise in an attempt to secure a licence. For this reason, the promises usually form the basis of the winner's licence conditions. Additionally it is common for a detailed business plan to be required which shows achievable forecasts for revenue and expenditure.

11.2.1.5   Spectrum trading

A common licence condition for block allocations specifies whether or not spectrum trading is permitted. If it is, the winning operator may subsequently transfer their licence to another party for instance if they wish to sell it or merge their licence holdings with those of another operator. If trading is prohibited, an operator is not at liberty to sell their licence and if they no longer require it, it must be returned to the government for resale.

11.2.2   Frequency bands

Allocations made for narrowband wireless local loop applications vary from country to country but in most cases they are in the 2 to 4 GHz band. Individual operators have been granted spectrum allocations of approximately 10 to 50 MHz within these bands, which is relatively small. There is increasing interest in using these frequencies for mid-band systems offering data services as well as voice. The advantage over higher frequencies is that these systems can work over greater distances and use less sophisticated and hence less expensive components. However, there is inherently less spectrum available and it is in demand for future wide area mobile systems.

Broadband systems require more spectrum and the allocations that have been made, or are being considered, lie within the 10 to 40 GHz band. Figure 11.1 indicates

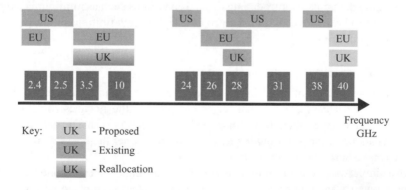

Figure 11.1   Main EU and US fixed wireless access spectrum allocations

the major allocations in the US and Europe. Shown are both issued and proposed licences, with the UK shown separately from Europe to illustrate that even within one region there are still national differences. It can be seen that a number of past licences issued for fixed wireless access in the UK are under consideration for reallocation, due to lack of use.

11.3   Radio channels

The performance of a digital radio system is expressed in terms of the bit error rate. This is related to the signal to noise ratio at the receiver. The noise is the sum of thermal noise added by the receiver, interference from other carriers, inter-symbol distortion arising from both multipath propagation and filters, and jitters, etc. In any radio system the choice of modulation method, coding and multiple access method is governed by the need for an adequate signal to noise ratio under the conditions imposed by the radio transmission path. With a given system design, the signal to noise ratio is limited by the received signal strength which is a function of path loss. This path loss is the reduction in signal level between the transmitter and receiver. The technical parameters for a system, such as transmit power, antenna gains and required signal to noise ratio define the maximum allowable path loss.

Total path loss between transmitter and receiver is determined by many factors. The following outlines those most significant for fixed wireless access systems.

11.3.1   Free space path loss

The radio waves launched for the transmitter spread out like ripples in a pond, only in three dimensions hence forming an expanding sphere. As the distance from the transmitter increases, the surface area of the sphere increases and power at any point on this surface, corresponding to the received signal strength, decreases. The received power density therefore decreases with the square of the distance from the transmitter. This reduction in received level is due to the geometric nature of the radio propagation and assumes the system is operating in a vacuum – it is known as free space path loss (FSPL).

The magnitude of the loss for a given system is also inversely proportional to the square of the signal wavelength. For a given link range, a doubling of operating frequency increases FSPL by four times.

11.3.2   Obstructions

The majority of energy in a radio signal is carried in an area surrounding its direct path, known as the first Fresnel zone. If this zone is completely obstructed by objects in the environment, no signal will arrive at the receiver. Partial obstruction leads to an loss in signal strength additional to that caused by free space path loss, but some power in the Fresnel zone passes the obstruction due to diffraction. For a given arrangement of obstruction and antenna positions, the size of the Fresnel zone and so received signal power reduces as the frequency of operation is increased.

Another mechanism by which radio signals can overcome obstacles is penetration. Usually considered in relation to buildings, penetration losses will occur when a signal passes through any object along its path. The majority of materials absorb huge amounts of the signal energy, the exact value increasing with material thickness and again signal frequency.

In practical systems operating at higher frequencies, above around 10 GHz, all links must be 'line of sight', that is a completely clear unobstructed path between transmitter and receiver must exist. If there is any obstruction the Fresnel zone is so small and the penetration losses so high that no useful signal power reaches the receiver. Thus this 'shadowing' imposed by tall buildings, trees, uneven terrain and so on, will result in it not being possible to reach every user location potentially within range. Very careful planning is therefore required to establish whether a new customer can be connected into the system. Much of this is done by site surveys, using binoculars and beacons, although increasingly more sophisticated techniques based on high resolution three dimensional data derived from aerial photography are being used as illustrated in Figure 11.2. Whichever technique is used, long term accuracy is limited by the changing nature of real environments – new buildings are erected, trees get bigger and grow leaves, and temporary structures come and go.

Surveys in the UK indicate that in downtown areas it is reasonable to expect to cover up to 70% of the buildings within a 2 km radius of an antenna mounted on the roof of a relatively high building. In practical deployments, however, this figure can vary enormously from perhaps 25% to over 90% depending on the exact terrain and buildings.

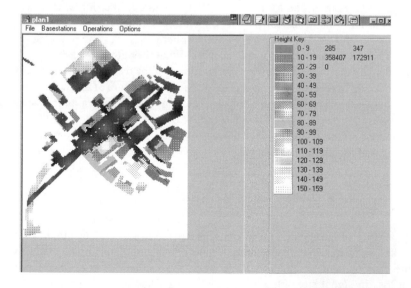

Figure 11.2   Example of 3D building data

As the frequency of operation decreases below 10 GHz, the requirement for line of sight becomes less absolute. By around 3 GHz systems will operate even when there are several buildings for the signal to pass through and around, but optimum range is still only achieved in line of sight conditions and therefore the coverage footprint is affected by the density and height of buildings.

11.3.3   Rain fade

At high frequencies, atmospheric water absorption of radio energy becomes significant. In most areas, this is dominated by rainfall and this has to be considered when planning networks or longer links will stop working during periods of heavy rain. The standard model used for this is provided by the ITU in the form of Recommendations P.837-1 and 838. These classify each region in the world in terms of the rainfall rate that is exceeded during the wettest part of the year, for instance for 1%, 0.1% or 0.01% of the time. When engineering a network, a required link availability figure is selected, for instance 99.99%. The fading that occurs due to the rainfall experienced in this region during the worst 0.01% of the time can then be calculated and allowance made for this when calculating the maximum workable link range.

In many countries the need to be able to cope with rain fading has an impact on the modulation methods that can be employed. Higher order modulation methods increase the bit rate that can be supported. However, because higher order modulation methods require a larger signal to noise ratio, it may be necessary to drop back to a lower scheme during rain fades in order to maintain a given link length at expense of reduced network capacity. During such fades priority would be given to delay sensitive services such as voice, but there is debate within the industry as to whether such protocols would be acceptable to the customer and few have been deployed.

Rain fade is the major contribution to a class of fading called slow fading. Other factors include foliage growth and atmospheric changes such as temperature inversions.

11.3.4   Gaseous absorption

In addition to loss caused by travelling through solid materials such as the walls of buildings, radio signals are also diminished by passing through air. One of the two main reasons for this, the interaction with water molecules, has already been discussed. The other major mechanism for loss is absorption of energy by the gaseous molecules within air, as shown in Figure 11.3. The combination of these two effects tends to limit the operation of practical fixed wireless systems to around 40 or 50 GHz.

11.3.5   Multipath

Due to the ability of radio waves to reflect from surfaces, in real environments which have a ground plane, buildings and other large objects, the receiver will see many versions of the transmitted signal. Such environments are said to exhibit multipath behaviour – there is more than one route through the channel. Initially the direct path

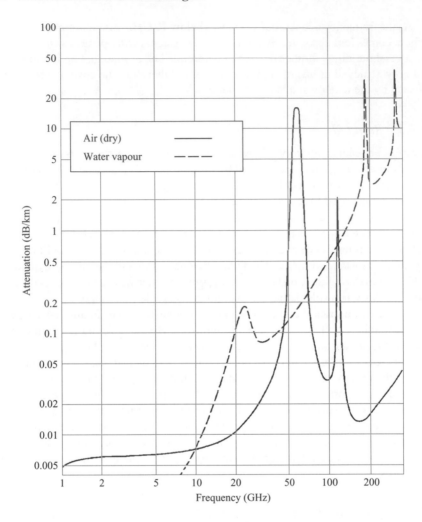

Figure 11.3   Attenuation caused by gaseous losses of dry air and water vapour

signal will arrive (presuming there is a line of sight) followed by other generally lower level signals delayed and phase shifted by their longer, reflected paths. To make sense of this multiplicity of inputs, the receiver must either select one and discard the others in such a way as to minimise the interference they introduce, or combine the signals constructively.

For mobile radio systems this problem is particularly acute, because as the user moves through the environment, the nature of the channel and the required behaviour of the receiver will be constantly changing, potentially very quickly. For fixed systems the user does not move but the environment may change (for instance if paths involve reflections from moving cars), affecting the multipath characteristics, albeit to a much lesser degree. How well a system can cope with multipath channels generally depends on the design of its channel equalisation, modulation and multiple access schemes.

Fast fading is a term used to cover rapid changes in propagation loss, the majority of which is caused by multipath effects.

11.4   Technologies

11.4.1   Modulation schemes

Modulation is the process by which the message to be transmitted is added to the radio carrier sine wave. In some systems the modulation method is fixed, in others it is changed to suit the required bit rate or transmission path conditions. There are three parameters of the carrier wave that can be changed to represent the data signal: amplitude, frequency and phase.

The most common form of amplitude modulation used for digital signal transmission is quadrature amplitude modulation (QAM). A series of symbols are transmitted, each representing one or more signal bits by the amplitude of the in-phase and quadrature components of the signal. The more discrete levels that the components are allowed to take, the greater the number of bits that can be encoded into each symbol and the greater the throughput rate of the modulation scheme. 4QAM allows each component to take one of two values, creating four possible combinations, or constellation points. This modulation scheme can therefore transmit two bits per symbol, with the four points representing 00, 01, 10 and 11. 16QAM and 64QAM can encode 4 and 6 bits per symbol respectively.

Phase modulation schemes, known as phase shift keying (PSK), are less commonly used in digital fixed wireless access systems. Here the relative phase of the carrier signal is adjusted to one of a range of allowed values, equally spaced between $\pm180°$. Typical choices are 2 (BPSK), 4 (QPSK) and 8 (8-PSK) constellation points transmitting 1, 2 or 3 bits per symbol.

Amplitude and phase schemes are described mathematically by linear expressions, whereas frequency shift keying (FSK) is exponential. The frequency of the carrier signal is deviated slightly from its nominal value, with each frequency representing a symbol. Minimum shift keying (MSK) is a commonly used variant which exhibits a constant envelope – the amplitude of the modulated signal never varies, even during transitions from one symbol to the next.

With all the schemes, increasing the number of constellation points increases the amount of information each symbol carries and so the throughput for a given symbol rate or transmitted bandwidth go up. This is measured by spectral efficiency which is the number of bits transmitted per Hz of spectrum consumed. Maximising this figure is essential as spectrum is such a limited and expensive resource. The limit on the modulation order that can be used is its noise immunity. As more constellation points are used, the gaps between them reduce and the amount of added noise that causes the received signal to be misinterpreted is reduced. In practical systems that can experience high levels of noise and interference, the choice of modulation scheme becomes a very fine trade-off.

11.4.2   Coding

Channel coding is employed for the detection and correction of errors caused by noise or interference. Error detection systems add extra check sum information to the original data which is used to determine if the received signal has been corrupted. Higher level protocols are then used to decide how to deal with the error, often by requesting retransmission. For error correction, more overhead information is added, leading to sufficient redundancy to be able to recover the original signal in the event of one or more bit errors occurring.

These schemes can be used to mitigate against errors caused by noise problems and allow higher order modulation schemes to be used. However, the increase in raw spectral efficiency achieved must be balanced against the decrease in user information throughput due the increasing requirements for overhead data. In practical systems there is physical limit to the useful data that can be correctly conveyed by the channel and this can be approached by careful balancing of the modulation and coding schemes. In fact so closely related are the two topics that the most successful implementations actually perform both operations at once using a process called coded modulation.

11.4.3   Multiple access

Various multiple access techniques can be used to combine individual signals in the radio spectrum to allow for simultaneous access by a number of users. The aim of all of them is the same – to allocate to each user a set of radio resources which are unique in one or more dimensions.

11.4.3.1   Frequency division multiple access (FDMA)

The total available spectrum in the network is divided into a large number of sub-bands, or channels. Each user is then allocated one or more of these channels depending on how much data they have to transmit. This allocation can be done on either a static or dynamic basis. In static systems a given user always uses the same channel, which means that the radio they use can be dedicated to just that small part of the total band. This allows a high performance and less expensive radio but considerable operational problems are caused by the extreme lack of flexibility and complex inventory holdings and network planning. More flexible systems use wideband radios able to retune and use any channel within the band. This has disadvantages in terms of radio complexity and cost, but enables, for instance, a user to change the bandwidth used by a simple management system reconfiguration, rather than the installation of a new radio unit. However, all but the most advanced systems are not flexible enough to enable dynamic bandwidth adjustment during a session.

11.4.3.2   Time division multiple access (TDMA)

All users of the system transmit over the same channel, which is usually much wider than in the case of FDMA systems. Each user is allocated one or more timeslots and when that timeslot is active they have exclusive access to the channel. The overall data rate offered to a user is the same as in the FDMA case as they are offered a higher data rate for less of the time. This makes the scheme well suited for the transmission of bursty data services and allows higher peak data rates. It is also easier to engineer radios that are agile in time rather than frequency and hence a TDMA system, which is inherently flexible, is likely to be less expensive than an FDMA system with similar flexibility.

11.4.3.3 Code division multiple access (CDMA)

This technique was originally developed for military applications as it allows communications to be made secure and resistant to jamming. CDMA has other properties that make it important as a multiple access scheme. There are two implementations, direct sequence and frequency hopping, which use different methods to achieve similar ends.

In direct sequence (DS-CDMA) systems, the data stream that is to be transmitted is first multiplied by a much higher speed bit stream known as the spreading code. The result is a high speed, large bandwidth signal for transmission, hence these systems are often called spread spectrum. On reception the signal is again multiplied by the known (and synchronised) spreading signal, a process that recovers the original data. The benefit of this process is that any interference and noise added to the signal during propagation is rejected by a ratio, the processing gain, equal to the ratio of original signal and spreading code speeds. If each user is allocated their own unique code, ideally orthogonal to all the others in use, the receiver can virtually eliminate the signals from all but its wanted user, forming a multiple access scheme.

Frequency hopping (FH-CDMA) systems transmit the original narrowband version of the data stream, but transmitter hops from one frequency to another during the transmission. In Fast FH, the dwell time on each frequency is less than the bit period of the data. For Slow FH, several bits are transmitted on each hop. In a similar way to DS-CDMA, each user's hopping pattern is controlled by a unique, orthogonal, pseudo-random code which is used to control the receiver. Narrowband interference will therefore only affect a small proportion of the data stream and unwanted signals of other FH users are largely rejected.

A important benefit of CDMA systems is their ability to compensate for multipath interference. By employing RAKE receivers the various delayed signals can be realigned in time and phase allowing them to be constructively combined. The RAKE is effectively an array of single CDMA receivers each using a version of the spreading code with its timing shifted to match a different incoming multipath signal component.

11.4.4 Channel compensation

Several techniques have been described that reduce the effects of a channel's transmission behaviour changing with time. Narrowband systems experience flat fading due to these changes, that is, the changing attenuation caused by the channel is equal across the whole band. This can be allowed for by designing in a fade margin where the system is effectively over engineered by a sufficient degree to cope with the worst expected fade (in a similar way to rain fade margins discussed earlier).

Wideband systems suffer from both this flat fading and also dispersion. This leads to frequency selective fading and is due to the effects of multipath. The effect on the receiver is that many time delayed versions of the signal arrive, causing inter-symbol interference. This is a condition where the received signals are so delayed relative to one another, that symbol n in one signal and $n + 1$ in another start to overlap. If the receiver simply adds all signals together (which a simple receiver does) the different symbols interfere and errors are generated.

CDMA uses RAKE receivers to minimise the effects of frequency selective fading. Another technique is channel equalisation. Here, predetermined 'training sequences' are transmitted in amongst the user data. Digital signal processing algorithms use the received signal and knowledge of what was originally transmitted to calculate a model of the channel. The inverse of this model can then be applied to future incoming signals to create an approximation of the signal transmitted. The channel estimation process must be repeated regularly so the model tracks the changing environment. The process is very complex and for rapidly changing and very wide channels is still a research topic.

11.4.4.1 Orthogonal frequency division multiplexing (OFDM)

Another technique used to compensate for the frequency selective fading experienced by wideband systems is OFDM. If the frequency response of a channel is viewed over a small enough window it will always be approximately level, even if it very uneven over the whole band. For instance, if the response decreases by 10 dB over 1 MHz, over 10 kHz it will only decrease by 0.01 dB, which for practical purposes is flat.

A solution to this fading problem is therefore to divide the wideband signal into many narrowband signals, each of which will experience only flat fading. This is exactly what multicarrier technologies such as OFDM do. The incoming bit stream is divided into many lower rate signals, 256 or 512 being common. Each of these streams is then modulated on its own narrowband carrier for transmission. At the receiver a corresponding set of demodulators and a recombination process are employed. The whole process requires large amounts of DSP resource and modern processors have only recently made this practical.

Due to the multiple carriers to be transmitted, the RF power amplifiers used in OFDM systems must either have higher linearity performance or higher power output. This means they will be more expensive than those used in other systems especially at higher frequencies where high performance amplifiers are hardest to build. OFDM systems also suffer from increased phase noise which again makes them very hard to design for high frequencies. The majority of currently available systems operate below 5 GHz.

11.5 Architectures and systems

11.5.1 Point-to-point (P-P)

A pair of dedicated radio transceivers at either end of a path forms a point-to-point system. Such systems can be deployed in core and backhaul networks as well as for access. The majority of commercially available systems offer high capacity and are expensive, aimed primarily at the backhaul market. One of the major applications in the UK is the connection of outlying base stations into the rest of a mobile operator's network. The number of links that can be supported from one site is strictly limited by the (relatively small) number of radio units that can be mounted on each mast. Economies of scale are largely absent as none of the deployed radio or spectrum resources can be shared by future links. These features lead to a very limited application in the access market and so this architecture will not be considered further.

11.5.2   Point-to-multipoint (PMP)

A much more promising architecture for access networks is point-to-multipoint as it allows resources to be shared between users. A central base station is established, which uses a single wide beamwidth antenna and multiple access techniques to communicate with many customers distributed over a large area. In this way, the expense incurred at one end of a link and the cost of the spectrum used can be shared across all the customers that are connected.

Depending on which multiple access and channel assignment methods are chosen, there may be a need for frequency planning to reduce interference between base stations (and customer units) in different cells. When frequency planning is required each cell operates on a different frequency from its neighbour.

To increase the number of channels that can be provided by a given base station a cell can be split up into sectors by using more directional antennas at the base station. Each of these then shares its capacity over a smaller geographical area, increasing overall system capacity. In this case different frequencies will be used in neighbouring sectors. The logical extreme of this process is for each sector antenna to be so directional it only serves one customer, leading to a point-to-point architecture with its complexity and flexibility problems. For this reason it is not practical to increase capacity by sub-dividing sectors indefinitely, with four or eight sectors per base station being the limit for most systems.

The number of frequency sets needed to provide ubiquitous coverage without excessive interference depends on the frequency reuse pattern which needs to be used with a given type of system. It will also depend on the frequency band used. Higher frequencies are more severely attenuated by obstacles and thus a tighter frequency reuse pattern can be used as interference received from other parts of the network is reduced. The reuse pattern can also be influenced by using antenna down tilt at the base station to again limit the propagation of signals into adjacent cells.

The frequency and cell planning has a huge impact on the raw spectral efficiency established by the modulation and coding schemes. A system that only allows each frequency to be used over 10% of the deployment geography offers a lower capacity than one with only half the raw spectral efficiency but a 50% frequency reuse factor. Cell size also plays an important role – a system that can cover an area with one cell using 100% of the available spectrum is at a great efficiency advantage over one requiring two cells each using only half the spectrum. For these reasons, selecting which system to deploy is much more complex than comparing raw spectral efficiency figures as these values are swamped by the various modifiers in practical deployments.

Instead an estimate of 'area' or 'network' spectral efficiency must be established and this can be very difficult without performing a detailed radio planning case study in a representative environment.

Two of the key attributes of PMP systems are (i) that the cost of the customer premises equipment (CPE) is not incurred until service is offered and (ii) the cost per home 'passed' (or 'covered' in radio terms) of the shared infrastructure can be very low compared with wired technologies. However, it is necessary to operate at high fill levels to get the cost per customer served down to competitive levels. This can be a problem for some of the smaller cell systems. The impact of fill and cell size can be partly offset if there is a high degree of modularity in the base station build. Manufacturers offerings differ considerably in the amount of modularity provided and thus the choice of system must be carefully matched to the target market.

To address the issue of limited coverage percentages due to line of sight, the base station coverage areas can be overlapped to increase the possibility that a new customer has line of site to at least one of them. Whilst this technique can increase coverage, it greatly increases the cost of building a network and the complexity of radio planning and ultimately suffers from the law of diminishing returns.

11.5.2.1 Customer premises equipment (CPE)

The radio specific parts of CPE will usually be externally mounted, having a wired feed to an internal interface unit, power supply and possibly battery back up depending on the services to offered. Increasingly lower frequency systems are attempting to employ a single integrated indoor unit combining the antenna with both radio and interface electronics, with the aim of reducing installation costs. This approach introduces additional propagation loss and reduces the working radius of a cell, with the resultant increased requirement for base stations offsetting the installation cost savings. However, the customer acceptance benefits may make this trade-off worthwhile.

Each customer can have their own radio unit or a multiple line version can be shared between customers. In the latter case service is provided to a unit mounted on the top of a multi-tenanted office or residential block, with wired feeds being used to interface units for individual customers. In most systems the CPE has a high gain directional antenna pointed at the base station. This antenna may contain integrated radio electronics or these may be mounted adjacent in a separate enclosure. The requirement for the antenna and radio electronics to be mounted in close proximity stems from the need to use low loss cable if large distances are to be spanned. Such cabling is large and inflexible and therefore leads to expensive and unattractive installations particularly unsuitable for residential applications.

11.5.2.2 Base station

The central base station antennas will usually be mounted as high as possible to lift them above obstructions and maximise the range of the cell. Ideal targets are existing tall buildings that have a convenient connection to the operator's backhaul network. When this is not possible, custom towers may be constructed and backhaul can make use of point-to-point radio links. Generally the radio electronics are too large to be mounted with the antennas and so are located with the network interface system, usually in a conventional equipment rack.

The requirement to provide sufficient radio power and capacity for all users means that base stations are much larger and more expensive than customer equipment. Modularity allows the lowest cost version to be installed to provide initial coverage of an area and then upgrades are made as customers take up increases in capacity requirements.

In many systems the traffic from a number of geographically clustered base stations is carried back to a controller unit which aggregates the traffic in preparation for transport over the core network. The controller may undertake other functions such as the co-ordination and synchronisation of its subordinates.

11.5.2.3 Narrowband analogue

The first PMP systems, designed in the early 1990s, provide one or more simple analogue telephony interfaces to each customer. They tend to operate in the lower frequency bands, below 1 GHz, and are often based on existing mobile technology such as TACS, engineered to provide the lowest possible cost per connection and very large cell sizes (50 km or more are possible).

The primary market for such deployments is developing markets, usually those without an existing wired PSTN access infrastructure. They are of limited interest in developed countries with the only likely application being the replacement of long rural copper connections.

11.5.2.4 Narrowband digital

There is a wide range of digital systems offering similar speeds to the PSTN, most of which operate below 4 GHz. Like analogue systems, some are based on mobile networks such as GSM or IS-95. The fixed access may be provided as a completely new network or by utilising an existing set of mobile base stations. Ranges are less than for analogue, but far larger than other digital systems, with 35 km being achievable.

Other systems are available that are based on cordless standards such as DECT. These necessarily use very low power levels and therefore offer limited link ranges of a few kilometres.

A large number of systems use proprietary solutions with typical systems offering voice and data services up to 128 kbit/s to each customer. Examples are available based on TDMA, FDMA and CDMA schemes. The range of such systems varies, but 5 km cell radii are common.

11.5.2.5 Wireless IP (WIP)

As a relatively new class of system that is usually described as mid-band, a WIP network can be viewed as a competitor to ADSL copper-based systems. Typically they offer peak data rates of up to 2 Mbit/s although this bandwidth is heavily contended between users. Operating between 2 and 10 GHz they are subject to some increase range limitations and require careful cell by cell planning. The first examples emerged in the USA, based on existing unlicensed 2.4 GHz wireless LAN technology, suitably adapted for fixed access use. Stricter licensing regimes in Europe made these systems less suitable and so many were modified to operate in licensed bands such as 3.5 GHz.

Successful competition against ADSL requires careful cost controls and it is with these systems that indoor customer units and transport schemes to overcome the resultant multipath effects are becoming increasingly common. The reduction in deployment expenditure significantly increases the cost competitiveness, but limits the number of customers that can be served with a given number of base stations. An operator may decide simply to accept a lower final penetration or quality of service or instead to offer a conventional solution (directional external antenna) to the outlying customers.

A particular limitation of early solutions was the lack of voice support due to their development from a pure data foundation. This is a particular issue for operators wishing to offer a full service package in competition to ADSL. Developments in voice over IP have lead to many systems being extended to offer voice, although this technology is still evolving.

Many proprietary systems have been developed, designed from the ground up to support a mix of voice and data. They have the potential to offer higher voice performance and a radio design tailored for fixed PMP use. The disadvantage is they cannot leverage the price reductions available to wireless LAN based systems due to the huge volumes their component parts achieve.

11.5.2.6   Local multipoint distribution system (LMDS)

LMDS was originally intended to provide local television services in America. Large amounts of suitable spectrum around 28 GHz were auctioned by the US government in 1998 and due to the flexibility in licensing regulations this spectrum was also allowed for use as a delivery channel for general telecommunications services. Focus has now shifted to the use of LMDS for providing data services such as IP access and leased lines. The use of the term has expanded beyond the US and is commonly used to refer to any high bandwidth point-to-multipoint fixed radio system and therefore no longer has a precise definition. The association with broadband means that it is a label rarely applied to systems operating below 10 GHz and current technology is limited to a maximum of around 40 GHz.

Spectrum is usually allocated in paired bands, one frequency used for the downstream traffic from the base station to the customer and one for upstream (this is known as frequency division duplexing or FDD). The amount of spectrum allocated to each operator varies enormously by country ranging for a pair of 28 MHz carriers in parts of Europe to more than 1 GHz in the US.

The frequencies at which they operate make these systems strictly line of sight and sensitive to heavy rainfall. To minimise these problems, the base station needs to be located as high as possible and customer antennas mounted at or above roof height. Even with careful planning cell radii are limited to between 1 and 3 km, depending on climate, topography and required service availability.

The majority of available systems are truly multiservice, usually using ATM for the transport of traffic over the air interface and back into the core network, although native IP based systems are now emerging. Rates offered to individual customer units range from 128 kbit/s to 25 Mbit/s, although 2 to 8 Mbit/s is more common. Due to their cost, customer radios are often shared between multiple tenants in the same building.

To maximise the use of the shared spectrum resource, it is desirable to employ a multiple access scheme that can allocate resources to users only when they have data to transmit. The system can then accept larger numbers of customers on the basis that they will not all be using it at once. For this reason TDMA is preferable to FDMA, with CDMA also being suitable but less commonly used due to the complexity of designing systems to operate over such wide bandwidths.

With most spectrum allocations being paired bands, the majority of systems employ FDD. When the services offered do not result in equal up and downstream traffic this leads to inefficiency. For heavily asymmetric applications such as Web browsing, it is common for a system to run out of downstream capacity while the upstream frequencies are still largely empty. Time division duplex (TDD) systems are designed to overcome this problem. In a similar way to TDMA, they consider all frequencies available to the system to be one logical block, which is divided into timeslots, and these are allocated to waiting transmissions, be they in the up or downstream direction. An issue with deploying this technology is that licence conditions imposed by the issuing government often mandate FDD operation.

The frequency of operation, speed, range of interfaces and quality of service offered by these systems means they are considerably more expensive than narrowband offerings. Typically they are targeted at small to medium enterprises, and with suitable in-building distribution, multi-tenanted offices or residential buildings. The economic viability in a given region is a function of the density of suitable customers and this limits ideal deployment locations to a relatively small number of urban centres.

11.5.3 Multipoint-to-multipoint (MP-MP)

The multipoint-to-multipoint architecture, more commonly known as 'mesh', is a relatively new development for wireless access, with few commercial offerings yet available. Systems will soon be launched in frequency bands extending from 2.4 GHz up to 40 GHz. The architecture gives rise to many potential benefits and will therefore become very important over the coming years, but specific implementations vary enormously and therefore capitalise on different aspects with varying amounts of success.

The essence of a mesh network is that customer equipment becomes an integral part of the network infrastructure by being able to relay data from other customers. This leads to the first advantage for systems operating at line of sight frequencies – the need for each customer to connect directly to the base station is removed. Rather than each outlying location having to be able to 'see' a specific unit connected to the backhaul network, the requirement is now to 'see' any other station in the system, so that traffic can be forwarded on towards the backhaul network. This means that coverage percentages depend on the number of units already installed, but in typical urban deployments the figure can quickly reach 90% with reasonable densities of customers.

11.5.3.1   Antenna system design

The type of antennas used is a fundamental variable in mesh networks and leads to the majority of practical differences between implementations. It is possible to classify the emerging designs into one of three categories based on the technology they employ.

Omni-directional

The most straightforward approach is the use an omni-directional antenna at each customer location, giving 360° coverage and hence allowing connection to other units in any direction. This allows for simple installation and operation as no antenna pointing or re-pointing is required – the node can just be installed and switched on. This approach has severe drawbacks in terms of link range, due to the resultant low antenna gains. Also, as each node is transmitting in all directions and receiving for all directions, interference levels are high, leading to low frequency reuse, poor spectral efficiency and so relatively low overall network capacity. PMP systems suffer from the same effect to some extent due to their wide angle base station antennas, however they benefit from directional customer antennas, therefore having higher potential capacity than omni-mesh systems.

Directional – static

The second potential benefit of mesh networks is increasing system capacity and a much better solution for achieving this is to use high gain directional antennas at both ends of each link. The trade-off over the omni-directional solution is increased complexity and/or limited flexibility. The simplest implementation uses static arrays of directional antennas at each node, one for each link that will be formed to another node. The result is a network of point-to-point links but without the need for one end of each link to be located at a common location with a connection to the core network. The chief disadvantage with this solution is its lack of flexibility – every time a change in the network configuration is required, such as a customer being added or removed, engineers are required to visit nodes and install, remove, or re-point existing antennas. This soon becomes an unacceptable overhead in larger or more dynamic networks.

Directional – steerable

Refining this solution further, it is possible to redirect the antennas remotely from the management system by fitting them with rotators. This removes the flexibility problems and retains the spectral efficiency advantages. As each connection between customers can be a narrow point-to-point connection, the generation of, and susceptibility to, interference associated with each link is much lower than omni-directional mesh or PMP systems. This allows frequencies to be reused much more often and can increase area spectral efficiency, and so network capacity, many times over. The main disadvantage of this option is the increased mechanical complexity of the network elements. The movement of antennas is slow relative to data rates and therefore the arrangement of links within such systems is considered to be quasi-static.

Another way of achieving the same result is the use of so called smart or adaptive phased array antennas. These devices are capable of being electronically rather than mechanically steered, forming high gain beams in any required direction without moving parts. This removes the mechanical complexity of the previous solution, and given suitable beam forming speed and management capability, allows dynamic network reconfiguration for instance on a packet by packet basis.

These remotely reconfigurable solutions introduce many other important degrees of design freedom which will have impacts on the performance of the system. For instance, a model similar to IP networks could be adopted, with traffic being packet switched across the network, using distributed routing algorithms. Instead, an ATM model could be used, with circuit connections being set up across the network under central management control.

Viewed in these terms, mesh networks appear similar to traditional wired networks and would seem to adhere to the same rules. However, there are some important differences that limit this analogy. In wired networks, the links between network elements are essentially a given, dictated by the fibre or copper cables that have been installed. With quasi-static mesh radio networks the elements can be physically connected in a vast number of different ways depending upon the arrangement of the individual antennas and so point-to-point links. This leads to a new degree of freedom or 'layer' in the network optimisation problem. The situation is somewhat different for dynamically reconfigurable smart antenna mesh systems as they can be viewed as having physical connections from each node to all others within range and therefore the analogy with conventional wired networks is stronger.

11.5.3.2 Deployment options

PMP networks have a very limited range of deployment options. Base stations are installed, the number of which is limited either by the maximum budget or minimum tolerable coverage and capacity. Service can then be sold to the customers with line of sight paths to these base stations. Coverage can only be increased by installing more base stations which is a very costly exercise and involves extensive network reconfiguration and customer equipment reinstallation unless the expansion was carefully planned from the outset. Capacity is also increased by the same route, or by introducing more sectors to the existing base stations, which has the same limitations.

Mesh offers many more options for deployment with the possibility of more attractive investment profiles. Networks can start with a collection of single nodes connected to the backhaul network. This looks very much like a base station network, but there are important economic differences. Most mesh systems employ the same type of node at these backhaul interconnection locations as they do for customer installations, so they are much lower cost than base stations. Additionally, as a network grows, coverage is provided by all the nodes within it not just these first nodes, therefore they can be located where access to the backhaul network is least expensive, rather than a location chosen for optimum coverage, as required in PMP. This can lead to a significant reduction in backhaul costs, which have proven to be a large issue in PMP network business cases.

Adopting this 'PMP base station' style approach leads to coverage that is limited initially to levels comparable with PMP networks. Coverage grows as customers are connected and hence increase the radio range of the network. However, relying on this 'organic' growth can be slow and unpredictable. If higher initial coverage is required a technique known as seeding can be employed. This involves installing an initial density of additional units, sufficient to provide the required coverage of the rollout area. These elements may or may not have actual customers connected and do not need additional connection to the backhaul network, provided that they can communicate with other elements and form a mesh. In practice, only a small number of these units are required to offer very high coverage and if planned carefully the cost has only a minor impact on the overall business case. Any future requirements for increased coverage can be met in a similar way, without the need for any re-engineering of the existing network.

A similar decoupling from installing additional expensive base stations and re-engineering the network is achieved for capacity increases in mesh networks. Due to the high spectral efficiency within the mesh itself, the capacity bottlenecks are likely to be the interconnection points with the backhaul network. As more capacity is required, more of the units in the mesh can be connected to the wired network, again at locations that minimise the cost of backhaul.

11.5.3.3   Limitations

There are many potential problems with multipoint-to-multipoint architectures that the better solutions promise to overcome. One example of these problems is controlling the maximum number of transit connections or hops that a given customer's connection must traverse on its way to the backhaul network, essential if an operator is to guarantee overall service availability and latency. Figure 11.4 shows that this is

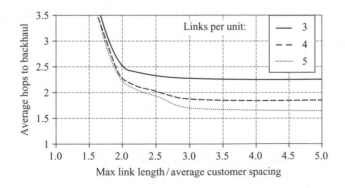

Figure 11.4   Average hop count versus link range

a function of the ratio between maximum radio link distance and the average distance between installed units. The practical implication is that for a system with a given maximum link range, there is a minimum target node density which must be achieved using a mix of customer and seed network elements.

Another issue is minimising the impact of customer churn or unit failure. In a carefully managed system, each customer will have more than one path through the mesh to the backhaul network. Hence the impact of any one customer disconnecting, or a node or link failing, is minimised. However, it has yet to be seen how often multiple failures will occur and what the effect of these will be. For residential customers who generally switch off equipment they are not using, this may be a considerable issue.

11.6 Applications and limitations

So is fixed wireless a useful access technology? The answer is undoubtedly yes but it is not the correct solution for all network operators nor all customers. It is very difficult to construct realistic yet positive business cases for fixed wireless, a problem common to all access technologies for which radio presents no magic solutions. A solid understanding of the market requirements and a carefully targeted deployment is needed if the project is to be a success.

Point-to-point systems are already extensively deployed within core and backhaul networks. They have limited applications for access due to their high cost and inflexible nature. Most likely target customers are those requiring high data rates at short notice in areas where an operator does not own and cannot lease wired infrastructure. Here a point-to-point link can be established to the operator's nearest point of presence, with a wired link being installed in the longer term.

For true access networks, rather than one-off connections, point-to-multipoint and multipoint-to-multipoint are much more suitable architectures. They are also suitable for backhaul networks requiring a relatively high density of connections, but their ability to reduce cost by sharing resources makes them much more commonly deployed for last mile customer connections. Systems are available to provide rural areas with basic telephony services or corporate customers with a diverse portfolio of high speed network solutions. Whatever the access requirement there is likely to be a fixed wireless system to address it.

The question of when to chose radio access over wired alternatives still remains. An operator may not have free choice in the matter if suitable spectrum is not available when they require it. Assuming this hurdle is passed, fixed wireless is best adopted when playing to its strengths – deployment speed and low initial investment. With wired access networks wide scale coverage can be established by the very costly and time consuming installation of cables to all potential customers, many of whom may never take service resulting in wasted investment. The alternative is wait for a customer order and then lay the required cabling to minimise unnecessary investment, however, this results in long order fulfilment times and the cost of cabling for even one customer is still high.

In contrast, wireless allows a whole city to be covered by establishing a handful of base stations which can be done very quickly. Depending on the development of the operator's core network, little or no digging of the streets to install cable is required. A given customer order can then be serviced very rapidly by installing radio equipment on their property and making simple network management changes – again no cable laying is required. This is the key benefit for fixed wireless and makes it very attractive for new operators without an established access network.

There is of course another side to this coin. While the initial investment in civil works required to obtain coverage is removed, base stations are generally expensive items. The complexities of radio propagation means that to achieve high levels of coverage and service availability many more are required than initial estimates might suggest. Additionally for the majority of systems the customer equipment is both expensive and complex to install.

Possibly the most important consideration when deciding between access technologies are the long time requirements of the network. As the requirement for capacity increases over time, more sophisticated electronics can be used at either end of the access link, be that copper, fibre or radio, to increase its performance. This approach has its limits in the fundamental bandwidths available over the transport medium and this is where radio loses out. Both copper cables and radio frequency allocations offer limited bandwidth; however, when copper is exhausted, it is always possible to run more cables in parallel. The same option does not exist with radio, when all the useful radio frequencies are used (a situation rapidly being reached) another electromagnetic spectrum cannot simply be laid alongside. A move to optical fibre is more likely than parallel copper cables and this is a technology that theoretically offers almost limitless bandwidth. The result is that whilst fixed wireless access networks can be engineered to offer expedient solutions to today's and tomorrow's requirements, they will lose out to wired networks on capacity grounds in the long term.

To date there have been many successful radio access networks deployed around the world. However, there have also been many failures. Most of these can be traced to poor management practice and the disregard for basic economic principles, rather than specific issues with the technology. Like all things though there are good and bad solutions available. If the correct system is being deployed by the correct operator, in the correct location, to serve the correct customers, fixed wireless access is and will remain an important access technology option.

11.7   Further reading

BORGNE, M.: 'Comparison of high-level modulation schemes for high capacity digital radio', *IEEE Transactions on Communications*, COM-33, pp. 442–449

FEHER, K.: 'Digital communications – microwave applications' (Prentice-Hall, 1981)

GLOVER, I. A. and GRANT, P. M.: 'Digital communications' (Prentice-Hall, 1998)

ITU-R Recommendation PN.837-1: 'Characteristics of precipitation for propagation modelling' (ITU, 1992–1994)

ITU-R Recommendation PN.838: 'Specific attenuation model for rain for use in prediction methods' (ITU, 1992)

PAHLAVAN, K. and LEVESQUE, A. H.: 'Wireless Information networks' (John Wiley & Sons Inc., 1995)

PROAKIS, J. G.: 'Digital Communications', 3rd Edition (McGraw-Hill, 1995)

Chapter 12

Wireless LANs

M. Begley and A. Sago

12.1   Introduction

The concept of a local area network or LAN has existed for many decades, and it forms the basis of all in-building office and retail data networks in the world today. Interconnection of personal computers, servers and peripherals is the rationale for such a network. Also, many homes are now equipped with a LAN to cater for multiple home computers and peripherals. We can also think of the LAN as an access network to wide area locations or the Internet. The LAN wiring can be replaced by a wireless medium using wireless LAN technology to give local mobility for users, sometimes known as nomadic access. This capability has existed since the mid-1980s in some niche applications, but recently there has been a resurgence of interest in implementing wireless LAN networks across a wider range of applications. The architecture and protocols in current use on wireless LAN networks are discussed in this chapter along with a view of current and future trends in the markets and technologies that constitute this field. The important areas of spectrum availability, regulation and security are also covered.

12.2   Network architectures

Wireless technology can be used to replace fixed wiring in many parts of a local area network, or to provide mobility on an existing network. Common architectures are described below.

12.2.1   Infrastructure

A typical small, medium or large enterprise would deploy wireless LANs as part of a wired network infrastructure, to enhance network coverage or provide mobility to employees within a building or complex of buildings. All users would need a client card, and be in range of an access point (AP) in order to communicate on the network. Client side equipment is widely available in the credit-card-size PC card or Compact Flash (CF) format or as a USB 'dongle', covering use in laptops, PDAs or desktops. Increasingly wireless LAN connectivity is being built into laptops and PDAs at the point of manufacture, rather than as an optional add-on. Access points typically occupy a footprint no greater than an A5 sheet of paper and are wall or ceiling mounted, or may be hidden behind ceiling panels or in cable risers. Cell size will depend on the building construction and the required capacity per cell which in turn depends on the number of users, but 10–30 m cells are typical indoors. An overlapping cell architecture will provide for continuous coverage and allow users to move freely throughout the building, using the access point roaming feature to have their network connections handed off from one cell to another as they move location.

12.2.2   Public

Cells of wireless LAN coverage at public 'hot spots' such as airport departure lounges, hotel receptions and shopping malls provide entertainment, information and Internet access services to members of the public. They can also enable a business traveller to obtain a high bandwidth connection back to their business office LAN services, such as intranet and email, using a laptop or PDA that they already carry with them. On the wireless side, an operator simply needs to provide good coverage and capacity through wireless LAN access points at the locations to be covered (Figure 12.1). Beyond the wireless architecture there is a lot of work to be done regarding provision

Gateway to wide
area network

Figure 12.1   Typical public location

of the service to the customer, providing and controlling access, ensuring accurate accounting and billing, and backhauling to the customer's corporate LAN or to the Internet.

Public access wireless LAN is gaining popularity, with the first operational systems appearing worldwide. Regulatory constraints in many European countries to the provision of public services in the wireless LAN spectrum have been withdrawn, and most major cities have hotspot services available at a number of locations. Depending on the business model for the hotspot operator, the service may be free at the point of delivery (a cross subsidied or promotional service), or the cost may be bundled into other services, e.g., so called 'surf and sip' services at coffee shops, or there may be a straightforward charge based on time used or data volume downloaded. The pioneering service in the UK is BT Openzone which became operational on 1st August 2002 and provides subscription services for business users with a number of flat rate or bundled minutes tariffs. It also offers scratch card access for any user through newsagents, hotel receptions and the like, and instant sign up services. These are where the user pays by credit card for 1 hour or 24 hours usage obtained over the Internet in advance, or by logging on and paying at the hotspot itself. BT Openzone has a target to provide 4000 public hotspots in the UK by the Summer of 2004.

12.2.3   Community wireless LANs

The public access wireless LAN model involves a commercial service run by a network operator for profit. The long incubation period for the development of wireless LAN technology and services led to a significant amateur market for experimentation with wireless LAN, and this has now spawned not-for-profit community wireless LAN configurations. Since a user can buy a decent client card for less than £50, an enthusiast can operate a private wireless LAN network from his house or a convenient central building, and either offer the service for free or make a small charge to cover the cost of the connection to the Internet. Some community wireless LAN networks have grown to become commercial operations themselves, providing multiple access point coverage across a line of villages, for instance. Although it is possible to provide such services in a legal fashion, some community networks are clearly operating with illegal power levels through the use of amplifiers and non-standard antennas for their access points. Thus the coverage provided seems very good in comparison to public access services from mainstream network operators. High power operation may not cause too many problems whilst usage of the wireless LAN bands remains low, but as congestion becomes an issue it may be that the appropriate agencies would start to take enforcement action which could see some networks going off the air. Notwithstanding the legalities, a few community wireless LAN networks have already folded for cost reasons but there are many hundreds still in operation across the UK.

An interesting twist on the community wireless LAN network is the use of a mesh topology. For all systems described so far, the access point is the central node and all client devices communicate directly with it, so typically need to be within a range of a few tens of metres. If a mesh capability is added to every client device, or to at least some of the client devices, then data can 'hop' from the access point through

Figure 12.2   Ad hoc networking example

intermediate client devices acting as relays until it reaches the intended destination. Community networks operating such a topology sometimes offer a discounted price for users who are prepared to let their client device act as a relay for traffic for other people. Mesh networks are in their infancy and community networks are a 'playground' for experimentation at this stage, but there are tremendous commercial applications which could harness the power of the mesh concept for cost effective delivery of services by network operators. BT has undertaken some trials of mesh technology in order to understand the potential for new types of self forming and self healing networks.

12.2.4   Ad hoc

For simple file sharing and small networks it is not necessary for client devices to access the network through a central access point. An ad hoc network can be set up by allowing client devices to communicate directly with each other, and an example network is shown in Figure 12.2. Typically Windows file sharing would be used so that users could have access to each other's files, or one device could be set up as a file server. This architecture would suit a small office situation, or an ad hoc network at a meeting which is only required for an hour or two to provide shared file access.

The peer to peer market may develop significantly in communities of shared interest, such as the college campus or local residential environment. A combination of this architecture with file sharing applications and key seed nodes for external connectivity has the potential to change the accepted business paradigm.

12.2.5   Home

Home networking is developing rapidly and the ease of use of wireless LAN in this context has seen a substantial uptake in this technology. The residential gateway is making inroads into this market by providing both connectivity to the Internet and a local area network in a single unit (Figure 12.3). In its simplest form the home network simply has client devices communicating in ad hoc mode.

Wireless LAN offers tetherless data connectivity for laptop and PDA users. In the future this may increasingly extend to video and voice.

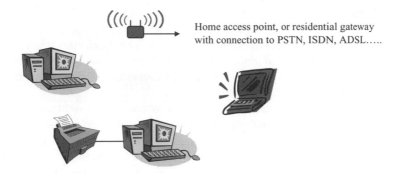

Figure 12.3   Home networking options

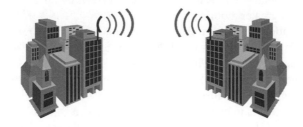

Figure 12.4   Point-to-point wireless connection

12.2.6   Bridge wireless LAN

Wireless LAN may also be used for building to building application so that businesses may connect remote offices or temporary buildings without the need to provide expensive cable infrastructure.

Networks on two sites can be interconnected using a point-to-point topology (Figure 12.4). Users on each site can be on the same subnet or different subnets and the wireless LAN equipment can be set up to provide the appropriate routing. Optical wireless technology may be employed where buildings have good line of sight visibility, using modulation of a laser beam to transfer high bit rate information over several kilometres. Optical wireless is not considered in this chapter, which concentrates on radio technology for wireless LANs.

A point-to-point architecture may not strictly fall under the heading of a 'Wireless LAN', but in practice it is a popular niche application for wireless LAN equipment, and many radio equipment manufacturers provide options to facilitate the point-to-point case. The main advantage to the company operating the network would be avoidance of the need to cable between buildings or to pay a leased line rental. Throughput may actually be slightly improved compared to a 2 Mbit/s leased line solution, but there will be a reduced quality of service in terms of link availability due to the use of a shared medium, namely the licence-exempt radio spectrum. To cover distances over approximately 100 m, directional antennas are typically required.

12.3   Market

The potential market for nomadic wireless LAN access has been strengthened by the growth of laptops and personal mobility products. Demand for access has never been higher and shows no sign of slowing. This presents service providers with incredible opportunities and challenges.

There is a vision to use wireless LAN as a low cost access network to provide nomadic broadband service combined with low speed ubiquitous cellular coverage. It is no longer sufficient to have Internet access at the office or at home. Increasingly, users expect Internet connectivity and secure access to corporate network in public places such as airports, train stations, hotels and conference centres. However, in the past a lack of open standards has hindered enterprises who wanted to deploy wireless infrastructure for nomadic users.

Wireless LAN technology sales are currently driven by an organisation's desire to give employees greater mobility.

While initial adoption of wireless LAN systems is mainly in retail, warehousing and manufacturing, future growth will be driven by other market segments. These include enterprise, small office/home office, telecommunications/ISP and public access environment. Voice over IP, wireless peripheral connectivity, wireless Internet and higher throughput are maturing services.

A number of barriers to wireless LAN acceptance persist that pose real threats to the future of wireless LAN market development. The take up of wireless LAN has been slowed by multiple technologies and standards. Other factors that have hindered the market include concerns over security and seamless access to services.

Security of wireless LANs has been highlighted as being of major importance, especially among corporate users. The security model and implementation of several products have been questioned, casting doubt on the use of the built-in security, Wired Equivalent Privacy or WEP. Even wireless features, such as spread spectrum or frequency hopping, are limited in terms of security as the same consumer wireless LAN equipment may be used in monitoring. Thus a determined hacker could intercept traffic in minutes using a £30 PC card and tools downloaded for free from the Internet. The higher layer IPSec approach has been the most favoured security solution for wireless LANs since the failings of WEP became understood, and a corporate Virtual Private Network (VPN) solution can form an end-to-end secure IPSec tunnel through the wireless network from user device to corporate gateway.

The successor to WEP is being developed in the standards groups and it is expected that it will be fully implemented in new wireless LAN chipsets during 2004. Due to limitations through the use of the closely controlled Advanced Encryption Standard (AES) it is not possible to provide a software upgrade to existing equipment. In the interim, a subset of the future standard is already available though Wireless Protected Access or WPA. This is both forward and backward compatible and is designed to run on existing wireless LAN devices as a software download. It uses the Temporal Key Integrity Protocol (TKIP) for encryption and employs 802.1x authentication with one of the standard Extensible Authentication Protocol (EAP) types available today.

Effectively WPA contains all the security features that are to be delivered in 2004, without the AES encryption feature, and thus satisfies the security concerns of the vast majority of users when implemented within a network.

12.3.1 Seamless access

Wireless LAN may be viewed as a low mobility with high bandwidth technology. It is a complement to cellular technology which is characterised by high mobility with low bandwidth. These features are shaping the market for deployment and there are three distinct sectors; namely home, office and public.

The bandwidth the user experiences in the office may be replicated at home and whilst travelling. However connectivity is not overlapping and ubiquitous coverage may only be provided by cellular. There is a concerted drive in the industry for the development of seamless access so that the user is technology neutral and maintains his connectivity without his active intervention.

12.3.2 The Bluetooth alternative

There has been some attempt to position Bluetooth technology as a wireless LAN alternative by several vendors. There is some overlap in the market between Bluetooth and wireless LAN, however Bluetooth is primarily a cable replacement technology and not necessarily suited as a network technology.

Power consumption is a key issue, and is often quoted as 'low' for Bluetooth devices. Radio devices (of any type) consume their greatest amount of power when transmitting. An example can be seen in the talk times for mobile handsets that tend to be in the order of 1–5 hours, whereas standby times can extend into several days. Where devices are battery powered and may be used away from a charger for extended periods, maximising these usage times will contribute greatly to a good user experience. Efficient radio design can help to minimise power drawn while transmitting, but there are fundamental limits where a certain radio transmitter power must be maintained. There is more scope for being creative with the periods where the unit is not transmitting, and is simply waiting in receive mode for the next time it needs to use the channel. This leads to the science of power management in radio and protocol design, which in Bluetooth is addressed by moving between various connection states and connection modes. Bluetooth devices can be in a connection state operating in one of four connection modes, Active, Sniff, Hold or Park, or in a non-connection state of Standby.

Power management in IEEE 802.11 wireless LAN is less complex than in Bluetooth. There are only two modes, Active or Power Save, and two states, Awake or Doze. For wireless LAN, designed essentially as wireless Ethernet, devices in a network require the illusion that they are all still connected all of the time. Bluetooth, designed essentially for point to point links as a cable replacement technology, can afford to be 'off air' for greater periods of time. In addition, in many cases Bluetooth trades off extended range to reduce power consumption by only offering low transmit powers in comparison to wireless LAN. Where both technologies are used in a LAN situation then it may be that any advantages of Bluetooth power management are lost.

Interoperability has always been a great problem with Bluetooth devices. With all devices currently on the market conforming to Bluetooth 1.1 or greater, interoperability has greatly improved recently. For end users, there remains a confusion over the features and capabilities of Bluetooth devices, which occurs through the use of Bluetooth profiles. It may seem obvious that a Bluetooth headset and a mobile phone will both conform to the Headset Profile and therefore interoperate. If the user then buys a Bluetooth PC card for a laptop, will it be able to talk to the Bluetooth phone to make a data connection over the mobile network? Does the card use the Serial Port Profile or the Dial Up Networking Profile (or both)? Are they implemented on the phone? It can be seen that display of a Bluetooth compatibility logo does not necessarily provide interoperability.

Bluetooth 1.2 (available by the end of 2003) is now bringing further features such as improved QoS support, extended voice modes and faster connection. The majority of these favour usage in the Personal Area Network (PAN) rather than for wireless LAN. At some time in the future Bluetooth 2.0 could bring higher rate support (up to 12 Mbit/s) and other improvements such as a full QoS solution.

Bluetooth deployment is expected to reach more than 30 million end user devices by 2003. This compares to only 4.3 million deployed wireless LAN products. There are few applications currently developed although it should be noted that almost all Bluetooth connectivity on cellular handsets is for a headset profile.

Bluetooth was not originally intended to become a network technology; it has limited ability to do handoffs between access nodes which is an essential feature to ensure mobility. Several manufacturers have solved this for data communication, but real time mobile voice communication is still problematic.

Industry may market Bluetooth's clear advantages as a cable replacement technology rather than focus on it as a wireless LAN alternative. Bluetooth's future is likely to be based on applications that take advantage of cable replacement capability embedded in millions of handheld devices. However it still needs to address interoperability, security, and the need for compelling applications.

12.4　Spectrum

This section will consider three frequency bands for wireless LAN systems, which cover most current systems as well as those expected within the next two years. This information is presented as background to assist in understanding the reasoning behind the equipment specifications for the different bands, and is necessarily a brief summary of a complex area. It should also be noted that it reflects the situation in the UK at the time of writing, which is subject to change over time.

12.4.1　DECT band: 1.880–1.900 GHz

This spectrum is set aside exclusively for DECT (Digital Enhanced Cordless Telecommunications) systems. Through the use of dynamic channel assignment in the DECT

Table 12.1   *Examples of applications permitted in the 2.4 GHz ISM band (UK only) [2]*

Service/technology	Operating range	Permitted power
Wireless LANs and Bluetooth devices	2400–2483.5 MHz	100 mW EIRP
Short range devices	2445–2455 MHz	500 mW EIRP
Short range devices	2400–2450 MHz	10 mW ERP
Electronic news gathering/outside broadcast	2390–2510 MHz	10 kW EIRP
Radio fixed access	2400–2483.5 MHz	100 mW EIRP
Government use	2400–2450 MHz	Various
Deregulated emissions, e.g. microwave ovens	2400–2450 MHz	Various

specification, unco-ordinated systems can co-exist amicably and no radio frequency planning is required.

12.4.2   2.4 GHz ISM band: 2.400–2.4835 GHz

The 2.4 GHz ISM band covers 2.400 GHz to 2.500 GHz and is designated worldwide for licence-exempt industrial, scientific and medical (ISM) use. The range 2.400–2.4835 GHz is defined for certain radio services within Europe, and this smaller band is generally referred to as the 2.4 GHz band. Products used as wireless LANs in Europe in the 2.4 GHz band must conform to ETSI (European Telecommunications Standards Institute) standard EN 300 328 [1], which mainly specifies radio parameters such as modulation methods, power levels, and spurious emissions. It is the responsibility of the product vendor to supply devices that operate with the correct radio parameters for the country of use. The European Norm (EN) provides a reference set of parameters and limits for conformance testing, and in doing so guards the consumer against products that either cause or are susceptible to interference when in use. This last point is very important, since this 2.4 GHz band is also open to a great number of other applications operating at various agreed power levels, most of which are licence-exempt and therefore not registered or controlled centrally by the RA (see Table 12.1). Therefore the wireless LAN protocol needs to be robust enough to cope with high levels of local interference, and the use of spread spectrum technology is mandatory from a regulatory point of view, and highly desirable from a user point of view in order to give a reasonable (if not guaranteed) quality of service.

12.4.3   The 5 GHz radio LAN bands: 5.150–5.250 GHz, 5.250–5.350 GHz, 5.470–5.725 GHz

In Europe and most parts of the world these three bands have been identified as appropriate for wireless LANs, and either have been or will be allocated for use by that service (Figure 12.5). However they also have current allocations worldwide to other services such as meteorological and military radar, radionavigation, satellite feeder links and the earth exploration satellite service. In order for these services to share

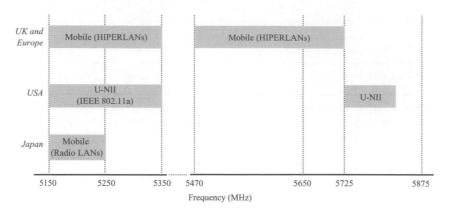

Figure 12.5   5 GHz wireless LAN frequencies

with wireless LANs, complex sharing rules are being developed within Europe as a result of simulations and studies, including power limits, transmit power control and dynamic frequency selection. These rules provide for acceptable cells sizes and data rates for future wireless LAN systems while offering protection for the other services from interference. Both HIPERLAN/2 and IEEE 802.11a will operate in these bands.

12.4.4   Development of spectrum rules and standards

Allocation of spectrum is agreed internationally by the ITU, ultimately at the World Radiocommunications Conference (WRC) held every three years or so, based on proposals from national administrations. Within Europe the use of radio spectrum is harmonised within the Conference of European Postal and Telecommunications Administrations (CEPT). Telecommunications standards are developed by groups of experts working in the ITU and a number of other organisations around the world, who share information through formal liaisons. Standards for equipment and rules for spectrum use were developed and agreed for regions such as Europe, Japan, US and Canada and co-ordinated at a global level for the 2.4 GHz band in the 1990s. This made it possible to use economies of scale to give volume production of licence-exempt wireless LAN equipment and so provide a whole new level of freedom for computer users, paving the way for today's wireless LAN systems. Much work has recently been undertaken to help define the spectrum sharing rules for the 5 GHz band which will cater for expansion of demand for wireless LAN services in the future.

12.5   Technology

12.5.1   DECT

DECT was developed by ETSI as a standard across Europe and originally received approval in 1992. To encourage global usage the word 'Europe' in the title was

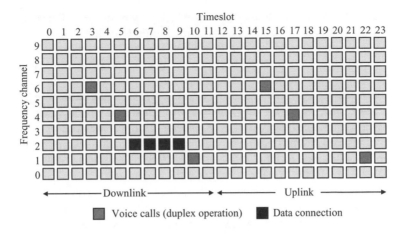

Figure 12.6   Frequency/time snapshot of typical DECT activity

changed to 'Enhanced' in the mid 1990s. Although one of the first products to market was a wireless LAN, long term success was in the field of cordless telephony, with sales reaching 28 million terminals in 2000 and over 75 million DECT terminals now deployed in 110 countries worldwide [2, 3]. The development of profiles for data networking took many years, and in advance of such standardisation a subsidiary of Olivetti called Sixtel introduced a proprietary data product in 1993 using DECT, called Net[3] [4, 5]. The form factor was a half length ISA card to fit inside the PC, connected (by a cable) to a box measuring $150 \times 170 \times 35$ mm containing the main radio transceiver and antennas.

The DECT spectrum allocation is divided into ten frequencies each with 24 time slots, typically using twelve frequency/time slot pairs in an MC-TDMA/TDD structure (Figure 12.6). Typically, for the voice service coded with 32 kbit/s ADPCM, one frequency/time slot pair carries a single voice call. Net[3] concatenated up to 12 time slots in each direction to carry data for one wireless LAN user. The air interface data rate of DECT as standardised at the time was 1.152 Mbit/s, but testing of this equipment at BT Labs could only achieve a maximum of 124.7 kbit/s for a single user. Multiple users shared this throughput equitably in a typical single cell system, for example three users communicating simultaneously would each receive one third of this value. Deficiencies in the implementation included a further performance degradation through the use of a connection oriented protocol. This had a set-up time of 30 ms before the start of the data transfer, and a default timeout of 1.1 s, configurable between 300 ms and 600 s, after which the connection was released if no data was sent. A further problem for UK users was regulatory rather than technical, in that it was deemed necessary at the time to protect users of voice DECT systems (both public and private) from bandwidth hungry data systems, and the UK RA required each installation to be licensed (at a cost of £200) so that the spread of systems could be monitored. Although a pioneering idea with some technical merit, Net[3] had a low takeup with the corporate market into which it was sold and the product was eventually discontinued following restructuring within Olivetti.

DECT standardisation has now produced a mature family of data profiles under the DECT Packet Radio Service (DPRS) as well as enhancements to the voice capabilities. The volume sales of voice products have helped to bring data pricing down to a level within the reach of homes and small businesses. This should be a more amenable marketplace when considering the available throughput, compared to the corporate market which tends to expect wired LAN speeds. Products are now small PC cards or attachments to serial or USB ports, compared to the bulky devices of just a few years ago. In addition to its DECT voice products, BT currently has data products in its portfolio under the On-Air and Airway banners which allow users to connect PCs and cordless voice handsets in a single system. These range from products integrating a DECT 56 k modem for home Internet access over PSTN, to an all-in-one DECT internal telephone system and wireless computer network suitable for SOHO and small business use, supporting connection to four PCs, four PSTN lines and two ISDN channels. The 242 kbit/s throughput on the data network for this product is achieved using 12 duplex DECT channels, and is automatically reduced when voice calls are in progress simultaneously [6].

12.5.2   IEEE 802.11

The IEEE produces standards for Local and Metropolitan Area Networks in the 802 series, such as IEEE 802.3 which defines the CSMA/CD protocol used for Ethernet [7]. The IEEE 802.11 committee was set up to develop conformance specifications and functional standards for wireless LANs. The vast majority of wireless LAN systems in use today conform to an IEEE 802.11 specification [8]. A common medium access control layer is defined across a number of different physical layers, including infrared. Since the number of implementations of infrared wireless LANs is very small, they will not be considered further in this paper.

12.5.2.1   Physical layer characteristics

This section will consider the radio physical layers for 2.4 GHz frequency hopping and direct sequence spread spectrum, and for the high rate addition to the direct sequence specification called 802.11b [9]. The protocol at the air interface needs to be very resilient to cope with high levels of interference, that will occur from other wireless LAN systems and from the other users of the band. Other wireless LAN systems in the vicinity may well employ a carrier sense mechanism to ensure politeness when attempting to gain access to the medium, but other sources will be less friendly. Refer to Table 12.1 for a reminder of the other users of this band.

Frequency hopping spread spectrum systems operate with an air interface data rate of 1 Mbit/s using 2-GFSK or 2 Mbit/s using 4-GFSK. A 2 Mbit/s device must always be capable of decoding and transmitting 1 Mbit/s data, for backwards compatibility with early devices. The principle of a frequency hopping system is shown in the sequence in Figure 12.7.

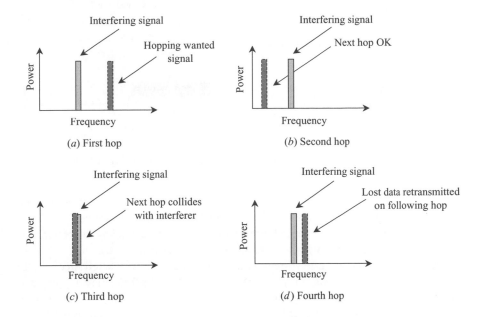

Figure 12.7   Interference rejection with frequency hopping

All units in a network follow the same pseudo random hopping pattern, changing frequency several times a second. The detail of the hopping mechanism is not defined in the standard but is left to local regulation. For Europe, under EN 300 328, a dwell time is defined of <0.4 s on any one frequency, using >20 hops and occupying each channel of the hopping sequence at least once during a period not exceeding four times the product of the dwell time per hop and the number of channels. Practical systems hop over more channels and much more frequently than these minimum values. For a point source of interference into a frequency hopping wireless LAN, there will come a point at which the wireless LAN transmission is swamped by the interference. The 'lost' data is simply retransmitted on the next hop, causing a slight reduction in throughput. For an 802.11 frequency hopping system employing 79 frequencies and providing a user throughput of 500 kbit/s, the reduction is 1/79, giving a throughput during such interference of 493.7 kbit/s, which would be a negligible degradation from a user perspective.

Direct sequence wireless LANs operate at 1 or 2 Mbit/s with the same modulation schemes and backwards compatibility requirements as for frequency hopping systems. High rate 802.11b systems add 5.5 Mbit/s and 11 Mbit/s operation using higher order modulation schemes, and are able to adapt to any of the four rates according to circumstances, such as local propagation conditions or capabilities of other devices that wish to communicate. The direct sequence spread spectrum method employed has similarities to CDMA, in that a spreading code is used with the result that every 1 Mbit/s of information occupies many MHz of spectrum but at a reduced power level. However, direct sequence in 802.11 employs only a low chipping rate of

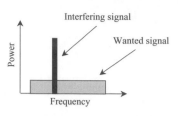
(*a*) Transmitted direct sequence spectrum

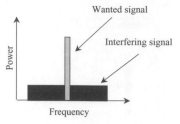
(*b*) Effective received spectrum after despreading

Figure 12.8   Interference rejection with direct sequence spread spectrum

11 MHz, since the objective is to mitigate against interference rather than to provide several orthogonal transmission channels as in CDMA. When the spreading code is reapplied at the receiver, the wanted information is recovered and any interference is uncorrelated and acts as low level noise at the receiver (Figure 12.8).

Both frequency hopping and direct sequence products were marketed successfully for several years in the 1990s, with the addition of the 2 Mbit/s frequency hopping and then 5.5 and 11 Mbit/s direct sequence enhancements to give improved performance. It is this latter 11 Mbit/s standard, 802.11b, which is in predominant usage today. Recently the 802.11g standard, compatible with 802.11b networks, has brought speed up to 54 Mbit/s.

12.5.2.2   Medium access control layer characteristics

The 802.11 family, currently consisting of the original 802.11, and the additions through 802.11b, 802.11g and 802.11a, all share the same MAC layer. Access to the radio channel is gained using a collision avoidance (CA) protocol. On a wired Ethernet 802.3 LAN [9], collision detect (CD) technology is employed to recover from two stations transmitting at the same time. Stations sense the signal on the wire as they transmit, and quickly detect when simultaneous transmission is occurring. The protocol requires both to cease output, and employ a pseudo random backoff algorithm for retransmission, so that subsequent attempts by both parties to transmit are unlikely to occur at the same point in time. For a radio system, CD is not an option because it would require a station to have both its receiver and transmitter active at the same time, and, if receiver destruction did not result, the transmission would in any case swamp any signal received from the other party. The answer chosen for the IEEE 802.11 family of standards is carrier sense multiple access/collision avoidance or CSMA/CA. Initially the station performs a carrier sense/clear channel assessment (CS/CCA) and if the radio channel is sensed as idle then it begins transmission immediately. If, on the other hand, the medium is busy, the station waits for the end of the current transmission and then enters a contention period. If the channel is still idle at the end of this period, then the station has chosen the shortest delay and has gained access to the channel in advance of any competing contenders. This procedure occurs for every packet using random contention delays, ensuring equal access to the channel for all stations.

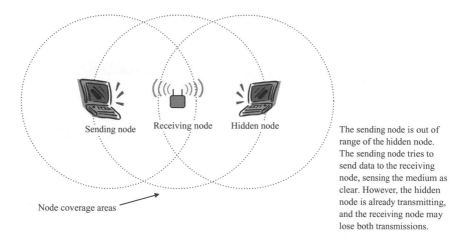

Node coverage areas

The sending node is out of range of the hidden node. The sending node tries to send data to the receiving node, sensing the medium as clear. However, the hidden node is already transmitting, and the receiving node may lose both transmissions.

Figure 12.9   The hidden node problem

In a radio network, use of CSMA/CA sense does not guarantee a free medium. Apart from the small possibility of choosing the same contention delay as another station, there may also be another device, out of range of the sending station, which is nevertheless within range of the target and causing interference. This is the classic hidden node problem, and is illustrated in Figure 12.9.

This problem is overcome in IEEE 802.11 using optional RTS/CTS (request to send/clear to send) handshaking, also known as virtual carrier sense. When a station has a packet to send, it first sends an RTS and waits for a CTS from the target node. This will show that the target node is free to receive the data packet. As a bonus, the hidden node hears the CTS sent from the target and, since both RTS and CTS contain a field indicating the length of the following data packet, the hidden node refrains from any transmissions for the required amount of time. Of course, RTS packets can still collide, but the time lost is much less than if data packets collide and efficiency is improved in a busy network with many access points (Figure 12.10).

As well as the possibility of interference, the nature of the radio medium means that the characteristics of the environment (such as shape and size of room and people moving around in the vicinity) will produce unwanted effects on the signal known as fading. Measures can be taken at the physical layer to mitigate against these effects, but there could still be the occasional loss of data. Higher layer protocols used over a radio medium may not cope well with data loss, and recovery must take place within the wireless LAN medium access layer. This is simply achieved by returning a short positive acknowledgement (ACK) packet for every transmitted MAC data packet. The absence of an ACK indicates that the previous data packet should be resent, either because the packet was lost or because the ACK itself was lost.

CSMA/CA is a distributed control channel access mechanism, called the distributed co-ordination function in IEEE 802.11 terms, with each node controlling its own access to the radio channel. The bursty nature of LAN traffic and the mobility

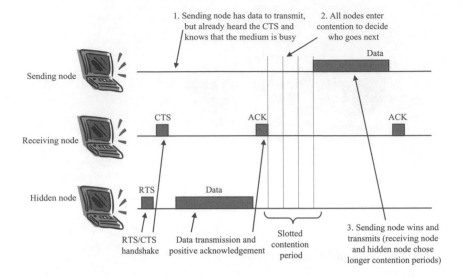

1. Sending node has data to transmit, but already heard the CTS and knows that the medium is busy

2. All nodes enter contention to decide who goes next

Data

Sending node

Receiving node

CTS   ACK   ACK

Hidden node

RTS   Data

RTS/CTS handshake

Data transmission and positive acknowledgement

Slotted contention period

3. Sending node wins and transmits (receiving node and hidden node chose longer contention periods)

Figure 12.10   IEEE 802.11 RTS/CTS mechanism

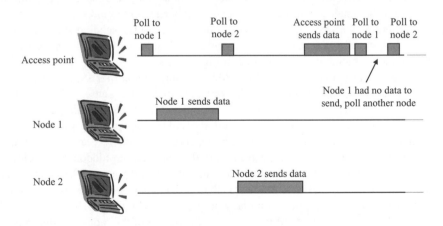

Poll to node 1   Poll to node 2   Access point sends data   Poll to node 1   Poll to node 2

Access point

Node 1 had no data to send, poll another node

Node 1 sends data

Node 1

Node 2 sends data

Node 2

Figure 12.11   IEEE 802.11 polling mechanism

of terminals to move from one cell to another means that a TDMA solution, with timeslots allocated by a central access point in a cell, would result in inefficiencies in most wireless LAN networks. However IEEE 802.11 does provide for a polling channel access mechanism called the point co-ordination function. The access point sends a poll packet to indicate that a particular station is free to transmit. On reception that station simply sends what it has to transmit. The access point has control over the channel, but unlike full TDMA it is possible for stations to send variable amounts of data and for the access point to quickly send out a poll to another station if the first one had no data to send (Figure 12.11). Polling is a useful feature, but the majority of IEEE 802.11 wireless LANs on the market use the CSMA/CA distributed mode in their implementations.

12.5.2.3 Interoperability

Interoperability of wireless LAN equipment is often the initial aim of standardisation activities. Although the multiple physical layers of the original IEEE 802.11 prevented interoperability within the standard as a whole, this has been achieved to a great extent in 802.11b. On top of the mandatory features of the standard, individual vendors will naturally provide optional or proprietary features which will differentiate and enhance their product, while at the same time restricting the choice of vendor when the time comes for additional infrastructure or client cards. Examples of this can be found in proprietary protocol extensions in some access points to increase data throughputs, and the various access server security solutions, which only operate with wireless LAN access points from the same vendor. Nevertheless, vendors are aware that success in the market place is through volume selling which will be greatly enhanced by good interoperability, achieved by close company co-operation. To address this, the Wi-Fi Alliance (previously the Wireless Ethernet Compatibility Alliance) was set up in August 1999. Products from member companies are put through basic interoperability tests to ensure that a client card from one manufacturer will communicate with an access point or client card from another. Successful products are awarded a 'wireless fidelity' or WiFi sticker, and this simple step has contributed greatly to the recent dramatic take-up in the acceptability of wireless LANs. In late 2003, 205 chipset manufacturers, test equipment and wireless LAN vendors and resellers were listed as members of the global Wi-Fi Alliance, with 915 products having received Wi-Fi certification since testing began in March 2000.

As a final caveat on interoperability, certain areas have not been standardised and are therefore not part of the WECA WiFi testing. For instance, handover between access points is not tested. If you operate in a network which has access points from more than one vendor, it may be that your client device will not be able to roam onto and associate with a second access point that is much closer and offering better performance, without first completely losing contact with the first access point (and consequently losing the connection to the network).

12.6 Technology futures

12.6.1 HIPERLAN/2

The future is not all rosy for 2.4 GHz band wireless LANs. It has already been mentioned that a number of existing unco-ordinated services and technologies share the spectrum with wireless LANs. A concentration of wireless LAN or other usage in one particular area could create spectrum crowding and interference, with a reduced throughput for all users. The UK RA, who have responsibility for the radio spectrum, perceive this as a potential problem in the future, particularly in large commercial areas [1]. Migration of wireless LANs to technologies such as HIPERLAN/2 or

Table 12.2   Comparison of main features of IEEE 802.11 family with HIPERLAN/2

Feature	802.11	802.11b	802.11a	HIPERLAN/2
Frequency band	2.4 GHz	2.4 GHz	5 GHz	5 GHz
Maximim air interface data rate	2 Mbit/s	11 Mbit/s	54 Mbit/s	54 Mbit/s
Media sharing mechanism	CSMA/CA	CSMA/CA	CSMA/CA	TDMA/TDD
Connectivity	Connectionless	Connectionless	Connectionless	Connection oriented
QoS support	Limited support through the PCF	Limited support through the PCF	Limited support through the PCF	ATM 802.1p RSVP DiffServ
Convergence layer support	Ethernet	Ethernet	Ethernet	Ethernet IP ATM UMTS IEEE 1394

802.11a could be the answer, and some of the key features of this technology are compared with the 802.11 family in Table 12.2.

There are a number of advantages to HIPERLAN/2, which offers backbone network independence by allowing for interoperability with Ethernet, IEEE 1394 Firewire, ATM and third-generation mobile systems. There are other advantages in quality of service (QoS) and multiple antenna support will enhance external deployment. However the marketplace is currently favouring IEEE 802.11a solutions because of the perceived compatibility with the installed base of 802.11b systems. Nevertheless HIPERLAN/2 systems are available in certain markets and for certain applications.

HIPERLAN/2 was developed by the ETSI BRAN committee as a wireless multimedia standard. HIPERLAN/1 [10] was developed in 1996 but did not make it to the marketplace, since at the time it offered insufficient benefit compared with the emerging 11 Mbit/s 2.4 GHz wireless LANs. HIPERLAN/2 is an all new design for the 5 GHz band [11, 12], offering a number of air interface data rates up to 54 Mbit/s, mobility and QoS for applications such as multimedia, voice over Internet protocol (VoIP) and real-time video. The physical and data link control (DLC) layers are specified, with convergence layers defined above the DLC for cell based services such as ATM, and for packet based services such as Ethernet, Internet protocol (IP), IEEE 1394 and UMTS. At the physical layer the radio uses orthogonal frequency division multiplexing (OFDM) technology, which is well known in Digital Audio Broadcasting (DAB) and asymmetric digital subscriber line (ADSL). OFDM does not use frequency hopping or a spreading code, but it simultaneously modulates multiple narrowband sub-carriers in a channel. In HIPERLAN/2 there are 48 sub-carriers carrying data (plus 4 pilot sub-carriers) in a 20 MHz channel using BPSK, QPSK, 16QAM or 64QAM modulation to deliver the required data rate from 6 to 54 Mbit/s. All the sub-carriers are used for one transmission link between a mobile station and an access point. OFDM has been chosen because it offers good protection against multipath propagation, which affects the performance of any radio link. Although the technique has been known for some time, it is only recent advances in silicon and processing power (to undertake the necessary fast-Fourier transform signal processing) that have enabled OFDM transceivers to be realised practically and at reasonable cost.

In contrast to the situation with 802.11 equipment, in a HIPERLAN/2 network, control plane signalling is used to set up a path between the client node and the access point before any data is transmitted. Essentially a connection-oriented protocol is used, with a TDMA/TDD structure over the air interface. This enables QoS parameters and fair access to the channel to be negotiated, controlled from the access point. Ad hoc networking is partially supported through provision of a direct mode for client to client communication, but both clients need also to be in communication with an access point in order that the radio resource can be allocated.

As mentioned in the regulation and spectrum section of this chapter, HIPERLAN/2 operates in 455 MHz of spectrum in a band at 5 GHz, where sharing is with particular services such as satellite uplinks, space science exploration and radar. International studies have examined the various sharing issues over a three year period which has resulted in a primary mobile frequency allocation at the ITU-R World Radiocommunications Conference in 2003.

12.6.2   IEEE 802.11a and MMAC HiSWANa

The US IEEE 802 committee have developed the 802.11a standard [13] for the 5 GHz band, broadly in parallel with HIPERLAN/2. (Although the idea was thought of first, hence the 'a' suffix, the 802.11b work at 2.4 GHz was completed first.) In Japan, the wireless multimedia body MMAC has developed the HiSWANa standard, essentially for the same frequencies. Although the physical layers are closely aligned for HIPERLAN/2 and 802.11a, HiSWANa has some operational frequency differences and all three standards have differences at the MAC layer. Also, the regulatory situation is quite different in the three regions for the 5 GHz band. Nevertheless a Globalisation Study Group and a Joint Task Force have been established to investigate whether and how these standards could coexist, interoperate, or merge to create a single global market for 5 GHz systems.

The IEEE 802.11 committee has a number of task groups set up to investigate wireless issues, such as enabling support of QoS parameters and improving security.

12.6.3   Ultra Wide Band

This is a development which is generating considerable interest within the wireless community. Ultra Wide Band (UWB) has its origins in military radar applications

[14], but the benefits of using UWB to provide multi-user communication have long been recognised since it provides good performance in multipath environments such as inside buildings [15]. Recent advances in silicon technology are now enabling practical communication systems to be built. The principle is that pulsed rather than sinusoidal transmissions are used to convey information at a high data rate. The transmission is effectively multiple narrowband signals over a very wide bandwidth, perhaps as much as 1.5 GHz. Pulse position modulation is employed with a low power level (typically 50 μW) but a high processing gain, so that the data is recovered even in the presence of high power levels from other services in one part of the spectrum. Of course, although the power level may be low, the pulses affect existing equipment in many frequency bands, with interference into the satellite Global Positioning System (GPS) being a major concern. The nature of ultra wide band transmissions poses a challenge for the regulatory authorities which is currently being considered in the US and Europe. When current regulations are written for signals that fit into clearly defined frequency bands and occupy a few MHz of bandwidth at most, how is it possible to regulate transmissions that cover several bands, with an ultra wide bandwidth by definition?

12.6.4   Bluetooth

Bluetooth could be considered as a wireless LAN technology. Although it has been the subject of much hype over several years, initial products are finally beginning to appear. The Bluetooth vision is to cut the cord between devices, providing short range radio links to interconnect equipment. The idea is to provide Bluetooth functionality in all consumer and business equipment, eventually as a built-in capability rather than an add on accessory. Transmission is in the 2.4 GHz ISM band using GFSK modulation and frequency hopping spread spectrum at 1600 hops per second with a 1 Mbit/s signalling rate [16]. Thus Bluetooth networks are a future source of interference for 802.11 wireless LANs.

With a Bluetooth-enabled PC and either a Bluetooth-enabled server, a Bluetooth access point or a Bluetooth access server, it would be possible to set up a local area network (Figure 12.12). The Bluetooth standard does provide a LAN access profile,

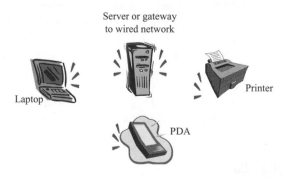

Figure 12.12   Bluetooth networking example

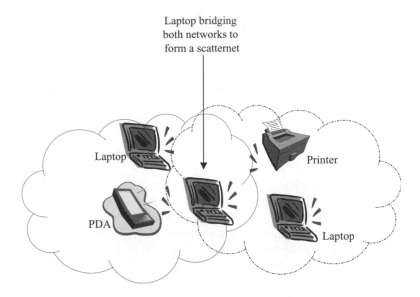

Laptop bridging
both networks to
form a scatternet

Laptop

Printer

PDA

Laptop

Figure 12.13 Bluetooth scatternet example

which allows the transfer of data at a maximum rate of 721 kbit/s [17]. It should also be borne in mind that a Bluetooth network (called a personal area network or PAN) can only contain eight devices according to the standard (early implementations only allow two devices). Although it is possible to have one Bluetooth client in two PANs, acting as a bridge to form a scatternet (Figure 12.13), it is likely that such implementations are stretching the capabilities of the Bluetooth specification to its limits. Whilst it might suit a home user or some ad hoc networking, it is unlikely that Bluetooth technology is sufficient to satisfy the needs of today's corporate users if we think only in terms of a traditional wireless LAN solution. Nevertheless, Bluetooth technology has the potential to enhance or alter our interactions with devices and therefore our working practices, in ways that will only become apparent as applications and services start to harness the power of the Bluetooth vision.

12.7 Conclusions

Typical architectures for wireless networking have been presented, which introduce the concept of wireless LANs and provide a grounding for the subsequent technical analysis. An established worldwide market is already apparent and there is a strong undercurrent to suggest that this will continue to rise rapidly. DECT and IEEE 802.11 have been identified as currently available technologies suitable for use in the implementation of wireless LANs, but IEEE 802.11b is clearly the current winner, being applicable across a wide number of scenarios, with a mature feature set and good interworking between equipment from different manufacturers.

Five prospective technologies have been examined that are each likely to have some suitability across different applications as wireless LAN requirements and markets mature. It has been noted that spectrum and regulatory issues are as important as technical excellence when it comes to the development of new technologies. The wireless LAN concept is capable of delivering broadband access now for nomadic users in a local area and provides a stepping stone to future high bandwidth mobile services.

12.8 References

1   European Telecommunications Standards Institute, ETSI EN 300 238: 'Electro-magnetic compatibility and Radio spectrum Matters (ERM); Wideband Transmission systems; data transmission equipment operating in the 2.4 GHz ISM band and using spread spectrum modulation techniques; Part 1: Technical characteristics and test conditions', http://www.etsi.org, July 2000

2   European Telecommunications Standards Institute: 'Annual Report and Activity Report 2000', http://www.etsi.org, April 2001

3   DECT Forum: 'DECT Sales Continue to Grow', http://www.dectweb.com/DECTForum, December 2000

4   Sixtel: 'LAN NET3: Installation and Management Guide (Instalazione e Gestione)', Sixtel S.p.A, April 1993

5   Sago, A: 'The Olivetti Net^3 Wireless LAN: Product Description and Test Results', Technical Report 359:T:TEP:0183, BT Laboratories internal publication, 31st January 1994

6   BT product webpage, http://www.bt.com/on-air

7   Institution of Electrical and Electronic Engineers, IEEE Std 802.3 – 2000: 'Information technology – Telecommunications and information exchange between systems – Local and metropolitan area networks – Specific requirements – Part 3: Carrier Sense Multiple Access with Collision Detection (CSMA/CD) Access Method and Physical Layer Specifications', http://www.standards.ieee.org, 2000

8   Institution of Electrical and Electronic Engineers, IEEE Std 802.11 – 1999 Edition, 'IEEE Standard for Information technology – Telecommunications and information exchange between systems – Local and metropolitan area networks – Specific requirements: Part 11: Wireless LAN Medium Access Control (MAC) and Physical Layer (PHY) Specifications', http://www.standards.ieee.org, 1999

9   Institution of Electrical and Electronic Engineers, IEEE 802.11b – 1999: 'Supplement to IEEE Standard for Information technology – Telecommunications and information exchange between systems – Local and metropolitan area networks – Specific requirements: Part 11: Wireless LAN Medium Access Control (MAC) and Physical Layer (PHY) specifications: Higher-Speed Physical Layer Extension in the 2.4 GHz Band', http://www.standards.ieee.org, 1999

10 European Telecommunications Standards Institute, EN 300 652 V1.2.1: 'Broadband Radio Access Networks (BRAN); HIgh PErformance Radio Local Area Network (HIPERLAN) Type 1: Functional specification', July 1998

11   JOHNSSON, M.: 'HIPERLAN/2 – The Broadband Radio Transmission Technology Operating in the 5 GHz Frequency Band', HIPERLAN/2 Global Forum, 1999

12   KHUN-JUSH, J., MALMGREN, G., SCHRAMM, P. and TORSNER, J.: 'HIPERLAN type 2 for broadband wireless communication', Ericsson Review, 2002, 2

13   Institution of Electrical and Electronic Engineers, IEEE Std 802.11a and errata: 'Supplement to IEEE Standard for Information technology – Telecommunications and information exchange between systems – Local and metropolitan area networks – Specific requirements: Part 11: Wireless LAN Medium Access Control (MAC) and Physical Layer (PHY) specifications: High-speed Physical Layer in the 5 GHz Band', http://www.standards.ieee.org, 1999

14   ENGLER, Jr, H. F.: 'Technical Issues in Ultra-Wideband Radar Systems,' in TAYLOR, J. D. (Ed.): 'Introduction to Ultra-Wideband Radar Systems' CRC Press, 1995, available on http://www.uwb.org

15   WIN, M. Z. and SCHOLTZ, R. A.: 'Impulse Radio: How It Works,' *IEEE Communications Letters*, 1998, **2** (1) available on http://www.uwb.org

16   Bluetooth Special Interest Group: 'Specification of the Bluetooth System: Core', Version 1.1, http://www.bluetooth.com, February 2001

17   Bluetooth Special Interest Group: 'Specification of the Bluetooth System: Profiles', Version 1.1, http://www.bluetooth.com, February 2001

Chapter 13

Satellite access services

I. Rose, A. Fidler, G. Hernandez, T. Pell and D. Bryant

13.1 Introduction

The provision of end user services, including broadband, via satellite is now a reality. Technological advances over the last 3–4 years have contributed to a significant reduction in costs for delivering a wide range of services to the end customer. A number of unique features that satellite brings to the market, such as wide area coverage, the ability to broadcast and multicast and distance independent cost, have made the use of satellite a more attractive proposition for a large number of existing and emergent network providers. Furthermore, satellite has established itself as the dominant transport technology in delivering direct to home digital TV services in the UK and elsewhere. Costs for space segment, while still high, could start falling and demand is still buoyant for certain types of capacity and services. This is despite some notable failures in certain market sectors.

This chapter addresses the access capabilities of satellite. It does not go into depth on the more traditional services that satellite have been associated with, except as an initial positioning statement. It then goes on to describe the options for access, performance of IP services over a satellite path, discusses emerging technology that will impact on future service delivery, presents information on regulatory and spectrum aspects and closes with an brief overview of standards activities.

13.2 Background

Historically, two-way communications satellites have been used to provide trunk connections, particularly to countries that are difficult to reach, the so-called thin routes. There are still several 10s of countries where satellites provide the only reliable means of connection. The services provided are normally un-switched circuits or switched

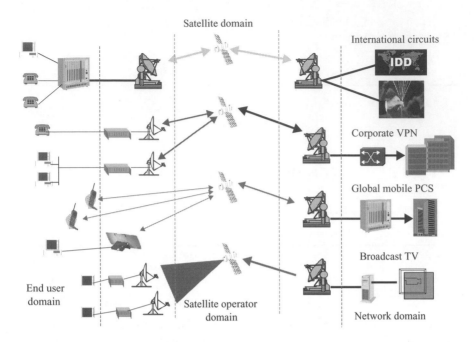

Figure 13.1   Traditional satellite service domain

(dialled) circuits that offer private leased services or public switched voice services respectively. Of all the satellite services, these traditional ones still generate the greatest revenue and profit to the service providers and satellite operators. However, this revenue is under pressure with the increased rollout of fibre optic networks by international cable operators where capacity is reduced to a commodity. The challenge here is that the migration rate to the access domain is not yet high enough to counter losses in trunk revenue. However, the rate is predicted to climb steeply as satellite operators and new service provider entrants focus more on last mile connectivity and the next generation of satellite systems is introduced [1,2]. A view of the traditional role of satellite is shown in Figure 13.1.

Apart from trunk communications, satellites have always been used for television transmission between countries and, over the past 15 years, directly to user premises where technology and higher frequencies (Ku band) have been used to reduce the size and cost of receiving equipment. TV is now transmitted digitally using the Digital Video Broadcast (DVB) ETSI standard. DVB, which is a layer-2 protocol, is a multiplex containing several programme channels that can be broadcast from a satellite with gross rates of between 35 Mbit/s and 45 Mbit/s within a 36 MHz transponder. Within the DVB multiplex, it is now common practice to replace some or all of the MPEG-coded video streams with IP datagrams using multiprotocol encapsulation (MPE). This enables a mix of TV and Internet services to be delivered directly to users with a 45–90 cm antenna (usually requiring no planning permission) and low-cost receiving equipment. It is this development in particular that is enabling satellite to move from the international domain to the access domain, and in particular become cost effective in delivering Internet services.

13.3   Satellite orbits and examples of use

There are three main orbits used for satellite systems – geostationary, medium and low earth orbits as shown in Figure 13.2.

In communications systems, low-earth orbit networks (LEOs) have a large number of satellites circling the earth in a number of planes, each plane containing a number of satellites, at altitudes between about 500 and 1500 km. Examples of satellite providers that have adopted LEO systems are Motorola's Iridium (low rate voice) and Globalstar (low rate voice and data). Initially broadband satellite access proponent Teledesic considered using a LEO architecture but is now focusing on a MEO approach. Another broadband proponent Alcatel's Skybridge is looking at both LEO and GEO approaches.

Business cases for LEO systems need careful consideration because of the large up-front costs (several billion dollars), complexity and terminal logistics associated with them. Furthermore, LEO business models are very sensitive to early year revenues and must take into consideration the fact that the entire constellation, which must be around 40–80 satellites to cover the Earth's surface, requires replacing every 7–10 years at a cost of several billion dollars. A prime example of this is Motorola's Iridium which filed for Chapter 11 bankruptcy and was then bought for a fraction of its cost, and one of the main factors in keeping it operational was its major use by the military.

With medium-earth orbit (MEO) systems, satellites are located at altitudes between 10,000 and 20,000 km using inclined circular orbits (GPS satellites are located at an altitude of 24,000 km and use these orbits). At the time of writing

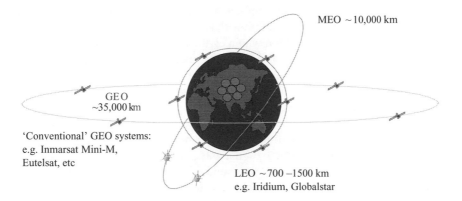

Figure 13.2   Satellite orbits

a MEO system is being constructed by ICO, a spin-off from Inmarsat. This was originally designed as a low-bandwidth voice system, consisting of 12 satellites. ICO is now going through a re-design phase to alter the service set to include broadband data as a result of the global expansion of GSM. It is due for service launch in 2003 and its market focus is now to target markets that are under-served by terrestrial communications services. It is perceived that its initial strategy will be to target existing markets, such as the maritime, transportation, oil and gas and construction industries and governmental agencies, where there is an established demand for satellite communications services. It is believed that ICO plans to eventually extend its focus to broader markets such as small and medium-sized businesses, small office and residential users. This system could provide DSL-like access in rural areas, but it remains to be seen what actually will emerge, and when.

Both LEOs and MEOs require the 'big bang', in other words all space segment must be deployed before the network can go live.

Geostationary satellites (GEOs) orbit the earth at an altitude of approximately 36,000 km and have a period equivalent to the earth's rotation (23 hr 56 min 4.1 s). Most GEO satellites are transparent, in other words they only convert frequency and amplify the signal for retransmission, although most of the broadband satellite proponents are designing for on-board processing to layer 2. At the time of writing there are at least two GEO systems under construction with satellites that are twice the mass and power of anything presently in orbit. One such system is Hughes Spaceway, another is WildBlue. Both of these systems will operate in Ka-band (20/30 GHz) and propose to offer 384 kbit/s uplink and a few Mbit/s downlink from <1 m terminals costing < $1000. Currently all commercially viable satellites are geostationary.

13.3.1   Attributes of satellite systems

Satellite systems should be viewed as a complementary technology to other forms of access technologies and used where its unique features give it a distinct advantage over other access technologies [1–3]. From the service provider perspective, the main attributes that satellites provide are:

- Wide area coverage. Current satellites have a variety of antenna types fitted to them that generate different footprint shapes and sizes. Footprint sizes typically range from the coverage of the whole earth as viewed from space (about 1/3 of the earth's surface) down to a shaped spot beam that covers much of Europe or North America (see Figure 13.3). All these coverage options are usually available on the same satellite, with selection between coverage being made on transparent satellites by the signal frequencies (i.e. at layer 1). It is spot beam coverage that is most relevant for access since they operate to terminal equipment of least size and cost. Increased power capabilities within spot beams of more modern satellites are now allowing smaller receive terminal antennas to be used. Future systems will have very narrow spot beams a few hundred miles across. Furthermore, future satellite systems may also use on-board switching at layer 2 and above to switch traffic between different beams and transponders within the same satellite. This introduces a new engineering challenge since on-board IP/ATM switches must be

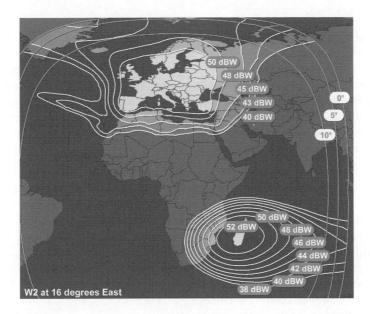

Figure 13.3   Typical shaped beam continental coverage (reproduced by permission of Eutelsat)

developed that can survive the harsh satellite environment as well as being light weight, have low power consumption and be intelligent to aid upgrades since they will have to exist in operation for 7–15 years, the life of the satellite. This contrasts considerably with conventional terrestrial IP/ATM switches, which can become obsolete after eighteen months but can be seamlessly and easily upgraded.

- High bandwidth. Transponder bandwidths on communications satellites are commonly 72 MHz, whilst TV broadcast satellites typically have 36 MHz bandwidths. Trials have shown that 72 MHz communication satellites can support the carriage of data rates up to STM-1 (155 Mbit/s). However, there is one problem with supporting these higher data rates, in that large earth stations (several metres plus) are typically required at both the transmit and receive ends. However, it is perceived that future satellites with narrow spot beams will be capable of delivering data rates of up to 100 Mbit/s to small 90 cm antennas associated with end user terminals. On the uplink side 90 cm end user terminals are typically capable of transmitting data rates between 384 kbit/s and 2 Mbit/s.
- Set up costs are distance independent, since the wide coverage area from a satellite means that it costs the same to receive the signal from anywhere within the satellite footprint.
- Fast access. Once the hub and network connection is in place, which is necessary for the first user, more users can be added in the time it takes to install the equipment, pending regulatory requirements.
- Increased reliability. Satellite links only require the end stations to be maintained and they are less prone to disabling through accidental or malicious damage.

These attributes combine to make satellite a flexible and fast way to connect users both in the core network and access network domains. They can be used as a strategic tool to bring services to market quickly and to bring in early revenue, ahead of the terrestrial network development. Furthermore, such strategic use could also help identify clusters or hot spots of demand so that the network operator can then rollout alternative technologies on a more commercially sound basis. Of course, it can also be used to meet un-served demand for outlying company and residential premises located in rural or remote areas.

Generally satellites are less suited to:

• Games and other applications that require fast interaction, however, proxies and PC settings can be used to automatically re-route this kind of traffic via terrestrial paths in hybrid satellite solutions. Furthermore it is worth noting that hybrid satellite Internet solutions only exhibit a delay of approximately 260 ms due to a single satellite hop.

• Download applications that require data rates in excess of 1 Mbits/s. The impact of the satellite latency on TCP handshaking and window mechanisms tends to limit downstream throughput, per TCP session, to about 1 Mbit/s for hybrid satellite systems using standard TCP/IP settings. One also needs to take into account, especially for hybrid satellite systems, that the narrowband return path can also impact on the maximum downstream data rate, due to slowing down the rate at which TCP acknowledgements are sent back to the Internet from the end user. Having said this, these limitations can be mitigated to some extent by using commercially available enhanced or 'satellite friendly' TCP stacks, performance enhancing proxies, TCP spoofing, by launching multiple TCP sessions or by using UDP instead of TCP and building in reliability at the application layer. These options are described more in Section 13.5.3.

• Voice with inadequate echo cancellation.

The applications listed above are affected by the GEO hop delay, however this delay is much reduced with LEO satellites. In fact a LEO satellite hop exhibits less delay than a fibre for distances greater than about 1/3 of the way around the earth, because the speed of light in a fibre is only about 0.6 of that in free space.

13.4   The move into the IP access domain

The satellite communication industry is currently entering a new chapter by migrating from its traditional trunk, broadcasting and VSAT roles into the new and challenging area of delivering broadband Internet and multimedia content directly to end users. With the four main growth areas being the provision of ISP links to Internet backbones, hybrid Internet access services, two-way Internet access services and content distribution and caching. Table 13.1 [4] shows the recent growth in these new areas in terms of market by transponder lease value, where the cost of a long term transponder lease (10 years) could be between €2M and €3M. It is clear that there has been significant growth in all four areas over the last four years. However, it is notable that

Table 13.1 *Growth in satellite Internet and multimedia services[1] (reproduced by permission of DTT Consulting)*

	Jan 1998	Jan 1999	Jan 2000	Apr 2001
ISP links to Internet backbone	70.5	210.4	746.0	1060.3
Hybrid Internet access services	38.4	58.9	106.5	127.2
Two-way Internet access services	0	0	10.8	88.8
Content distribution and caching	0	1.5	20	42.9
Total	108.9	270.8	883.3	1319.2

Split of IP over satellite market by transponder lease value (in US$m)

two-way Internet access services has shown the most significant year on year growth, from nothing in January 1999 to US$88 M by April 2001, due in part by demand for IP services but influenced by lower user terminal costs.

Prime early examples of this migration are Starband's operation of a two way satellite Internet service in the US with a customer base in excess of 30,000. Closer to home, BT Openworld [5] are offering a two way satellite service throughout the UK targeting SOHO/SME customers, whilst more recently BT Wholesale are trialing a hybrid satellite service targeted at consumer/SOHO markets. These satellite Internet operators are currently exploiting to some extent the unique ubiquitous coverage characteristics of satellite by being able to deliver broadband content to areas presently outside the reach of conventional terrestrial x-DSL networks. For example the BT Openworld Broadband Satellite 500 service provides ADSL equivalent services to customers in rural or remote areas, which are not currently served by ADSL or rate adaptive ADSL and already has customers as far from a main telephone exchange as the island of Unst in the Shetlands.

13.4.1 Services

Satellites are currently used to deliver the following services, an indication of required end user terminal size is also given for information:

- Direct-to-home broadcast TV and pay on-demand movies (staggercast) examples are Sky via SES Astra and Eurochannels via Eutelsat. Recently, Sky have enhanced their service offering with Sky+ a combined satellite receiver and a personal video recorder in one set top box.
- Hybrid satellite fast Internet services for consumer and SOHO markets – satellite downstream (Internet to user) and a narrowband (PSTN.ISDN) return path (user to Internet). Typically end users' terminals use 0.45 m–0.9 m diameter antennas.

---

[1] Internet via Satellite 2001 report by DTT Consulting.

Variable contended downstream satellite data rates, e.g. 256 kbit/s to 4 Mbit/s, and narrowband upstream data rates, system dependent.

- Two-way via satellite services for SOHO and SME markets, e.g. BT Openworld's Broadband Satellite service. Typical end user terminal size 0.9 m. Contended service downstream data rates typically from 512 kbit/s to 2 Mbit/s and upstream data rates from 64 kbit/s to 512 kbit/s system dependent.
- Two-way VSAT (very small aperture terminal) services for dedicated corporate LAN/intranet interconnections. Typical end user terminal size is 1 m–2 m. Various upstream and downstream data rate and contention options available dependent on system design.
- Mobile services, with bit-rates of between 10 bits/s (for paging) to N × 64 kbit/s. These are mainly via Inmarsat and Eutelsat services for land and maritime. A particular example is extension of GSM to cruise liners using Inmarsat B services to provide N × 64 kbit/s links to a small GSM base station on board. End user terminal sizes range from hand held to 1.4 m diameter.
- High bandwidth links – up to STM-1 (current satellite technology limit) between core networks and Internet nodes. Currently there is a growing market for ISP backbone connections via satellite, especially asymmetric ones with large bandwidth out from USA/UK to eastern ISPs and smaller bandwidth return paths. For example, BT presently provide an STM-1 out and 34 Mbit/s back (uncontended) service for an ISP in India and similar systems to Turkey, Singapore. Terminal sizes 2–6 m.
- Leased and switched circuits to countries that are not fibred. Data rates up to 100 Mbit/s (uncontended) both way. Terminal sizes 6–30 m, reducing high value market.
- Military use, very low rate secure data, voice and data up to about 1 Mbit/s.

13.4.2   IP encapsulation

IP to DVB gateways or encapsulators are used by the majority of hybrid and two way satellite Internet services to encapsulate IP traffic into MPEG transport streams, as per the DVB standard, for transmission over satellite access networks.

ETSI standard EN 301 192 [8] specifies the way in which IP packets are encapsulated into MPEG2 transport stream using multi-protocol encapsulation (MPE).

Since satellite is a broadcast media in that all customers receive the same information, a mechanism is needed to assist the satellite receiver's in filtering out traffic just destined for them only. This is commonly done by filtering on the satellite receiver's MAC address, but to do this the receiver MAC addresses must be mapped onto end user IP addresses at the satellite IP to DVB gateway. There are several ways of implementing this mapping function:

- *Static table:* the mappings are entered manually in a file. This requires static IP address assignment to the receivers and has obvious scalability limitations.
- *Broadcast MAC:* all IP addresses are mapped to the broadcast MAC address. This forces all receivers in the network to decapsulate all IP traffic on the PIDs (packet

identifiers) of interest and to filter at IP layer, which may have performance implications.

- *IP copy:* the destination MAC address is generated based on the IP address. This is a mechanism created by Europe Online and now supported by a number of gateway and receiver manufacturers. The receiver needs to be able to set its MAC address based on the IP address assigned to the terrestrial interface. To date this is the most common solution for hybrid satellite fast Internet solutions.
- *Dynamic table:* the mapping table is populated dynamically as receivers join the network. This relies on the receiver to send its IP and MAC addresses to the encapsulator (or to the entity managing the encapsulator) once an IP address has been assigned. Currently there is no defined standard to achieve this and only a few proprietary solutions are available.

13.5   Satellite Internet delivery systems

Current satellite Internet delivery systems can be summarised into two basic architectures – hybrid satellite systems and two-way satellite systems.

13.5.1   Hybrid satellite system

Hybrid satellite systems consist of a satellite downstream path in the Internet to user direction and a terrestrial return or upstream path in the user to Internet direction. Typically the terrestrial return path is achieved with either a PSTN dial-up, ISDN dial-up or leased line connection. This topology is inherently asymmetric since it typically consists of a broadband satellite connection in the downstream direction and a terrestrial narrowband connection in the upstream direction.

Hence downstream content is delivered via satellite, whilst all upstream information (e.g. Web page requests and TCP acknowledgements for received data) is sent terrestrially. For optimal TCP/IP performance, care should be taken to correctly dimension the terrestrial return path for the particular application service set required. TCP acknowledgements, unless the TCP stack is optimised, typically require about 20% of downstream bandwidth, so the bandwidth of the return channel is very important. Examples of applications that use this topology are satellite Internet services and corporate virtual private networks (VPNs). It should be noted that in addition to the delivery of the usual unicast Internet and data services this topology can support the bundled delivery of multicast and broadcast services, e.g. electronic newspaper delivery and local Web caching services, which can be off-line. Figure 13.4 shows the typical network topology for a hybrid satellite solution.

There are several design options available for integrating the terrestrial return path and associated dial access platforms into a hybrid satellite access network:

- A solution whereby the Internet service provider (ISP) undertakes the responsibility of routing downstream traffic to the satellite hub earth station via either a direct leased line or fibre connection, or across the Internet using tunnelling protocols

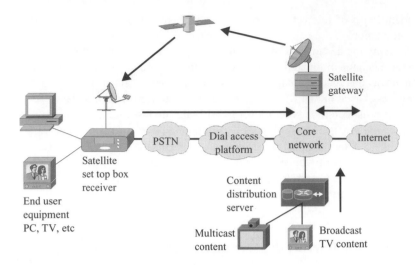

Figure 13.4   Hybrid satellite system

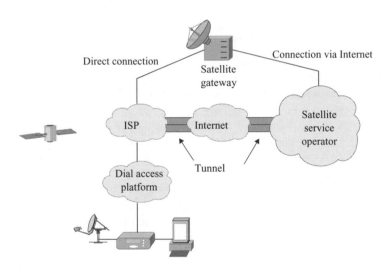

Figure 13.5   Direct connection solution

such as generic routing encapsulation (GRE) and point-to-point tunnelling protocol (PPTP). This principle is shown schematically in Figure 13.5.

With a tunnel solution, as default, all downstream traffic irrespective of its traffic type or protocol (HTTP, HTTPS, FTP, Telnet, etc) is routed downstream via the satellite path.

- A proxy solution whereby all Internet requests from end user PCs are routed to a designated proxy server off the Internet. The network is then configured to route

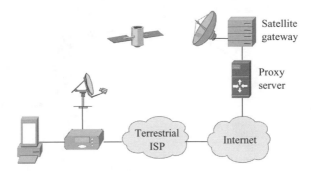

Figure 13.6   Proxy solution

all downstream traffic from this designated proxy server via the satellite hub earth station. With this solution the proxy option within the end user's Internet browser needs to be enabled. Some solutions also require the installation of customised local proxy software that resides on the user's PC. With this approach the end user can gain access to the service using any terrestrial ISP. This principle is shown in Figure 13.6.

For simplicity Figure 13.6 assumes collocation of the satellite gateway and proxy server hence it shows a direct connection between the two. In reality, if these units are not physically collocated, they could be connected via leased line, fibre link or across the Internet.

With a proxy solution only the protocols supported by the particularly proxy server being used will be routed via satellite, e.g. HTTP and HTTPS. All other unsupported traffic (e.g. Telnet, MP3 peer-to-peer services, gaming) will be automatically routed two-way via the terrestrial path. Proxy based solutions do not support VPNs whereas tunnelled solutions can be designed to do so.

13.5.2   Two-way satellite systems

In contrast two-way satellite systems consist of asymmetric satellite paths – a broadband downstream path for the delivery of the actual content and a smaller upstream path for the carriage of the user to Internet requests. As with hybrid satellite systems, the upstream path needs to be carefully dimensioned and contended to ensure that it does not limit the overall performance of the service.

The majority of solutions available today use the internationally recognised DVB standard for the downstream satellite carrier. However, there are presently two alternative options for the upstream carrier – the use of the new DVB Return Channel via satellite (DVB-RCS) [7] open standard or the use of proprietary solutions offered by a number of satellite service providers and suppliers. Figure 13.7 shows the typical network topology for a two-way satellite solution.

The principal service application area for two-way satellite solutions is broadband satellite Internet access and corporate VPNs. In addition to supporting point-to-point Internet applications, two-way solutions can also support the delivery of broadcast and

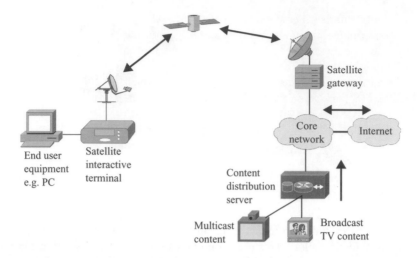

Figure 13.7   Two-way satellite network

multicast content. There is considerable interest in the delivery of broadband content to remote small office home office (SOHO) and small to medium enterprises (SMEs) via two-way satellite systems. For example, a number of ISPs and operators have now being offering these services in the UK for several months – e.g. BT Openworld, Tiscali, Armasika, Bridge Broadband, Crystal Data, Isonetric, net-work, Satweb).

13.5.3   DVB RCS

A European standard for Digital Video Broadcasting (DVB) with a return channel via satellite has been issued [6]. This standard provides a baseline specification for the provision of the interaction channel for GEO satellite interactive networks with fixed return channel satellite terminals (RCST). The standard, therefore, does not apply to services over the proposed new generation of LEO or MEO satellites intended for broadband applications, neither does it apply to mobile services. The standard may be applied to all frequency bands allocated to GEO satellite services. Key highlights from the DVB-RCS standard are:

- DVB-RCS is intended for use in a domestic environment, although most vendors are interpreting this as including SMEs and SOHOs, at least in the first stages of market exploitation.
- The forward link is DVB; the RCST must be able to receive digital signals conforming to the DVB standard [7].
- Forward link signalling, to manage the return link, is undertaken within the DVB structure. The forward link signalling is multiplexed with the user data onto a standard satellite digital video broadcast (DVB-S).

- The return link is a burst format, namely MF-TDMA. Many RCSTs can share the same bandwidth in both the time and frequency domain. The RCSTs retrieve the centre frequency, the start time and the duration of their transmit bursts by examining the forward link signalling.
- Traffic bursts within the return link can be defined as either ATM cells or MPEG2-TS packets and there are two terminal types – Type A and Type B to accommodate this.
- The RCST symbol rate range is currently specified from 128 ksymbols/s to 4 Msymbols/s, but it should be noted that a revised standard [8] is likely to be ratified that extends the allowable symbol rate range from zero to infinity. QPSK modulation is employed with a range of forward error correction (FEC) options. If ratified, the revised standard may promote a new generation of lower speed, lower cost products.
- One key aspect of the ETSI standard is the DVB forward channel format. This allows the possibility of mixing other services within the multiplex, so using the satellite transponder more efficiently. DVB also offers the user the possibility of more easily receiving other services than may be available from the same orbital location, for example DTV.

13.5.4   Hybrid versus two-way satellite Internet solutions

The main advantages of hybrid satellite Internet solutions are:

- Low cost CPE since these are largely the same as existing mass market TVRO components.
- The ability to self-install satellite antenna and associated indoor CPE.

The main disadvantages with hybrid satellite Internet solutions are often focused around the use of narrowband return paths, for example:

- Not having an 'always-on upstream connection' and hence having to dial up for Internet access.
- The call blocking and contended nature of narrowband dial access platforms.
- The additional cost to ISPs of bundling anytime ports with a satellite Internet offering to emulate an always on service.

Although hybrid satellite Internet solutions cannot provide an always-on pull Internet service analogous with x-DSL offerings, they can provide 'pseudo always-on' in the downstream path in that they can be designed to receive off-line content such as e-mail, pre-fetch files, multicast/broadcast feeds and news ticker feeds. In fact, DVB-compliant set top boxes will be emerging on the market in the next 1–2 years that will have mass storage devices and interactive play out facilities and will enable true multimedia in the home, including dial-up IP services [9].

In contrast the main advantages of two-way satellite Internet solutions are:

- Their ability to offer a true always-on solution, similar to that of terrestrial x-DSL solutions.

- Their self contained nature and non-reliance on other Telco platforms such as dial access. Hence their ability to be rolled out quickly to target areas once the hub earth station is installed and operational.

The main disadvantages of two-way satellite Internet solutions are:

- Current high cost CPE compared to hybrid solutions.
- Requirement for the terminal to be licensed to transmit by the Radiocommunications Agency in the UK.
- Requirement for all satellite antennas and associated indoor CPE to be professionally installed.
- Requirement for larger CPE antennas.
- Requirement for additional satellite space segment in the upstream direction and careful dimensioning of this to ensure that, when heavily contended, the overall service performance is not degraded.

The requirement for planning consent affects both types of solutions, for instance on listed buildings, within national parks, where the end-user has a dish already installed, etc. Although a General Development Order allows certain sizes of antennas to be installed without the need for planning permission, this is interpreted and implemented differently across the UK by the hundreds of planning authorities that exist.

13.5.5   IP performance over satellite

The two most widely used transport layer protocols are UDP and TCP. UDP is an unreliable transport mechanism that does not provide any error checking, retransmission or congestion avoidance facilities. UDP works therefore seamlessly over satellite and the only characteristic of satellite networks that may have an influence on the performance of UDP-based applications is the latency of the satellite link and the very rare occurrence of packet loss. The typical end-to-end network latency of a one-way satellite system is approximately 300 ms, and a two-way satellite system approximately 700 ms[2].

On the other hand, TCP is a reliable transport protocol that provides facilities like error checking, retransmission and congestion avoidance. It is well documented and understood that the maximum throughput per TCP session is limited by two key parameters: the TCP receive window size on the end-user PC and the link round trip time. This limitation is due to the guaranteed service nature of TCP which limits the transmitting server to only transmitting a certain window of N packets before it needs to receive a packet acknowledgement back from the receiving client:

Maximum throughput per TCP session = Receive Window Size/Round Trip Time
Assuming a round trip delay of approximately 300 ms and a default receive window size of 8192 bytes this gives a maximum throughput per TCP session of approximately 210 kbit/s, based on the hybrid solution.

---

[2] These are typical values experienced when pinging IP addresses or Web sites over a satellite access network.

However, it should be noted that this restriction is for a single TCP session and that in theory higher throughputs up to the defined return channel limit may be possible by launching multiple sessions.

While TCP works over satellite links, there are a number of well-understood mechanisms that can be used to improve the performance of the protocol over satellite networks. RFC 2488 [10] describes a number of performance-enhancing techniques that have been standardised by the IETF.

The techniques recommended in RFC 2488 include:

- The use of large TCP windows to increase the throughput of individual TCP connections. For example, by specifying a larger default receive window size at the client's PC TCP/IP stack. This assumes that a service provider's cache can tune its send window size to that advertised by the client. In principle, this should be possible by altering the receive window size in the customer's PC during or following installation (e.g. via a client set-up CD). At the client side, operating systems such as Window95© only supports a maximum receive window size of 65 kbytes, however, in theory Windows 98 and 2000 should support a maximum receive window size of 5 Gbytes based on RFC 1323.
- The use of fast retransmit/fast recovery algorithms for better congestion handling.
- The use of selective acknowledgements to avoid unnecessary retransmissions in case of lost segments.

In addition to the standard techniques proposed in RFC 2488 there are a large number of alternative TCP enhancements that are currently being investigated. The items being researched have been recently compiled and published as informational RFC 2760 [11]. These include additional techniques to mitigate the effect of the slow start algorithm, techniques for speedier recovery from loss situations, TCP/IP header compression techniques and the use of multiple TCP connections.

Another research area outlined in RFC 2760 aims at improving TCP performance over highly asymmetric networks. This is of particular interest because the topology of most satellite networks used to deliver broadband applications presents a bandwidth asymmetry: a very high data rate in the downstream direction together with a low-speed upstream link.

TCP can tolerate a certain amount of bandwidth asymmetry. However, if the network asymmetry is too large TCP performance may be reduced, for example through there being insufficient return channel capacity for the TCP acknowledgements (TCP acknowledgements, unless the TCP stack is optimised, typically require about 20% of received bandwidth).

This is due to the need for the receiving client to send back acknowledgements to the server for all packets received and for the server to be able to send a certain window of N packets before requiring the reception of an acknowledgement packet.

$$Maximum\ TCP\ throughput = \frac{return\ channel\ data\ rate \times maximum\ segment\ size}{ACK\ packet\ size + PPP\ overhead}$$

Using a maximum segment size of 1500 bytes, for an example, this approximate formula indicates a maximum TCP (all sessions) throughput for a typical PSTN connection of approximately 1 Mbit/s. This is based on the following assumptions:

- a PSTN return channel rate of 33.6 kbit/s;
- a maximum segment size of 1500 bytes;
- an ACK packet size of 40 bytes (maximum transfer unit = maximum segment size + header (ACK));
- a PPP overhead of 7 bytes.

Although many of the techniques already mentioned can be effectively used to improve this situation, there are several mechanisms (like selective ACK, ACK congestion control and ACK filtering) specifically aimed at reducing the limiting effect that a low-speed upstream link can have on the data flow over the faster downstream link.

Even though a lot of emphasis is placed on improving the performance of transport layer protocols over satellite links, it is important not to overlook the improvements that can be made to the performance of application layer protocols. Some application layer protocols are particularly 'chatty' or use inefficient headers. In this case, application layer performance enhancing proxies (PEPs) can be effectively used to improve protocol performance over satellite links. PEPs is a generic term for a device in the path intended to improve performance. PEPs are used to overcome certain link characteristics that have an adverse effect on protocol performance. In theory, they can operate at any layer in the protocol stack but they are usually found at the transport or the application layer. Although several transport and application layer PEPs are currently available, there is no standard implementation. A survey of PEPs (not only for satellite networks but also for wireless WAN and LAN environments) is available as informational RFC 3135 [12]. For example, one form of PEPs are TCP spoofers that send timely acknowledgements back to hosts while packets are in transit over the satellite link.

In other cases, performance improvement can be achieved by making an effective use of features already present in the protocol. An example of this approach can be found with HTTP, where the use of persistent connections and request pipelining (both available with HTTP/1.1 but not with HTTP/1.0) can lead to a significantly improved Web browsing experience.

13.5.6   Bandwidth management

Another performance aspect that must be considered when developing both hybrid and two-way satellite Internet services is that of bandwidth management, contention ratios and the number of users that can be supported within a given satellite transponder capacity. Generally the calculation of satellite contention ratios is more complex than its terrestrial counterparts since one has to factor in dial access contention as well as the satellite component for hybrid systems and multi-frequency TDMA in the return path for two-way satellite systems.

13.5.7 Security aspects

Security of user information is essential if the use of satellite communications is to become widespread in the delivery of broadband content and applications to residential, SOHO and SME communities. In line with terrestrial networks, security within satellite networks can be applied at various levels within the OSI seven layer model, primarily being applied at either the application layer (e.g. SSL), the network layer (e.g. IPSec) or at the physical layer (e.g. smart cards and conditional access).

Although the majority of terrestrial based Internet traffic is sent unencrypted, it is generally perceived by the general public as being secure, mainly due to the fact that terrestrial solutions involve the end users having their own dedicated line between their premises and the local exchange. In contrast, with a radio delivery mechanism (such as satellite or wireless LANs), eavesdropping is seen as a major concern since the user information is broadcast over an open medium and with the right equipment it can be relatively easy to eavesdrop on unencrypted data.

Looking closer at the satellite Internet scenarios discussed within this chapter, the downstream satellite carrier can generally be considered as the least secure component in the network due to its broadcast nature, e.g. all users (both authorised and rogue) are capable of receiving unencrypted downstream content. With hybrid satellite Internet topologies, the upstream path can be deemed relatively secure due to its terrestrial nature. With two-way satellite Internet topologies, the upstream path can also be deemed relatively secure due to the frequency hopping nature of both proprietary and DVB-RCS access schemes used.

Although radio is inherently insecure as an access medium, secure services can be delivered effectively via satellite by employing the following precautionary measures.

13.5.7.1 Firewalls

As the satellite-based Internet topologies can be considered as 'always open to attack' on downstream satellite carriers, even when a terrestrial link is used, a firewall is recommended at the customer site to prevent unauthorised access into the customer's network. This recommendation aligns with the advice given to terrestrial based Internet users (both residential and business) that use access networks such as ADSL.

13.5.7.2 Conditional access

Smart card technology, similar to that employed by satellite broadcast TV operators could be used to encrypt sensitive streams and content. A number of these systems are on the market that encrypt user data (either at the physical layer or by IP payload scrambling) and employ public key/digital certificate mechanisms to allow decryption. However, the key distribution and management of these systems is not currently perceived to be scalable to the size required for mass-market penetration, but could be applied to small or medium sized VPNs.

13.5.7.3 VPN security

For corporate and SME environments, a secure hybrid or two-way satellite service can be offered by implementing a VPN solution to enable remote access to a company intranet. This process involves setting up a secure tunnel between the VPN server, located at the corporate or SME hub and the remote worker's PC. Standard terrestrial VPN client and server software can be used, since the VPN software will just see the satellite link as a very long leased line and thus should operate in its normal way.

VPN solutions are typically based on tunnelling protocols such as L2TP, PPTP and GRE, which are then encrypted using various security and encryption protocols. It is also becoming popular to instigate secure VPNs using just IPSec without the extra tunnelling protocol, the main driver being to reduce the size of the overhead introduced when tunnelling.

IPSec is a suite of open standards, defined by the IETF to provide network layer security to IP for both private and public network infrastructures. These services include confidentiality, authentication, integrity and compression. Tests carried out by BTExact at Adastral Park have revealed that IPSec can operate fully over satellite access networks, but that this is heavily dependent on where exactly IPSec is implemented within the network. For example, the use of performance enhancing proxies that spoof TCP can cause problems for IPSec enabled networks, dependent on the order that IPSec and the proxy enhancement occur. The reason being that performance-enhancing proxies generally cause IPSec to think that packets have been intercepted or tampered with, which results in the IPSec client software discarding them. However, this problem can be easily overcome by performing the PEP function before the IPSec operation. Furthermore, IPSec can work either in a transport mode, where the IP payload is encrypted but the IP header information is left untouched, or in a tunnel mode, where the IP address and payload are hidden by the IPSec header. Presently it is perceived that the IPSec transport mode may be a more attractive solution for satellite networks where multicast operation is preferable to unicast operation.

13.5.7.4   Application security

Application security within the Internet is primarily implemented using secure servers, which use techniques such as secure socket layer (SSL) WWW sites and secure e-mail. Tests carried out by BTExact at Adastral Park have revealed that these services work perfectly well over hybrid and two-way satellite networks, as the security is implemented above the transport layer and is therefore considered part of the user data as far as the satellite networks are concerned. However, these secure services require the service provider to invest time and possibly technology in setting up these services.

13.6   Commercial positioning

A recent report by Fletcher Advisory entitled 'Small firms unlock broadband's potential' [13] defines broadband as 'technologies offering data speeds of 512 kbit/s and above downstream combined with always-on' and narrowband as 'having Internet access through a standard modem with speeds between 28.8 kbit/s or

56.6 kbit/s'. Based on these definitions two-way satellite Internet services can be clearly referred to as broadband services since they are always-on and can exhibit downloads speeds of up to 512 kbit/s. However, the definition of hybrid satellite Internet solutions is less clear cut since they can exhibit download speeds of up to and in excess of 512 kbit/s but cannot be really classed as always-on due to their use of terrestrial return paths. Having said that, one can argue that hybrid satellite Internet solutions can support some always-on services in that the downstream satellite path is always-on and can be used to receive off-line services such as e-mail and multimedia streaming without the requirement for a terrestrial return path.

Another area open to debate is that of the positioning of hybrid and two-way satellite Internet services against various market sectors, e.g. consumer, SOHO, SME and enterprise customers. One way to position these services is to look at the results of market surveys showing end-user key uses of broadband and the importance of high-speed uploading and downloading.

For example, the Fletcher report shows that for small businesses 41% of people surveyed see always-on as the most valued feature of broadband, as shown in Figure 13.8.

It is also interesting to note that the same report shows that main uses of broadband identified by small users are instant e-mail, always available tool, download of large files and upload of large files as shown in Figure 13.9.

The relative importance attached to high speed uploads and downloads is shown in Figure 13.10.

This result is interesting since it shows that the majority of small businesses see high speed uploads being equally important as high speed downloads which tends to

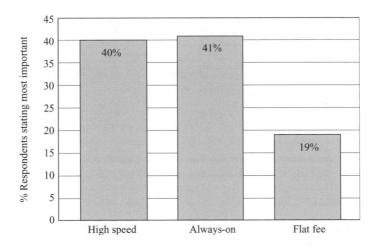

Figure 13.8   Most valued feature of broadband (reproduced by permission of Fletcher Advisory)

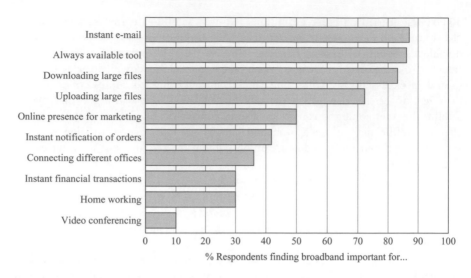

Figure 13.9   Main uses of broadband (reproduced by permission of Fletcher Advisory)

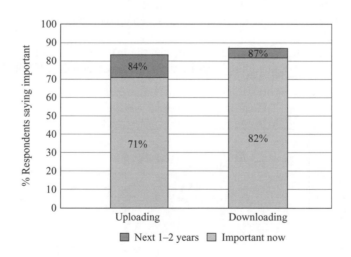

Figure 13.10   Importance of high speed uploads and downloads (reproduced by permission of Fletcher Advisory)

favour the two-way satellite Internet solution. This, combined with the requirement for always-on connectivity, tends to suggest that two-way satellite Internet solutions are more suited for high end SOHO, SME and enterprise markets, whilst hybrid solutions are more suited to low end SOHO and consumer markets.

13.7 Regulatory/licensing

13.7.1 Regulatory considerations

Most of the present satellite services use frequency bands between 1 and 20 GHz. Those most likely to be used for access networks in the UK are the 11/12/14 GHz bands and 20/30 GHz bands. The ITU Radio Regulations (RR) provide the framework in which these frequency bands may be used for satellite services and set forth the appropriate procedures for the advance publication, notification and co-ordination of satellite systems and procedures for earth station co-ordination vis-à-vis terrestrial microwave links.

The use of satellite spectrum within the United Kingdom is regulated by the UK Radiocommunications Agency (RA) under the powers invested by the Wireless Telegraphy Act 1949 (WT Act '49) and subsequent amendments, together with the WT Act 1998. The RA lays down more specific guidelines about the use of spectrum in particular bands (e.g. by giving a national priority to one service over another). Each country, in which satellite service providers operate earth stations, has its own licensing arrangements and national planning procedures have to be adhered to.

In most countries, the fixed satellite service (FSS) is allocated on a co-primary status with (typically) terrestrial fixed service (FS) links. Permanent earth stations at a fixed location require co-ordination with fixed service terrestrial links in accordance with the Radio Regulations. The required co-ordination area (as defined by ITU RR Appendix 7) may extend several hundred kilometres and hence may also require international co-ordination (especially for coastal sites such as Goonhilly Earth Station on the Lizard Peninsular in South-West England). Sharing between the two services has worked very successfully in the past in the situation where the fixed link service is used for trunk transmission networks and the satellite services operate via a limited number of Earth stations. Potential difficulties have occurred sharing with high densities of terrestrial links and this has made it difficult to get earth station clearance in some areas[3].

In the spectrum allocated in the UK between 10.7 and 14.5 GHz, only the frequency bands 12.5–12.75 GHz (downlink) and 14.0–14.25 GHz (uplink) are not shared with terrestrial fixed link services and this is the only pair of bands in which VSAT networks are currently allowed to operate. The RA have made extensive use of the 14.25–14.5 GHz up-link band for fixed link services. Due to the impact on scarce satellite uplink spectrum, both the EC and CEPT have recommended that no further fixed links be deployed in this band and it should become principally an FSS band in Europe.

In recent years we have seen moves by non-GSO satellite operators (e.g. low earth orbit systems) to operate in bands used for GSO satellite services. Alcatel's Skybridge low earth orbit network (and others) succeeded in being able to operate in these bands by meeting the requirements of RR 22.2 whereby the GSO FSS services are

---

[3] Sharing with point-to-multipoint fixed wireless access networks is a particular problem.

protected by ensuring that the NGSO networks employ satellite diversity. Procedures are in place to protect GSO services from short-term interference bursts from NGSO satellites and by enforcing their compliance with 'hard' power limits in Articles 21 and 22 of the Radio Regulations, agreed at WRC-2000.

The RA are obliged to ensure that the transmissions from UK licensed earth stations do not cause interference to the space stations of other Administrations. This could prove difficult to monitor with the current plans by some satellite proponents to deploy large numbers of small user terminals (typically 30 cm or less) and it is for this reason that the RA advocates a continuing conservative approach to licensing, at least for the time being. This is particularly pertinent in the 20/30 GHz bands and above, as spectrum for high density fixed satellite service (HDFSS) networks is currently being actively identified in the ITU in these bands. Sharing has been recognised as difficult when either or both networks are deployed ubiquitously. The approach of some countries is to use band segmentation to avoid the complexities of terrestrial frequency sharing (and to further sub-divide the satellite spectrum into NGSO and GSO in some bands). The fixed link services in these bands tend to be for short-haul customer access applications or for cellular base station interconnections, rather than the traditional trunk networks. In the 20 GHz downlink band this is probably going to constrain ubiquitously deployed terminals to 19.7 – 20.2 GHz, although 17.3–17.7 GHz is also being studied for HDFSS. Potentially, only co-ordinated gateway earth station terminals would be able to be deployed in the 17.7–19.7 GHz band. The situation in the 30 GHz uplink band is not quite complementary. The exclusive satellite band 29.5–30.0 GHz will certainly be used for HDFSS applications, but the band 27.5–29.5 GHz has been segmented both in the USA and Europe so that the FS and FSS have different priorities in different parts of the band. The global downturn in the telecommunications business has caused many of the potential HDFSS operators to defer their plans, although acquiring the spectrum and ensuring the regulatory procedures are in place is an important early consideration.

13.7.2 Licensing

All transmitting radio equipment in the UK is licensed by the RA for monitoring and policing purposes, either via individual licences (e.g. permanent earth stations, terrestrial fixed links, radar), or a network licence (as in the case of VSATs, public mobile telecommunications, fixed wireless access, etc) or a specific exemption order within a statutory instrument. Hence licensing is required for all two-way satellite systems.

Last year, the UK Radiocommunications Agency (RA) consulted earth station licensees and has now developed a new pricing regime for permanent earth stations (PES, i.e. the traditional larger type of earth stations), with a view to extending similar principles to other new products. The new PES administrative pricing structure is now related to spectrum use. In many cases there has been a fee reduction, particularly for the case where antennas share a site. The new pricing algorithm reflects the amount of bandwidth and power used.

In addition to licensing, the UK has a national site clearance procedure, which was established to ensure that the possibility of interference should be minimised. This ensures protection of other equipment that may be adversely affected by radio emissions. The site clearance procedure requires up to 28 days to process an application for presentation to radio users on the UK National Frequency Assignment Panel (NFAP). The NFAP meets regularly on a three-weekly basis to review, register and clear all licence applications that use a transmitter power exceeding an e.i.r.p of 17 dBW. The consequence of this is that, for permanent earth stations, the normal time between submission of an application and granting of a licence is between 3 and 6 weeks, even in the exclusive satellite bands.

In order to facilitate the rapid deployment of VSAT networks, the RA has negotiated a rapid clearance concession for VSAT networks where the e.i.r.p of the remote terminals is less than 45 dBW. Approval for the location of compliant sites can normally be provided well within a week at the present time.

The only exception to the 3–6 week site clearance is the use of transportable earth stations (TESs) in the 14–14.5 GHz band, where a same day (normally 15–20 minutes) turnaround is provided. In connection with this, the RA has implemented a new automatic Web-based tool, eFLATCO, that will further simplify the process for obtaining authorisations for TESs in line with government eBusiness initiatives.

Recognising that new licence products are required to encourage the take-up of 'ubiquitous' terminal services within the UK, the RA have consulted with operators, the CAA and MoD about the remaining types of earth station, notably those using small terminals (e.g. SITs, SUTs and VSATs). This is expected to result in a suitably responsive licensing regime to cater for this new environment, including site clearance procedures where authority to operate may be able to be obtained at point-of-sale (in exclusive bands only). A new Fees Order was passed through Parliament in July 2002 and this new licensing regime is currently being prototyped and under development.

In Europe, a more open market approach to the deployment of satellite services has been ongoing for many years. There have been a number of significant developments regarding authorisation of terminals, the most significant probably being the EC R&TTE Directive coupled with 97/13/EC on the common framework for general authorisations which remove many of the artificial barriers to marketing satellite terminals in Europe. Another success has been the one stop shop (OSS) arrangement within ECTRA/ERC for satellite terminal licensing[4].

CEPT ERC has taken some 35 satellite-specific decisions to facilitate market access. The UK has implemented 26 of these Decisions[5]. The most significant ERC Decisions from a satellite terminal network viewpoint are (00) 03,04 and 05 on 'Exemption from Individual Licensing of SITs, SUTs and VSATs' but the UK has not signed up to these particular decisions because the RA insist on terminal

---

[4] In parallel with the work within the EU, a number of satellite-related standards were developed within ETSI and the essential requirements of those standards have been published as ENs in support of the RTTE Directive. Those of current interest include 301 428 (14 GHz VSAT) and 301 459 (SIT & SUT).

[5] JPT SAT 'Comprehensive Satellite Initiative Report' – source SAP RWG.

registration. With this constraint, it makes it difficult for operators to provide pan-European services. However, it is hoped that the current discussions within the RA's Satellite Consultative Committee will lead to these services being facilitated in a satisfactory manner in the UK.

13.7.3   Planning issues

The issue that satellite customers face with local planning approval is determining whether or not they actually require consent to install a satellite antenna on their property. Generally they need to seek planning permission if they fall into one of the following categories:

- the building is in a conservation area or a national park;
- it is a listed building;
- the satellite antenna to be located is on a street-facing wall in Northern Ireland;
- there is already an existing satellite antenna on the premises.

Costs associated with planning approval can be from £80 upwards. A further issue is the time taken for planning permission to be granted which increases the time taken to have the service installed.

Obtaining planning permission can take at least six to eight weeks, during which time it is necessary to check with a customer how their planning application is progressing before proceeding to the installation phase.

The fact that there are different planning guidelines and interpretations of those guidelines across the UK increases the complexity of providing individual advice to customers about their requirements for planning approval. Furthermore, due to differences in planning guidelines across the UK, a separate set of planning questions is required for each country within the UK. There is a need to consult with the large number of planning authorities on the issues associated with obtaining planning approval, and have a set of consistent and user-friendly planning guidelines for installation of a second antenna.

13.8   Conclusions

In conclusion, satellite systems are now being increasingly used for the delivery of IP services in the access network. Wide area coverage and fast and flexible deployment attributes now make satellite an excellent access solution for delivering broadband Internet and multimedia services to rural users presently outside the reach of terrestrial x-DSL coverage. Furthermore, the unique natural broadcasting capability of satellite can also be used to enhance the range of Internet and multimedia services offered to end users, for example, multicast and broadcast TV content.

Today the majority of satellite IP services use the ETSI DVB standard to encapsulate IP packets into MPEG2 transport streams for transmission over satellite. Two key service options exist: hybrid satellite Internet systems, which are generally more suitable for consumer; and low-end SOHO markets; and two way satellite

Internet solutions, which are more suited to high-end SOHO, SME and enterprise customers.

Future satellite systems will be even more suited to the access network, probably incorporating layer-2 switching and narrow spot beams, enabling even more bandwidth to be delivered to end users at a lower cost. The time frame for the next generation systems is 2004–2007.

13.9   References

1 WAKELING J. F. and DOBBIE, W. H.: 'Satellite access services', *BTTJ*, 1998, **16**(4), Local Access Technologies
2 FITCH M. and FIDLER, A.: 'An overview of satellite access networks', *BTTJ*, 2000, **18**(3), Carrier scale IP Networks
3 FIDLER, A. *et al.*: 'Satellite a new opportunity for broadband applications', *BTTJ*, 2002, **20**(1)
4 Internet via Satellite 2001 report, DTT Consulting
5 ROSE, I. G. *et al.*: 'BT Openworld satellite broadband from concept to launch', IEE Personal Broadband Satellite Colloquium, 22nd January 2002
6 ETSI EN 301 790 V1.2.2
7 DVB-RCS 333, Rev 5.3, 22 February 2002
8 EN 300 421: TR 101 202: ETS 300 802: EN 300 468: EN301 192: ETR 154
9 WAKELING, J. F.: 'Can satellite move into the access space?', *Future Comms*, 2001
10 ALLMAN, M., GLOVER, D. and SANCHEZ, L.: 'Enhancing TCP over satellite using standard mechanisms', *BCP28, RFC 2488*, January 1999
11 ALLMAN, M. *et al.*: 'Ongoing TCP research related to satellites', *RFC 2760*, February 2000
12 BORDER, J. *et al.*: 'Performance Enhancing Proxies intended to mitigate link-related degradations, *RFC 3135*, June 2001
13 Fletcher Advisory Report, 'Small firms unlock broadband potential', 2001

Chapter 14

UMTS

R. Mostafavi

14.1 Introduction

The introduction of cellular telephony in the UK in 1985 opened a window of opportunity to people in the business sector to communicate on the move. The system was based on analogue technology suited to speech and, although the coverage was not very good, the gradual increase in demand made it financially viable for the operators to expand the coverage area. There were a number of problems with this cellular system, including interference, low security and lack of common standards, even within European countries. The operators, in spite of low traffic, kept the quality of service acceptable by charging premium rates. Mobile telephone users could sometimes hear other users and it was possible to eavesdrop on users with cheap scanners. The result of non-standard cellular telephones even within European countries made roaming impossible. Higher call tariffs and the prices of handsets kept the mobile telephones beyond the reach of most people. A common cellular mobile system, at least within Europe, was needed.

To harmonise cellular mobile service within Europe a new cellular system, GSM (global system for mobile communications), was developed. This is a second generation system and is still in use across Europe and many other parts of the world. Unlike the first generation system it is digital, and as well as good speech quality, it can also provide data at 9.6 kbit/s, or at 40 kbit/s [1] using general packet radio service (GPRS). Recently some mobile handsets have become available with WAP (wireless application protocol) capability to access limited World Wide Web type information on a mobile telephone.

The second generation systems, i.e. GSM (Europe), PDC (Japan), PCS (Americas), are very good for voice wireless communications. Although the above systems are different, roaming is possible with multimode handsets. The explosion of Internet and data communications during the past five years made the second generation systems inadequate for data communications which is becoming at least as important as voice communications. The need for evolution is apparent from the current trends. The new system should be able to cope with data, as well as being backward compatible. The compatibility with the second generation is required because of the massive investment by the operators in the second generation systems.

Speed limitations of data transport and true World Wide Web capability of second generation mobile systems showed the inadequacy of the current cellular mobile systems for the modern communications needs. The following are some of the expectations from a third generation mobile system:

- High quality speech as well as multimedia.
- High data bit rates (up to 2 Mbit/s) to enhance access to information on public and private networks.
- Global availability and roaming.

The International Telecommunication Union's (ITU) original goal was to determine a single third generation system. This goal was unachievable in full because so many operators had heterogeneous interests. In 1992 a decision was made by World Administrative Radio Conference (WARC) to allocate a band of frequencies around 2 GHz for third generation mobile systems (IMT2000). The WARC allocated 220 MHz bands of frequencies for third generation systems, including 2×30 MHz also available for the satellite component. In Europe, 1920–1980 MHz (uplink) and 2110–2170 MHz (downlink) has been identified to accommodate FDD technology and 1900–1920 plus 2010–2025 MHz for TDD technology. These bands of FDD frequencies were also acceptable in Japan and Korea. Figure 14.1 shows the spectrum allocation in Europe, Japan, Korea and the USA.

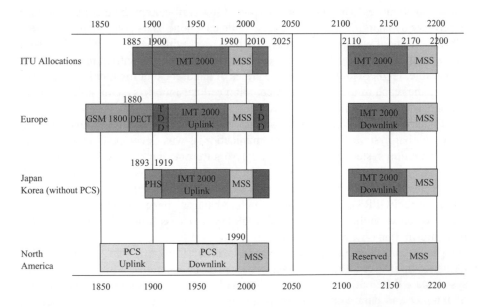

Figure 14.1   Spectrum allocation in Europe, Japan, Korea and the USA

As Figure 14.1 shows, a common frequency (IMT2000) is allocated for Europe, Japan, Korea. In the USA. the lower band (1920–1980 MHz) is occupied by the second generation system (PCS). The above common frequency arrangement will be possible for the USA, if a part of the second generation frequency band is made available for third generation systems. Without this arrangement, roaming will not be possible in the USA without multi-band handsets.

The ITU World Radiocommunications Conference 2000 addressed the issue of additional spectrum requirements for IMT2000 required to meet long term traffic projections and identified the bands 1710–1885 MHz, 2500–2690 MHz and existing bands in the range 802–960 MHz used for second generation mobile systems. The conference also identified mobile satellite service bands in the 1–3 GHz range for the satellite component of IMT2000 and took decisions to enable the optional use of High Altitude Platform Stations (HAPS) to deliver IMT2000 services in the original WARC-92 bands.

14.2 Standards and emerging technologies

The aim of International Mobile Telecommunications for thc ycar 2000 (IMT2000) was to have a single 3G standard. This was not achievable because of the difficulty of choosing a single system acceptable to all. Instead of a single system it was decided on a group of standards, providing a minimum specification should be met for each member of the group. Table 14.1 shows the agreed five technologies.

W-CDMA and TD/CDMA are evolved from GSM and are developed by the European Technical Standards Institute (ETSI) and collectively known as UMTS (Universal Mobile Telecommunications System). CDMA 2000 was evolved from cdmaOne technology and developed by Qualcomm Incorporated. The system was adopted in many countries including Korea and the USA. UWC-136 is a group of mutually compatible TDMA standards developed by the Universal Wireless Communications Consortium and used in the USA. Digital European (Digital Enhanced) Cordless Telephony (Telecommunications) is also developed by the European Technical Standards Institute. The method of multiple access used by the majority of the proposals was code division multiple access (CDMA).

Table 14.1   IMT 2000 standards

Name of the standard	Common name	Duplex method
IMT DS (Direct sequence)	W-CDMA	FDD
IMT MC (Multicarrier)	Cdma2000	FDD
IMT TC (Time code)	TD/CDMA	TDD
IMT SC (Single carrier)	UWC-136	FDD
IMT FT (Frequency time)	DECT	TDD

14.2.1   CDMA

In frequency division multiple access (FDMA) the available bandwidth is divided into a band of frequencies (channels). For the entire duration of the communication a channel is allocated to a user. In time division multiple access (TDMA) the available spectrum is divided into time slots and each user is allowed to access the whole band in a particular time slot. In code division multiple access the transmitted signal is multiplied by a high bit rate (chip rate) code. The result of this multiplication is the spread of the transmitted information. The codes for individual users are orthogonal to each other, therefore there is no correlation between them and hence the channels are separated from each other by unrelated codes. The customers have access to all frequency band all the time.

The advantage of this method is that it is spectrally very efficient. The signal can be detected even though it is below the noise level. The chip rate for wide band CDMA (W-CDMA) is 3.84 Mchip/s and the bandwidth of the spread data fits within a nominal 5 MHz channel. Figure 14.2 shows the principle of CDMA with a 2 chip spreading code.

14.2.1.1   W-CDMA

Wideband code division multiple access (W-CDMA) is called wide band because its bandwidth (5 MHz) is wider than the bandwidth used in IS-95 (1.25 MHz), which is the name of the standard that describes the cdmaOne technology in the USA. It uses code division multiplexing and therefore inherited all the benefits of CDMA. Theoretically the data rate is 2 Mbit/s but practically it is variable and limited to 384 kbit/s (depending on the radius of the cell). The operation of W-CDMA requires the accurate control of uplink power. This power is controlled by the network which ensures the maximum CDMA system capacity by allowing the minimum power (uplink) transmitted all the time. Minimising the uplink power results in prolonging the available talk time in a handset. In W-CDMA the capacity and the cell size are inversely proportional. The reason is that the down link power is shared by all users in a cell. When the

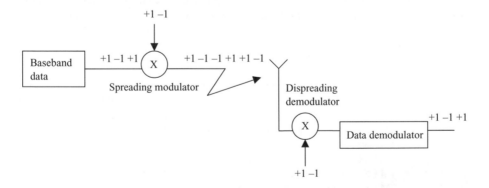

Figure 14.2   The principle of CDMA

capacity increases, i.e. the number of users increases in a cell, the maximum available power to individual users decreases and as a result the cell size shrinks. In case of data communication the effect is more drastic, i.e. the higher the data rate is, the closer to the base station the user should be. This makes the coverage planning of 3G (UMTS) system very complicated. 2G (GSM) planners distributed the base stations according to the density of the population. The high data rate of UMTS encourages more data communication and this is the reason for the complexity. From a capacity point of view one data user could be equivalent to fifty voice users (depending on the type of service).

14.2.1.2   TD/CDMA

TD/CDMA has all the advantages of CDMA. Unlike W-CDMA, which uses two different carriers, i.e. one for uplink and one for downlink, TD/CDMA uses only one carrier frequency for uplink and downlink. This can be done because TD/CDMA uses time division duplexing. Uplink and downlink is carried out at two different time slots and there are 15 time slots in a 10 ms frame. These time slots can be allocated for uplink and downlink dynamically. Figures 14.3 and 14.4 show some time slot arrangements for uplink and downlink.

As the figures show, the time arrangement is dynamic and depends on need. This characteristic of TD/CDMA is very useful for data communication where downlink may need more time slots than uplink. IMT 2000 has allocated two bands of frequencies 1900–1920 and 2010–2025 for time division duplexing. The disadvantage of TDD is the synchronization problem. In two neighbouring stations, if the timing of the transmit and receive time periods are different, the base station to base station interference becomes serious [2].

14.2.2   UMTS network architecture

In Europe the 2G (GSM) operators have spent vast sums of money on the network infrastructure and therefore the compatibility of UMTS with GSM is essential. Figure 14.5 shows the UMTS network architecture [3]. In Figure 14.5 two GSM base station subsystems (BSSs) and their counterparts (RNSs) in UMTS are shown. This

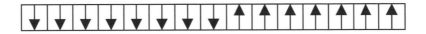

Figure 14.3   The symmetric allocation of time for uplink and downlink

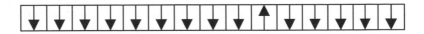

Figure 14.4   The asymmetric allocation of time for uplink and downlink

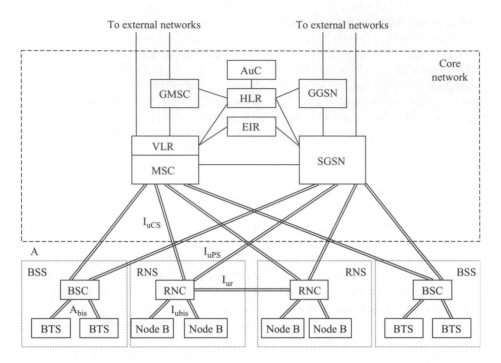

Figure 14.5   The UMTS network architecture

Key: I_{uCS}: Interface between an RNC and MSC, I_{uPS}: Interface between an RNC or BSC and SGSN, I_{ur}: Interface between RNCs, A_{bis}: Interface between BSC and BTS, I_{ubis}: Interface between RNC and Node B, AuC: Authentication Centre, HLR: Home Location Register, VLR: Visitor Location Register, EIR: Equipment Identity Register, MSC: Mobile Switching Centre, SGSN: Serving GPRS Support node, GGSN: Gateway GPRS Support Node, =====: Traffic and Signalling, _____: Signalling only.

figure is important as it shows the evolutionary path from GSM (Phase 2+) to UMTS. This compatibility is very important especially at the early states of UMTS where the customers can use the coverage of GSM for voice communications with UMTS handsets.

14.3   Launch of the world's first 3G system

Nokia made the world's first 3G W-CDMA voice call on the commercial 3GPP (3rd Generation Partnership Program) system on 17th August 2001 [4]. Within Europe, licences have been awarded in most countries in the European community [5] and several operators are planning to launch services in 2004.

14.4   Services and applications

Traffic in UMTS can be divided into four classes according to the sensitivity to delay:

- voice;
- streaming;
- interactive;
- background.

Traffic in the voice class is very sensitive to delay and is symmetric or nearly symmetric. Speech service over circuit switched bearers is the best known application for this class. In the streaming class the traffic is very asymmetric and can withstand more delay than the voice class. In this class the data is processed continuously and at a constant rate, for example the client browser can start displaying the data before the completion of the transmission. In the interactive class, the requesting media, i.e. a human or a machine, requires data from remote equipment, for example a server. The request in this case could be Web browsing or information from a database. The background class includes activities such as email delivery, SMS (short message service) or downloading of databases and immediate action is not required. The delay for this class of traffic depends on the data and can be from seconds to several minutes. UMTS, although initially to be used for voice and low data rate communication (depending on the distance from the base station), is designed to cope with all of the above traffic types.

14.4.1   UMTS concept terminals

There are a number of compromises that have to be considered when designing UMTS handsets; they will support a variety of services, for example, video telephony, which requires a relatively large screen. However, the handset should also retain its portability. The other constraints for the screen are the resolution and the power consumption. The power consumption and battery life must be such that the benefits of portability are not lost through a need for frequent charging. Higher resolution devices are more expensive, so the affordability for the end-user must also be considered. Therefore, there are conflicting interests and trade-offs that need to be made. Figure 14.6 shows pictures of some concept terminals from Nokia [6].

14.5   Conclusions

The introduction of cellular telephony in the 1980s revolutionised our communications. People in business no longer needed to be in a particular location to conduct their business. The first generation of cellular systems had limited use and tariffs and handsets were very expensive. The coverage was not very consistent and interference between the neighbouring handsets was common. The inadequacy of the second generation system became apparent as data communications became at least as important

Figure 14.6   UMTS concept terminals from Nokia

as voice. The operators decided not to lose out in data communications, which led to the birth of the third generation cellular mobile systems. The deployment of the third generation system (UMTS in Europe) is in the teething stage and momentarily delayed. The initial use will be for voice and low bit data but eventually will support up to 384 kbit/s (current version). This rate is less than that proposed initially (2 Mbit/s), but already researchers are turning attention to the further evolution of IMT2000 and development of systems beyond. UMTS has been made compatible with the GSM system so that it can use the coverage of GSM for voice; therefore GSM is here to stay for many years to come.

14.6   References

1   FRIEL, C.: 'IBTE: Telecommunications Engineering: A Structured Information Programme', Issue 36, 2001
2   CHUANG, J. C.: 'Autonomous time synchronization among radio ports in wireless personal communication', J. Proc. IEEE Vehicular Technology Conference, pp 700–705, May 1993

3 STEELE, R., LEE, C. and COULD, P.: 'GSM, cdmaOne and 3G Systems', John Wiley and Sons, Ltd, 2001, p. 425.
4 http://press.nokia.com/PR/200108/831208_5.html
5 http://www.cellular-news.com/3G/
6 http://www.nokia.com/3g/downloads_photos.htm

Chapter 15

TV, voice and broadband IP over cable TV networks

J. Tassel

15.1   Introduction

13.3 million homes have cable in the UK and 87% of them are broadband enabled. In this chapter we describe the key components of a cable TV system that enable these homes to receive voice, TV and broadband (and thus more voice and more and more video applications) over this single media. This integration is the attraction for the end users.

The cable industry, as opposed to the DSL industry for example, has a much more standard-based approach that is closely driven by service providers and vendors alike. The notorious DOCSIS protocol, for example, revolutionised the cable industry by providing interoperability between vendors and delivering high speed data services over the cable TV infrastructure.

In this chapter, we describe the various components of a broadband and voice-enabled cable infrastructure and discuss the key standards in use and forthcoming in this market place.

15.2   How do cable networks work?

The very first type of cable networks were strategically placed antennas with long cables attached to them that could reach houses where an antenna would not be able to receive the signal [1]. Amplifiers were needed along these cables to boost the signal so that it could reach the end of the cable. The signal may, for example, traverse 40 amplifiers before it would get to a home, each contributing noise and signal distortion and adding failure points.

In the 1990s, fibre optic replaced part of this cabling, hence removing the majority of these amplifiers and thus associated noises. The path was then clear for data over

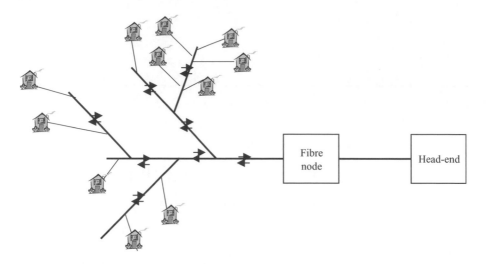

Figure 15.1   Illustration of an HFC network

cable networks. For economic reasons the fibre is only run up to a street location where coax is used to each individual home. Typically a fibre serves a neighbourhood of between 500 and 2000 homes. This mix of fibre and coax cable is known as HFC networks (hybrid fibre coax) and constitutes the majority of today's cable infrastructure. Figure 15.1 illustrates such a network; the fibre runs from the cable head-end (which aggregates many fibres) to a street location (a fibre node) where the coax cable then connects to the homes.

Cable TV channels are then delivered to each home over this infrastructure. A cable network consists of up to 750 MHz of bandwidth, each channel taking 6 or 8 MHz (Europe versus USA) of that bandwidth. These channels can then be used either for:

- Analogue TV: 1 TV channel per 6/8 MHz channel;
- Digital TV: up to 10 MPEG encoded TV channels (up to 1000 channels in total);
- Analogue voice: up to 240 lines per 6/8 MHz channel;
- Broadband IP data: 10 to 34 Mbit/s of bandwidth for 500–2000 users.

Not all frequencies are used for all type of cable services. Figure 15.2 illustrates the distribution of the frequencies for each service.

The cable network spectrum is divided so that voice, TV and data can be supported. The bulk of the spectrum is, of course, used for TV channels (analogue or digital), and the same channels can be used for analogue voice instead.

Lower frequencies are used for upstream data channels whereas higher frequencies (550 MHz and upwards) are used for downstream data. Fewer upstream channels are available in comparison to the downstream channels and many upstream channels are often allocated to each downstream channel. This follows the trend that residential users generate less upstream traffic than downstream traffic. DOCSIS

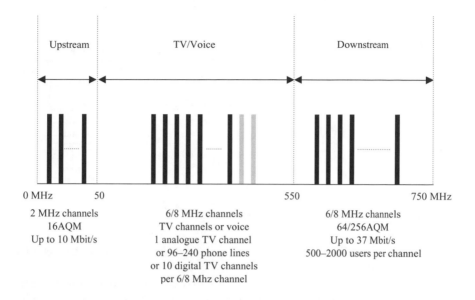

| Upstream | TV/Voice | Downstream |

0 MHz   50   550   750 MHz

2 MHz channels	6/8 MHz channels	6/8 MHz channels
16AQM	TV channels or voice	64/256AQM
Up to 10 Mbit/s	1 analogue TV channel	Up to 37 Mbit/s
	or 96–240 phone lines	500–2000 users per channel
	or 10 digital TV channels	
	per 6/8 Mhz channel	

Figure 15.2   Typical spectrum utilisation on cable networks

(Data over Cable Service Interface Specification) defines the standard used to provide high speed data services over an HFC network.

15.3   Voice over cable networks

Traditional PSTN-like services can also be supported over a cable TV network. Extra components are required in the infrastructure at the end-user premises to support voice services [2]. This section only discusses analogue voice services, although voice over IP services can also be offered using the data (DOCSIS) capabilities of the cable network. A later section describes the PacketCable specification that enables voice over DOCSIS.

Three new components are required to support analogue voice services over the TV cable infrastructure:

- NIU: network interface unit;
- HDT: host digital terminal;
- A telephony switch.

The location of these components is illustrated in Figure 15.3.

The network interface unit provides an interface to the RF network. This includes offering a twisted pair compatible connection in the home, analogue (PSTN) to digital conversion, packetisation of the voice traffic and dial tone functionality. The NIU may be powered by an external power source or via the coaxial cable.

The host digital terminal interfaces between the cable network and the telephone switch. The HDT therefore provides digital multiplexing; it takes the 64 Kbit/s

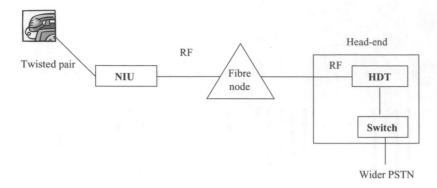

Figure 15.3   Analogue voice over HFC

channel from each NIU and delivers them to the switch and translates between RF and PSTN signalling.

The PSTN switch manages the individual PSTN calls and provide IN functions. Between 96 and 240 voice lines can fit into a single 6 MHz channel over the cable TV network.

15.4   Data over HFC networks (DOCSIS)

DOCSIS is the standard used to provide high speed data services over a cable TV network [3]. Cable TV standards for broadband are defined by the CableLabs consortium [4].

The HFC network is an inherently shared medium. As the signal passes between the optical node and the homes it passes through a small number of amplifiers that boost the electrical signal, using separate amplifiers for the upstream and downstream channels. As a result, while the signal may flow easily from the head-end to the homes and from the homes to the head-end the signal is very highly attenuated between homes making it impossible to communicate directly between stations. The best analogy to the HFC channel is probably a number of ground stations communicating via a satellite. All signals must pass through the satellite. Protocols like Ethernet, therefore, cannot be used for data transmission over an HFC network.

In the downstream direction the signal is broadcast from the head-end to each station attached to the system; a one-to-many channel. In the upstream direction all stations transmit to one receiver; a many-to-one channel. Due to the different characteristics of the upstream and downstream channels, separate physical specifications are required for each direction.

The key components to deliver data services over an HFC network are the CMTS (cable modem termination system) in the head-end and cable modems in the end-user premises (Figure 15.4). The CMTS and the CMs implement the DOCSIS standards so that IP traffic can flow over the HFC network. The CM [5] in the home decodes the RF traffic to pass data traffic to the CPEs. The CMTS puts the IP traffic

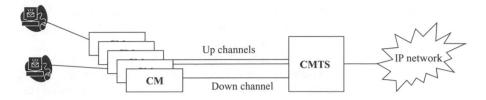

Figure 15.4   Data over HFC

into the RF network and manages the end-user IP connection by, for example, interfacing with a DHCP server. The CMTS also provides the interface to a wider IP network.

The MAC sub layer defines a single transmitter for each downstream channel – the CMTS. All cable modems listen to all frames transmitted on the downstream channel upon which they are registered and accept those where the destinations match the CM itself or CPEs attached to that CM. CMs can communicate with other CMs only through the CMTS. The upstream channel is characterised by many transmitters (CMs) and one receiver (the CMTS). Time in the upstream channel is slotted, providing for time division multiple access at regulated time ticks. The CMTS provides the time reference and controls the allowed usage for each interval. Intervals may be granted for transmissions by particular CMs, or for contention by all CMs. CMs may contend to request transmission time. CMs may also contend to transmit actual data. In both cases, collisions can occur and retries are used.

15.4.1   Downstream

On the downstream, because of the high quality of the channel, highly efficient modulation techniques can be used. A robust downstream transmitter can be obtained by using 64-QAM or 256-QAM modulation and including error correcting codes.

15.4.2   Upstream

The introduction of a reverse path to traditional cable plants is not easily achieved due to the noise funnelling effect of the tree-and-branch topology. In the upstream direction the signal propagates towards a single point, that is, the head-end node. The ingress noise introduced at the end of the plant (for example, within the home or in the drop cable) is collected from all the branches and amplified along the path to be finally recombined at the head-end node. Noise from external interferers is very high in the upstream region. This noise originates from sources such as amateur radio or emergency services; man-made machines, electrical motors, appliances, switching power supplies; or natural sources, such as lightning and solar activity. Thus, the upstream bandwidth is not only scarce but also noisy. The HFC plant mitigates this problem to some extent by eliminating the long strings of analogue amplifiers found in the traditional cable plant and by having each branch of the system serve far fewer households.

A more robust modulation technique is required in the upstream channel than in the downstream to overcome the additional noise problems. The IEEE802.14 Working Group specifies the use of either QPSK or 16-QAM modulation.

Bandwidth allocation in the upstream is based a reservation mechanism. A reservation is made with the head-end by transmitting a request on the upstream channel in a small slot assigned for this purpose (contention slot). Upon receiving the request the head-end sends a grant to the requesting station on the downstream channel, permitting it to transmit its data in the assigned slots.

Two important measures of the performance of a protocol are the throughput that can be achieved, that is, the fraction of the raw capacity provided by the physical channel that can be used to carry useful traffic, and the average delay associated with transmitting a frame of data for a given load. The throughput will be governed primarily by how much capacity is required to implement the reservation process. Allocate too many slots for reservations and capacity will be wasted, but the delay will be low. Allocate too few and the throughput will also be low since there is not enough reservation capacity to fill up the available data slots. An efficient system will dynamically adjust the capacity of the reservation channel to match the needs of the current traffic mix.

15.5   DOCSIS QoS

As the HFC network is a shared medium, the bandwidth available on one channel is shared between all the end-users on that channel. The more end-users on a single channel the less bandwidth they are each likely to get, and this is valid both for the downstream and the upstream directions. This has created some surprises, notably in the US market, where contention was not mentioned to the end-users who bought a cable TV Internet subscription and saw their actual throughput decrease as more end-users joined their segment.

With the promise of growing numbers of end-users and bandwidth hungry applications, the cable TV design community very early on focused on ensuring some bandwidth control was available. Solid QoS features are available since DOCSIS 1.1 [3]. DOCSIS 1.1 defines QoS mechanisms between the CMTS and the cable modems so that traffic can be differentiated over the HFC network.

Figure 15.5 illustrates the DOCSIS 1.1 specification. As shown, DOCSIS defines classifiers (downstream and upstream) and the notion of service flows. These classifiers and service flows are implemented and managed in DOCSIS 1.1 hardware and software. Traffic is classified into an appropriate service flow. The CMTS implement the downstream classification and enforces the QoS whereas the cable modems enforce the upstream classifiers. MAC layer mechanisms are used to manage these classifiers.

Classifiers can identify traffic based on information from a variety of layers:

- IP classification parameters: IP TOS, IP addresses;
- Transport parameters: TCP/UDP ports;

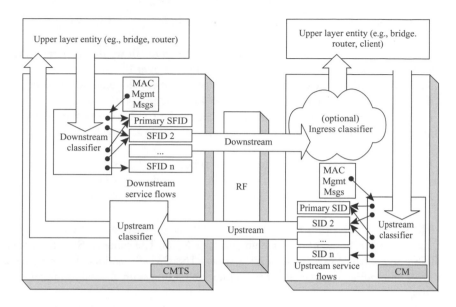

Figure 15.5   DOCSIS 1.1 functional architecture from the DOCSIS specification [3]

- LLC classification parameters: MAC addresses;
- IEEE 802.1P/Q parameters: priority range, VLAN ID.

(Note that a classifier may also use many attributes to identify traffic.)

This richness in classification means traffic from groups of users to individual users or even individual application sessions (i.e. a single file download) can get protected bandwidth over the shared access network.

Service flows then take care of giving the appropriate treatment to the classifier traffic. The QoS treatment can be one of the following:

- maximum bandwidth;
- minimum bandwidth;
- priority during contention;
- maximum burst;
- maximum downstream latency.

How these QoS treatments are implemented depends on whether they are applied to downstream or upstream traffic.

Upstream is based on scheduling appropriate time slots on the access medium for all the cable modems that share that medium to deliver the specified QoS requirements. Scheduling services are designed to improve the efficiency of the poll/grant process via which cable modems get transmit times in the upstream direction. By specifying a scheduling service and its associated QoS parameters, the CMTS can anticipate the throughput and latency needs of the upstream traffic and provide polls and/or grants at the appropriate times. The following services are available to meet the QoS mechanisms available:

- Unsolicited grant service (UGS): for real-time service flows that generate fixed size data packets on a periodic basis, such as voice over IP.
- Real-time polling service: for real-time service flows that generate variable size data packets on a periodic basis, such as MPEG video.
- Unsolicited grant service with activity detection: for UGS flows that may become inactive for substantial portions of time, such as Voice over IP with silence suppression.
- Non-real-time polling service: for non real-time service flows that require variable size data grants on a regular basis, such as high bandwidth FTP.
- Best effort (BE) service: for data traffic with no specific QoS requirements

The scheduling profiles above are not used in the downstream direction as there is a single point of contention management: the CMTS. The same QoS attributes are available to control downstream service flows. The CMTS vendors can implement the QoS enforcement at the IP layer and no interaction with the cable modems is required.

MAC layer messaging has been selected as the most appropriate mechanism to request classifiers and service flows to be setup. Both the CMTS and the CM can issue MAC requests. The CableLabs specification [6] details those messages and interactions. MAC requests may be issued at registration time (that is when the CM is switched on) or post-registration. Both the CM and the CMTS may perform admission control for the resources they manage. The CMTS implement some admission control such that the downstream may not be overbooked, for example. Some vendors, however, offer the facility to turn this functionality off.

An example of post-registration dynamic service flow setup is if an end-user decides to purchase a higher bandwidth package for a weekend. The end-user logs on with his default subscription package to a Web page where he can then select a different package, and a dynamic service flow with a different maximum bandwidth burst value is set up for the end-user to enjoy. This can also be used on a per session basis where an extra service flow would be provisioned for a given session, for example an individual file download may enjoy a higher maximum burst rate than the default end-user traffic.

It is also possible to change the end-user experience for a single session without using dynamic service flows. Static service flows may be set up on the access network so that traffic marked with a given TOS value (i.e. using the TOS field diffserv like) get one type of service and other traffic non-marked get another class of service. It is then up to some external entity to mark the traffic appropriately so that it gets classified by the CMTS into the appropriate service flow to get appropriate QoS treatment.

Note that DOCSIS does not impose how vendors implement the delivery of the QoS; this is vendor specific and each vendor can then provide its own value and specificity. DOCSIS only defines interfaces, as does the rest of the CableLabs standards.

15.6 PacketCable for voice and multimedia applications

There are two versions of the PacketCable specification: version 1.x is dedicated to voice applications and version 2, renamed 'PacketCable Multimedia', is dedicated to providing solutions to all type of multimedia applications. Note that PacketCable necessitates a DOCSIS 1.1 network infrastructure.

There are various versions (aka. 'Evolution of the same architecture') to the PacketCable 1.x specification:

- PacketCable version 1.0 is the architecture basis;
- Version 1.1 provides the add-ons to be able to provide a primary line-capable service;
- Version 1.2 enables inter-MSO (multiple service operator) operability via a managed backbone IP network that has sufficient bandwidth (i.e. QoS is not an issue);
- Version 1.3 defines the functional components, interfaces and a data model to perform subscriber provisioning on call management servers.

The architectural specification of PacketCable version 1 is illustrated in Figure 15.6. This specification defines interfaces for following key aspects that are needed to enable voice services over a data network:

- The call signalling to enable calls to start, stop and parties to join and leave calls (this is the role mainly of the CMS: call management server).
- Quality of service (QoS) interfacing. DOCSIS provides QoS control over the shard HFC network. PacketCable specifies the way in which these QoS mechanisms are used for voice services (this is the role on the CMTS together with the gate controller). DQoS (dynamic QoS) is the interface specification defined to achieve this.
- Codec to encode the voice signal (this is the role of the MTA: multimedia terminal adapter).

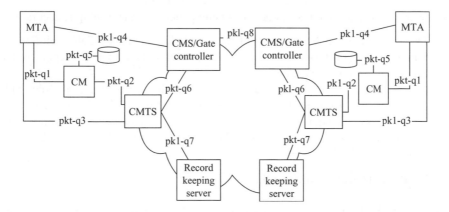

Figure 15.6   PacketCable 1.0 reference architecture from CableLabs [6]

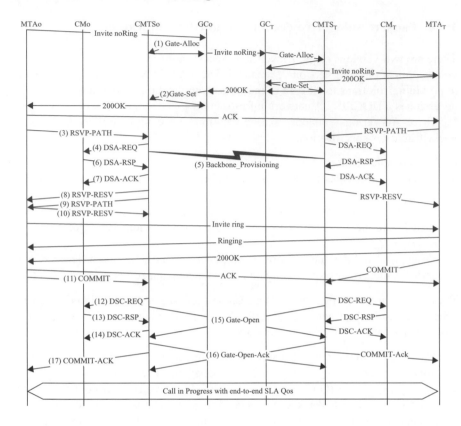

Figure 15.7   Example PacketCable interface interaction from CableLabs [6]

- Billing event message collection to ensure calls can be billed accurately (this is the role of the CMTS and RKS: record keeping server).
- PSTN (public switched telephone network) interconnection.
- Security interfaces necessary across all components so that billing is accurate.

These interfaces can be used to implement a single-zone PacketCable solution for residential Internet protocol (IP) voice services. Refer to the CableLabs report [6] for a detailed description of each interface.

For reference Figure 15.7 illustrates the use of each interface for an IP voice call. Note the gate establishment for resource control a priori of the ringing tone, and the RSVP message exchange for QoS establishment. (Backbone provisioning is not happening dynamically but is provisioned over a large period of time, months.)

15.7   PacketCable multimedia

PacketCable multimedia is looking at improving the version 1.x specification to support any type of multimedia applications. At the heart of PacketCable multimedia

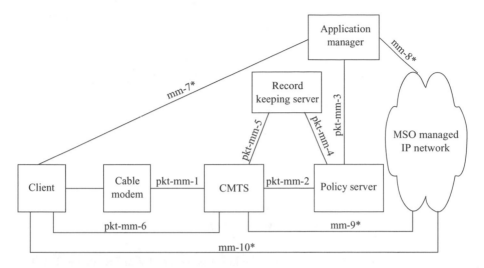

Figure 15.8   PacketCable multimedia reference architecture [from specification document]

is DOCSIS 1.1 and PacketCable DQoS used by PacketCable version 1. Figure 15.8 illustrates the key components of the PacketCable multimedia specification.

The new components provide the ability to integrate any application with the PacketCable QoS and control mechanisms. The following entities and interfacing between them has been defined to enable any application to make use of the QoS and record-keeping mechanisms of DOCSIS so that multimedia interactions can take place.

- Client is the end-user device or application.
- Application manager is an application server or third party server that has access to application messaging and is able to proxy QoS requests on behalf of a client that is QoS unaware. If the application is QoS unaware then the application manager must infer or define the particular QoS requirements of a session.
- Policy server does admission control and stores QoS allocations policies.
- Record keeping server is performing resource accounting.

With this specification the application can, for example, be a simple FTP server or a complex multimedia conferencing application and both can make use of the advanced control functions of PacketCable.

15.8   Conclusions

As we presented in this chapter, a single head-end can be used to provide TV, telephony and data services over a single HFC access network [2]. Figure 15.9 illustrates the different logical components that are required to support this set of services. Fundamentally each component has access to some spectrum on the shared HFC network

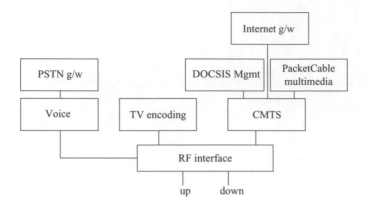

Figure 15.9   Head-end combining TV, voice and data

and interprets the signals as it needs (whether it be voice, TV or broadband IP). The CableLabs standards such as PacketCable are adding further software layers to provide more intelligent integrations with applications to deliver value added services such as multimedia conferencing over IP.

15.9   References

1   ABE, G.: 'Residential broadband', 1st or 2nd edition (Cisco Press), 2000
2   CICIORA, W.: 'Modern cable television technology: video, voice, and data communications', (Morgan Kaufmann), January 1999
3   DOCSIS interface specification document from CableLabs; Data-Over-Cable Service Interface Specifications, Radio Frequency Interface Specification, SP-RFIv1.1-I09–020830, 2003
4   CableLabs website: www.cablelabs.com
5   MAJETI, V.C.: 'Cable modem : technology', (available from http://www.cadant.com), November 1999
6   PacketCable technical report document, from CableLabs PacketCable™ 1.2 Architecture Framework Technical Report; PKT-TR- ARCH1.2-V01–001229, November 2002

Chapter 16

SDH in the access network

S. Christou and D. Austen

16.1 Introduction

An increasing number of bearer and transport technologies are used in the access network to deliver to customers an expanding and diverse range of products and services; synchronous digital hierarchy (SDH) is one of them. SDH was originally deployed in core networks, to take advantage of the improved reliability and network management for delivery of premium services. The technology has subsequently been used to provide efficient and reliable transport connectivity between broadband switch elements. Today, it represents a mature and stable technology used by many operators around the world. The resultant economies of scale, coupled with obsolescence of older technologies, are driving its increased use in the access domain.

This chapter describes how SDH technology is used to support the delivery of a broad portfolio of products and services. Experience from the BT network is used to describe and discuss the technology's generic functionality, deployment and potential.

SDH equipment has been present in customer sites since a decision was made to introduce the new technology into the network; however, until now it has been standard core network equipment rather than something specifically designed for the purpose. The first part of the chapter, Section 16.2, describes how SDH is deployed in the network and provides an overview of core networks and developments to date. The type of components found at customer sites is then described along with the ways in which these components can be connected back into the core network. Some examples of the early projects where BT installed SDH in customer sites conclude the section on the use of core equipment in the access network. Section 16.3, describes a new SDH-based technology, referred to as access SDH (ASDH), which has recently been introduced to initially support delivery of wide band services. A description of the technology and its components is followed by an overview of the solution and deployment designs. The potential to further exploit access SDH technology to support additional service delivery is then discussed.

16.2   Synchronous Digital Hierarchy

The definition of SDH standards during the latter part of the 1980s signalled the beginning of a new stage in the evolution of transmission networks. SDH with increased capacity, improved restoration and reconfiguration capability along with sophisticated network management would enable operators to deliver the more complex telecommunications services that customers were seeking.

The arrival of SDH came when operators were struggling to meet capacity requirements with plesiochronous digital hierarchy (PDH) technology in the core network and performance targets for private circuits to key customers. As a result, operators were quick to appreciate the potential and benefits of SDH and were moving forward rapidly with plans for new networks to enable them to offer services to these key clients.

A response from such an operator, BT, was to design a network, for national deployment that would deliver the benefits of SDH initially to such customers, through improved resilience in the core network and deployment of SDH equipment onto the customer site. The vision for SDH in the longer term was to support all the key service platforms and enable a migration from legacy PDH transmission.

The planned coverage of this new national network is illustrated in Figure 16.1, and is described in more detail later (SDH narrowband network). It was envisaged that the number of nodes served by the new network would be much reduced from the 6000 nodes served by PDH, and that customer nodes would be fibred back to SDH serving nodes as illustrated in Figure 16.2. The customer options are described in more detail later (SDH access configuration).

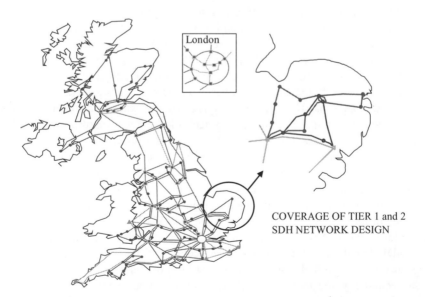

Figure 16.1   SDH network coverage (UK)

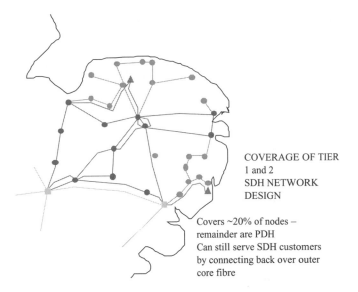

COVERAGE OF TIER
1 and 2
SDH NETWORK
DESIGN

Covers ~20% of nodes –
remainder are PDH
Can still serve SDH customers
by connecting back over outer
core fibre

Figure 16.2   SDH ring coverage and customer access (example)

16.2.1   Overview of core SDH network

16.2.1.1   1993–1999 – SDH narrowband network

The national SDH network deployed by BT is of a tiered design, comprising a high capacity mesh (2.5 Gbit/s point-to-point systems linking the major cities across the UK – tier 1, the inner core) and three layers of rings (super-cells, tier 2 and tier 3, the outer core). The tier 2 rings linked key towns at 622 Mbit/s initially and 2.5 Gbit/s later. This originally provided sufficient coverage to enable customers to have diverse access if required, i.e. to two separate serving sites. However, as customer demand increased, SDH coverage became necessary further out into the network and another tier of rings (tier 3) was introduced. The super-cell layer was employed to provide efficient delivery of the tier 2/tier 3 traffic into the tier 1 layer. This architecture is illustrated in Figure 16.3. Linkage to the international backhaul network (tier 0) is provided by connections from tier 1. The network management architecture that supports the BT SDH network is shown in Figure 16.4. These systems enable a high degree of automation for planning and building of the network and for the provision and maintenance of services. A brief description of the function for each component follows.

Service provision. SSD supports circuit design for service requests; it supports circuit routing across the different technologies used in the network.

PACS, the core system in the management of the SDH network, supports centralised planning and configuration and handles all service requests to provide, cease and re-arrange circuits.

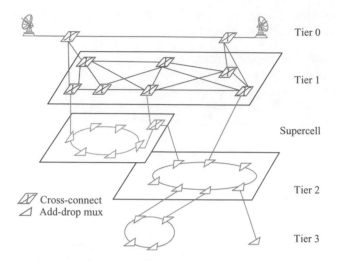

Figure 16.3   SDH narrowband architecture

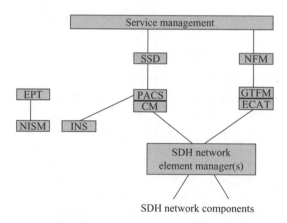

Figure 16.4   SDH network management architecture

CM provides the interface between PACS and suppliers' element managers. Requests from PACS for circuit provision, cessation and re-arranges are translated to requests for the establishment or deletion of cross connections in the components managed by the element manager.

Maintenance and repair. NFM collates fault data from across the BT network and provides for the handling of problems. The data it receives is manipulated to form associations that can help pinpoint the root of problems and faults in the network.

GTFM is a technology-specific fault management system. Algorithms are used to create associations on the data it receives from the network that can help with problem handling.

ECAT provides the fault interface between suppliers' element managers and GTFM for all alarm and event information.

Plan and build. EPT provides a visual front end for data held on NISM and INS. It is the tool used by planners to view plans and equipment layouts and supports the generation of job packs and instructions to the field teams for equipment installation.

INS stores all fibre and location data for the network; it is a collection of many systems that support the planning, assignment and maintenance activities for the network.

NISM provides physical inventory records for all technologies and components in the network and is updated when new equipment is installed.

16.2.1.2   1999–2001 – SDH broadband network

Towards the end of the 1990s a change was occurring with the type of circuit that customers were requesting, 2 Mbit/s demand was still strong from BT's customers and internal client platforms, but many customers were demanding increasing bandwidth and wanted it at a much lower price. The narrowband network could not deliver significant volumes of the higher bandwidth without danger of sterilisation and certainly could not deliver the price reduction required to remain competitive.

This led to plans to build an overlay ring network in tier 1. This new network would be VC4 (155 Mbit/s) granular and hence not have the flexibility to switch/act on individual 2 Mbit/s or 34 Mbit/s circuits. A new concept that arrived with this network was that of shared ring protection, rather than the 1 + 1 protection offered on the narrowband network. Modelling demonstrated that this network configuration would enable capacity delivery to customers at a significantly reduced price, whilst retaining almost all the performance improvements. Equipment, such as the Marconi MSH64, operating at 10 Gbit/s and with the capability to transport 155 Mbit/s, 622 Mbit/s and 2.5 Gbit/s circuits was available and deemed architecturally suitable. More recently this capability has been pushed out further into the network, targeted at nodes where demand for VC4 type circuits is growing. Management of the new overlay SDH network within BT is now based around the MV38 network manager from Marconi at the level below PACS/GTFM. The MV38 then controls a number of individual and supplier specific element managers (MV36).

16.2.1.3   2001 and beyond – the new core network

As demand for capacity increases it is envisaged that there will be a further deployment of broadband optimised SDH networks, limited investment in 2 Mbit/s granular SDH and significant investment in optical systems throughout the core network. This will lead to less low level granularity being available in the core network and a requirement for systems capable of bulking up capacity for transport across the access and outer core networks, ensuring efficient utilisation of the network capacity.

This requirement can be met now with the equipment described earlier, although this assumes there are no new access interface requirements, (e.g. Gbit/s Ethernet, ESCOM, etc.); these will all require development of new interfaces on the core SDH equipment. Work is currently in progress to assess the development required to provide the necessary customer interfaces on core equipment. Alternatively new access specific equipment, access SDH for example, is deployed to provide the grooming functionality required. Access SDH is described later.

16.2.2   Customer sited SDH components

Described briefly in this section are the most commonly used SDH components that can be found serving customer sites. These components are linked into either the narrowband SDH or broadband SDH networks described earlier and are used to deliver the full range of customer bandwidth requirements from 2 Mbit/s (specified to G703) up to 2.5 Gbit/s. The more specific capability offered by each component is described below and the options for connection into the core network are described in the SDH access configuration section later.

Add/drop multiplexor (STM1). A standard SDH muliplexor offering single stage multiplexing from 2–155 Mbit/s, capability to directly multiplex/de-multiplex up to 63 × 2 Mbit/s circuits. Protection (1 + 1) capability is available on the network side of the equipment to another ADM; this could be in the serving node or to a distant customer site using SNCP (sub-network connection protection). Tributary cards are also available at 34, 45, 140 and 155 Mbit/s. Protection is available on the customer side of the ADM on a 1 + 1 MSP (multiplex section protection) or 1:*n* basis dependent on customer requirements and capability.

Originally, this was the only ADM available to support SDH products and was deployed to customer sites irrespective of capacity required. Eventually VC-TS was introduced to serve smaller sites.

Add/drop multiplexor (STM4). A standard SDH muliplexor offering single stage multiplexing from 2–622 Mbit/s, capability to directly multiplex/de-multiplex up to 126 × 2 Mbit/s circuits (or up to 252 × 2 Mbit/s with newer equipment). 1 + 1 SNCP capability is available on the network side of the equipment to another ADM; this could be in the serving node or a distant customer site using SNCP. Tributary cards are also available at 34, 45, 140, 155 and 622 Mbit/s. Protection is available on the customer side of the ADM on a 1 + 1 MSP or 1:*n* basis dependent on customer requirements and capability.

Add/drop multiplexor (STM16). Standard SDH muliplexor offering single stage multiplexing from 2 Mbit/s–2.5 Gbit/s, capability to multiplex/de-multiplex up to 252 × 2 Mbit/s circuits. 1 + 1 SNCP capability is available on the network side of the equipment to another ADM, this could be in the serving node or a distant customer site using SNCP. Tributary cards are also available at 34, 45, 140, 155 and 622 Mbit/s. Protection is available on the customer side of the ADM on a 1 + 1 MSP or 1:*n* basis dependent on customer requirements and capability.

Add/drop multiplexor (STM16-VC4). A 2.5 Gbit/s ADM (like the MSH51c) having minimum granularity at the VC4 level only. Supports up to 32 STM1 tributaries delivered via electrical or optical connection. An STM4 optical tributary connection is also available; this can deliver a 622 Mbit/s aggregated signal or a concatenated

622 Mbit/s signal (VC4-4c, comprising 4 × VC4s). The choice of 622 Mbit/s signal configuration is software controlled.

- Tributary cards can be protected on a 1 + 1 or 1:n basis, where n can be 2, 3 or 4.

Add/drop multiplexor (STM64-VC4). A 10 Gbit/s ADM (like the MSH64) having minimum granularity at the VC4 level only, this ADM is the same as ADMs now deployed within the BT core broadband SDH network. Supports up to 64 VC4 (STM1) tributaries delivered via electrical or optical connections. Optical tributaries are also available at STM4 (622 Mbit/s aggregate or VC4-4c concatenated) and STM 16 (2.5 Gbit/s aggregated or VC4-16c concatenated). The choice of signal configuration is software controlled.

- STM1 and STM4 tributary cards can be protected on a 1 + 1 or 1:n basis (n can be 2, 3, or 4).
- STM16 cards can only be protected on a 1 + 1 basis.

Virtual container transport system (VC-TS). The VC-TS was developed to enable deployment of SDH functionality to smaller customer sites. The costs associated with deploying the core STM1 ADM for limited 2 Mbit/s requirements was restricting the deployment of SDH into customer premises. The VC-TS provides similar functionality to the core device, including 1 + 1 protection but has a limited capacity (maximum 14 × 2 Mbit/s, G.703 only). The VC-TS comprises a virtual container transport module (VC-TM) deployed at the customer site and a virtual container access module (VC-AM) deployed at the SDH serving node. Each VC-AM can support four VC-TMs, and up to eight VC-AMs can be supported on an STM4 ADM. The link between the customer site and SDH serving node is normally single fibre with a power budget of 14 dB, for links where this limit would be exceeded dual fibre can be used, increasing the power budget to 22 dB.

16.2.3   SDH access configuration

This section identifies the options for connecting the customer site SDH equipment back into the core SDH network.

The three most common SDH access network configurations are the single unprotected link, the protected link back to a single serving exchange and a diverse link back to two serving exchanges.

These access configurations are then plugged into the core network to give a number of different end-to-end scenarios and performance levels. These can range from totally diverse and protected from customer site to customer site to totally unprotected. The three most used core protection mechanisms are unprotected, SNCP protected (1 + 1 customer/customer – only available across the narrowband SDH network) and SPRing protected (where protection capacity is shared).

At the core node there are two commonly used methods for connecting customer sited ADM equipment into the core network: direct connection from customer ADM feed to core ADM equipment or connection via an ADM grooming function.

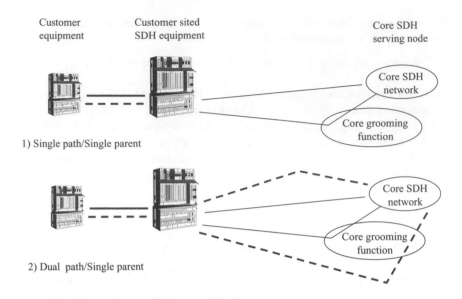

Figure 16.5   Customer SDH access options (1)

The grooming function enables termination of high bandwidth customer line rates and enhanced core network utilisation. Figure 16.5 illustrates the options where access to a single core SDH site is required.

Configuration 1 comprises $1 + 1$ multiplex section protection or $1:n$ tributary card protection between the customer equipment and the customer sited SDH equipment. In this option the SDH equipment could be any of the components previously described (ADM @ STM1, STM4 or STM16, MSH51c, MSH64 or VC-TS). No protection is applied between the customer node and the core network serving node. Connection at the core node would normally be via the grooming function to the core equipment although direct connection may be considered for STM1, STM4, or STM16 ADMs and VC-TS. Connections from the MSH51c and MSH64 would be into the core broadband equipment, connections from the STM1, STM4, STM16 ADMs would be into the narrowband core equipment for circuits lower than 155 Mbit/s and into the core broadband equipment for circuits of 155 Mbit/s and above. Connections from VC-TS would be into the narrowband core equipment.

Configuration 2 comprises $1 + 1$ multiplex section protection or $1:n$ tributary card protection between the customer equipment and the customer sited SDH equipment. In this option the SDH equipment could be any of the components previously described (ADM @ STM1, STM4 or STM16, MSH51c, MSH64 or VC-TS). Protection is applied between the customer node and the core network serving node. Connection at the core node would normally be via the grooming function to the core equipment although direct connection may be considered for STM1, STM4, or STM16 ADMs and VC-TS. Connections from the MSH51c and MSH64 would be into the core

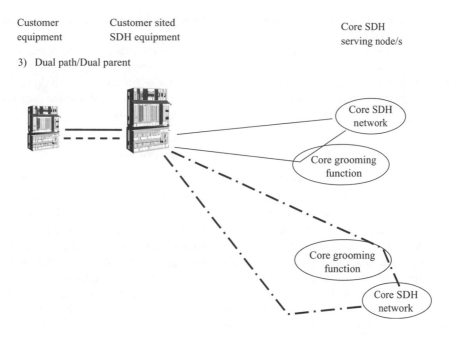

Customer   Customer sited   Core SDH
equipment   SDH equipment   serving node/s

3) Dual path/Dual parent

Figure 16.6   Customer SDH access options (2)

broadband equipment, connections from the STM1, STM4, STM16 ADMs would be into the narrowband core equipment for circuits lower than 155 Mbit/s and into the core broadband equipment for circuits of 155 Mbit/s and above. Connections from VC-TS would be into the narrowband core equipment. Figure 16.6 illustrates the options where access to two core SDH sites is required.

Configuration 3 comprises $1 + 1$ multiplex section protection or $1:n$ tributary card protection between the customer equipment and the customer sited SDH equipment. In this option the SDH equipment can only be the following components: STM1, STM4 or STM16 ADMs or VC-TS. With this option a second route is provided to an alternative serving node; protection can be applied at the customer sited SDH equipment to enable a completely diverse and protected circuit configuration to exist across the core and access networks to a distant ADM of the same customer. Connection at the core node would normally be via the grooming function to the core equipment although direct connection may be considered. Connections from the ADMs and VC-TS equipment would be into the narrowband core equipment for circuits lower than 155 Mbit/s and into the core broadband equipment for circuits of 155 Mbit/s and above.

Other configuration options. *Multiple customer ADMs* – In some instances a customer may have two ADMs connected in series, this can provide additional resilience to failure of the core components within a single ADM, e.g. switching fabric. In this configuration duplicate links can be provided between the customer equipment and customer sited SDH.

Customer rings – Occasionally customers can be provided with complete ring solutions, this is more likely in confined urban areas where a single customer has a high density of sites that require connectivity. These solutions involve linking the customer sites together via access fibre and providing connectivity into the core via gateway nodes (see Reuters 2000 example in Section 16.2.4.4).

16.2.4   Early projects using SDH in the access network

16.2.4.1   Initial SDH network (ISDH)

In order to test the new SDH technology a number of BT customers were approached with a view to participating in the first deployment of SDH. The initial deployment would be based upon the BT target SDH architecture but only a subset would be built to link two or three of the customers' sites. SMA1 equipment would be deployed at the customer site enabling the provision of a limited number of fully protected 2 Mbit/s circuits. A number of trial customers were identified and two networks designed, one from each of the suppliers identified through an SDH tender process. Ericsson would build a network linking sites in the Forest of Dean, Uxbridge and Marlow; GPT a network linking sites in Manchester and central London. This would provide BT with significant SDH coverage and enable the launch of SDH-based products after the initial testing was complete. Due to problems with supply of Ericsson equipment the two initial networks were combined and built by GPT during 1993, with a new initial deployment of Ericsson equipment designed to support the international backhaul network. Thus the sole supplier arrangement for the provision of BT's UK core network was born.

The fully protected 2 Mbit/s circuit configuration was successfully tested on the GPT network and was then progressed to a targeted national rollout as the MegaStream Genus product.

16.2.4.2   SuperJANET

The Joint Academic NETwork (JANET) is a private network used to link a number of university sites across the UK. Around the time that BT was setting up its new core SDH network these universities were looking to upgrade JANET, from a low bit rate PDH network to a higher rate network based on SDH. The functionality required could easily have been supported with PDH technology but by this time customers including the universities were increasingly aware that a new technology was available and they would accept nothing less. The superJANET requirement demanded the ability to provide a 155 Mbit/s links between each of five sites, namely Birmingham, Manchester, Oxford and two London colleges (Imperial and UCL). Each university site was equipped with an SMA1 provided with an STM tributary card and a 3 × 34 Mbit/s tributary card. The access link was terminated at the tier 1 serving node on another SMA1, two connections were then provided either onto a cross connect or into a pair of tier 1 line systems. Given five sites with potentially two links (unprotected) the maximum STM1 circuit requirement at any given time could only be five, and these were configured on demand via the NOU at Manchester. This requirement was provided on the ISDH network described above.

In later phases of superJANET the number of sites was increased and connections transferred onto cross connects at serving sites.

16.2.4.3 Project Tiger

One of BT's largest customer groups currently is the mobile operator; this has in fact been the case for some time and project Tiger was specifically set up to deliver SDH capacity to mobile operators in the mid 1990s. The key aim of the project was to offer SDH delivery to the major switching nodes. This normally comprised a four node ring to serve each switch site, employing two STM4 ADMs at the customer site providing up to 126 protected 2 Mbit/s circuits to each of the two serving nodes, i.e. 252×2 Mbit/s per customer site in total. The circuits were then routed via SDH core network (if available end-to-end) or PDH to the distant MOLO switch site. The prime benefit of this was the automatically protected access link from the customer switch site to the serving nodes.

16.2.4.4 Reuters 2000

BT was approached during 1995 and asked to design a customer specific network for Reuters. The customer requirement was for a number of 2 Mbit/s links between their key sites in London, with limited breakout to the rest of the UK. A number of customer specific rings were designed with SMA1 equipment deployed at each customer site. These rings served the majority of the customer requirements, however, breakout via two gateway nodes to the wider core SDH network was provided for those sites not reached by the customer specific rings. 1+1 protection was implemented on these circuits.

16.2.4.5 Euro 96

Although no SDH equipment was actually deployed at the customer site the Euro 96 project (in support of TV requirements for the 1996 European Football Championships) was an important step for the UK SDH network. Access fibre tails equipped with CODEC equipment were deployed from the nearest SDH serving node to individual football grounds. These PDH tails were then successfully back-hauled on SDH to the appropriate broadcast centre for transmission to a world wide audience; demonstrating expertise in deploying and managing this new type of network and its confidence in the technology.

16.3 Access SDH

Delivery of wide band and high band services utilises a variety of technologies over the access network. Until very recently, the majority of wide band services have been delivered to customer sites from BT serving exchanges using fibre and copper systems based on plesiochronous digital hierarchy (PDH) technology. Several types of systems have been used since the launch of MegaStream, the most recent being known as digital wideband serving section (DWSS), fibre based, and copper wideband serving section (CWSS). These systems are capable of delivering 2 Mbit/s services using either G.703, X.21 or I.421 interfaces dependent upon customer requirements.

The continuing use of such equipment did not address the changing expectations of customers who expected SDH technology to be used as the standard delivery method for wide band services. At the same time, market changes, driven by the Internet, regulation and competition, added more urgency to the ongoing commitment within BT to continually drive down the cost for providing these services to its customers. The most powerful approach in achieving this is to reduce the unit cost of the equipment used. Earlier studies indicated that such unit cost reductions could be achieved using synchronous digital hierarchy (SDH) technology more broadly in the access network; SDH had traditionally been used in access only for delivering premium wide band and high band services as discussed earlier. Other benefits were also identified, which included improved management capabilities from end-to-end SDH service delivery and a resolution to forthcoming PDH technology obsolescence. The introduction of access SDH equipment was therefore identified as the means to satisfy those requirements, meet customer expectations for SDH delivery and provide equipment cost reductions.

The key requirements for the supply of access SDH equipment and its associated management include:

- ease integration to the existing operational and business model;
- improve asset management;
- improve infrastructure utilisation;
- facilitate operational enhancements;
- support cost-effective migration to future bulk ATM-based service delivery.

The introduction of the access SDH technology in the BT network has been phased, driven primarily by business priorities, component availability and relevant system changes becoming operational:

- Stage 1 of the access SDH project is addressing the current service delivery requirements for wide band services and it has already been delivered to the field.
- Subsequent releases of stage 1 are looking to exploit the technology further to cost-effectively complement solutions for delivering either bulk or high bandwidth services (that require a customer-sited ADM).
- Stage 2 of the project will be addressing the evolutionary potential of the equipment to cost-effectively support new access architectures for the delivery of new and evolving data services.

Access SDH technology will eventually supersede the existing DWSS and CWSS systems for wide band service delivery; the deployment strategy is currently driven both by demand-levels at certain areas and by the need to asset-manage utilisation of existing equipment. As a result, older technology and components will continue to be used during the transition period to exploit the installed shelf capacity.

16.3.1   Technology overview

Access SDH is comprised of a range of components that can be used to support a number of different network designs (see Table 16.1) and architectures. Figure 16.7 shows the equipment and element management system (EMS), which forms the basis of the platform.

Table 16.1   ASDH capability overview

ASDH offers the following capability for service support: • SDH delivery of 2/34/45/140/155 Mbit/s over fibre and copper (2 Mbit/s only) • All wideband services, with the following customer presentations — $n \times 64$ kbit/s X.21 — 2 Mbit/s G.703 (75 Ω/120 Ω), G.704, I.421, X.21 — 34/45/140/155 Mbit/s G.703 75 Ω and STM-1 optical	Support for the following access network resilience types: • single access (single fibre single serving CU in exchange) • dual access (dual protected fibre single serving CU in exchange) • dual serving exchange access (only when CU is the NTE, with directly off the shelf OR via local NTEs)

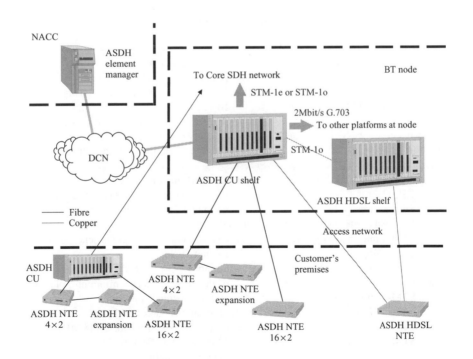

Figure 16.7   Access SDH components

- Consolidation unit (CU) – This is the ASDH ADM; it terminates NTEs and provides connections to the core SDH network equipment via STM-1 interfaces and to other platforms at a BT node via PDH break-out. It can also be located on customer's premises. It also provides:
  — the connection to the element manager for itself and its dependent NTEs;
  — synchronisation of terminated NTEs, based upon synchronisation derived from its link with the core SDH network or a synchronisation supply unit (SSU) (if present in BT node).
- ASDH HDSL shelf-exchange based shelf, which houses 3 port HDSL line cards (LTEs) and connects to the CU using an STM-1 link.
- Copper and fibre NTEs (fed from respective LTEs).
- Access SDH element manager.

16.3.1.1   Consolidation unit (CU)

The ASDH CU is an SDH add drop mux designed to also support line cards, and line terminating equipment (LTEs), that can terminate traffic from customer located NTEs. Table 16.2 provides an overview of the component's characteristics and Figure 16.8 depicts its actual construction.

Table 16.2   ASDH CU ADM features overview

Traffic interfaces supported:
- 2/34/45/140 Mbit/s and STM-1e G.703 75 Ω
- STM-1o and STM-4 (slots 1–4 only)
- Fibre and copper NTE interfaces

Protection:
- Unit protection:
  — 21 × 2 Mbit/s cards (1:3)
  — PSU (1:2)
  — Switch cards (1:1)
  — Fibre NTE interface cards (1:1)
  — STM-n interface cards (1:1)
  — 1 × 34/45/140 Mbit/s interface cards (1:1)
- MSP
  — STM-n interfaces
  — Fibre NTE interfaces
- SNC-P
  — STM-n interfaces (only in slots 1–4)

Network management
- Management of all dependent NTEs.
- Connection with element manager, via the SDH DCN

Interface for craft terminal or laptop PC

Software downloadable

Test facilities – loopbacks and embedded BER tester

Physical construction:
- Rack mountable. A 2.2 M tall rack can contain up to two CUs plus fibre distribution trays.

Power:
- Consumption (W) = 500 (depends on actual tributary load)
- Supply = 3–5 –48 Vdc feeds (one to each of the three PSUs as a minimum, plus two if HDSL is being supported)

Synchronisation sources (either):
- External 2.048 Mbit/s or 2.048 MHz
- 2 Mbit/s or STM-n interfaces
- Internal oscillator

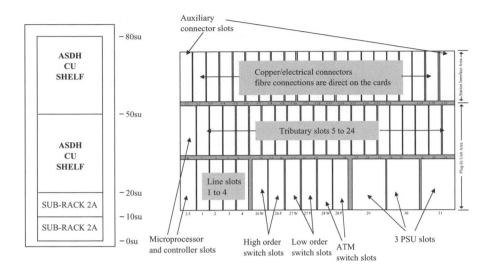

Figure 16.8   ASDH CU rack and shelf layout

The CU rack and shelf layout are shown in Figure 16.8. Note the fact that a standard rack in an exchange can accommodate two CUs which increases significantly the customer-port-density compared with the other existing technologies.

The total traffic capacity of the CU varies, depending on the choice of high order cross-connect SDH switch component used. It can be 16 (slots 1–4 have 4 × AU4 capacity) or 20 (slots 1–4 have 8 × AU4 capacity) ports where one port equates to an STM-1 payload (i.e. 63 × 2 Mbit/s or 3 × 34/45 Mbit/s or 1 × STM-1, etc). The total break-out capacity of a CU is 126 × 2 Mbit/s G.703.

Support for cell-based (ATM) services can be added via an in-service upgrade, which includes the addition of an ATM switch module (see switch slots in Figure 16.8) and a firmware upgrade via software download from the element manager.

16.3.1.2   The HDSL shelf

The DSL CU shelf and backplane provides the required connectivity for ASDH to support the delivery of wide band services using access network copper using 2-pair HDSL technology. HDSL shelf(s) can connect to 'parent' CUs via an STM-1 interface. The component offers a configurable HDSL core transport frame supporting both SDH and PDH mapping (i.e. TU-12 and 2 Mbit/s) – ASDH uses only the TU-12 option. The shelf backplane provides connections for a rack alarm bus, house keeping and user channel interfaces, clock/sync connection, an RS232 local terminal interface and a LAN management interface (ASDH uses the DCC in the STM-1 connection to the CU for management). A distributed power architecture design is also used which helps keep start-up costs low and can be exploited for thermal performance (potentially, 3 shelves can be installed in a standard exchange rack again significantly increasing port density compared with existing HDSL systems). BT policy is to install a maximum of two HDSL shelves per exchange rack.

Each shelf can support up to 128 customer interfaces, using 3-port copper LTEs. There are 17 slots in a shelf with the option of using up to 2 network cards for connection to a 'parent' CU (worker and protect if required) supporting various options:

1. 1 network interface card and up to16 line cards supporting up to 48×2 Mbit/s[1].
2. 2 network interface card and up to15 line cards supporting up to 45×2 Mbit/s.
3. 1 maintenance interface card and up to 16 line cards supporting 48×2 Mbit/s.

16.3.1.3   Access SDH NTEs

Access SDH offers a 2-pair copper NTE and a range of fibre NTEs. All fibre NTEs are single fibre working (SFW), without the need for an external splitter. The range of NTEs available are:

Copper NTE
- 1×2 Mbit/s copper HDSL (2B1Q) NTE with no expansion ports (line powered)

Fibre 4×2 NTEs
- 4×2 Mbit/s base NTE, unprotected fibre, with 2 TUG2 expansion ports and 1 TUG3 expansion port
- 4×2 Mbit/s base NTE, protected (MSP) fibre, with 2 TUG2 expansion ports and 1 TUG3 expansion port
- 4×2 Mbit/s expansion NTE with no fibre connectors and 1 TUG2 expansion port

Fibre 16×2 NTEs
- 16×2 Mbit/s base NTE, unprotected fibre, with 3 TUG3 expansion ports
- 16×2 Mbit/s base NTE, protected (MSP) fibre, with 3 TUG3 expansion ports
- 16×2 Mbit/s expansion NTE with no fibre connectors and 1 TUG3 expansion port

Fibre 34/45 NTEs
- $1 \times 34/45$ Mbit/s (software configurable) base NTE, unprotected fibre, with 3 TUG3 expansion ports
- $1 \times 34/45$ Mbit/s (software configurable) base NTE, protected (MSP) fibre, with 3 TUG3 expansion ports
- $1 \times 34/45$ Mbit/s (software configurable) expansion NTE with no fibre connectors and 1 TUG3 expansion port

The CU can be deployed as an NTE located on a customer's premises, with either STM-1 or STM-4 fibre connections to the serving exchange(s). Unprotected or protected (dual access-MSP and dual parent-SNC-P) resilience between customer's

---

[1] This is the preferred option for deployment in the network.

premises and BT serving node(s) is possible. Customer interfaces can either be presented directly off the CU (at 2/34/45/140 Mbit/s G.703 75Ω or STM-1e/o) or alternatively, via locally connected NTEs (e.g. for X21 service delivery).

All fibre network interfaces, LTE to NTEs, are operating at STM-1 line rate. A TUG2 expansion port can accommodate a 4 × 2 expansion NTE and a TUG3 expansion port can accommodate either a 16 × 2 or a 34/45 expansion NTE. These expansion ports offer the ability to increase the traffic drop at a customer site in a seamless manner to meet successive service requirements.

The 34/45 NTE is remotely configured, via the element manager, to the required mode of operation. It receives 34 Mbit/s or 45 Mbit/s PDH data (depending on configuration) from the customer's equipment. These are mapped into an SDH VC-3, processed to be part of an STM-1 frame and transmitted upstream to the CU. The complimentary process is carried out in the downstream direction and the respective PDH data is transmitted to the customer's equipment.

A battery backup kit is also provided for use with the NTEs, where required. A single battery powers a single HDSL or 4 × 2-type NTE for at least five hours in the event of power failure. Two batteries are required to power a 16 × 2 or 34/45 NTE-type for five hours. The battery backup kit can be mounted externally, using clips, on top of the NTEs. The information in Table 16.3 presents an overview of the physical and interface characteristics of the ASDH NTEs.

Table 16.3   ASDH NTE features

Common NTE characteristics:
- Status LEDs, front and back
- Loopbacks
- Monitor point (DIN 41616 socket), front and back configured to required customer interface using a craft terminal
- External alarm interface (9 way D-type socket)
- Interface for craft terminal (RJ45 socket, RS232C, 9600 bit/s, VT100), front and back
- Software downloadable
- External 2 Mbit/s G.703 120 Ω synchronisation port (RJ45 socket)
- Physical construction:
  - Mounting = rack or wall mounting
  - Dimensions 4 × 2 – (mm) = 44(H) × 450(W) × 240(D), Weight (kg) = <5
  - Dimensions 16 × 2 – (mm) = 113(H) × 450(W) × 240(D), Weight (kg) = <10
  - Dimensions 34/45 – (mm) = 44(H) × 450(W) × 240(D), Weight (kg) = <5

N × 2 NTE interface characteristics:
- each 2 Mbit/s interface supports the following presentations:
  - 2 Mbit/s G.703 75 Ω (2 × female BNC)
  - 2 Mbit/s G.703 120 Ω (RJ45 socket)
  - 2 Mbit/s G.704 (RJ45 socket)
  - 2 Mbit/s I.421 (RJ45 socket)
  - 2 Mbit/s X.21 (15 way D-type socket)
  - n × 64 kbit/s X.21

and hence the following services: Transparent 2M, N × 64 kbit/s structured lease lines (where N = 1 to 31), I421, 2.048 Mbit/s X.21 and N × 64 kbit/s X.21 (where N = 1 to 31).

34/45 Mbit NTE interface characteristics:
- One configurable 34 (HDB3) or 45 (B3ZS) Mbit/s PDH interface G.703 75 Ω

16.3.1.4   Access SDH LTEs (line terminating equipment)

There are two types of LTEs offered with access SDH: a single port, LTE1, and a three-port, LTE3, card. The LTE1 unit provides a 1310 nm (Tx-Transmit)/1550 nm (Rx-Receive) single fibre STM-1 optical NTE interface. Within the STM-1 the whole VC-4 can be used for the customer traffic payload. The LTE3 provides three STM-1, single fibre, optical NTE interfaces. Within each of the three STM-1s only one TUG-3 is used for the payload, the other two are unused (this is due to the fact that each tributary slot on the CU where an LTE is used can accommodate one VC-4 (max) of traffic). Tables 16.4 and 16.5 present the potential number of 2 Mbit/s interfaces that can be delivered to a customer site using the available combinations of ASDH LTEs and fibre NTEs. It is worth noting that a large number of service interfaces (to a combined total payload of an STM-1 when the LTE1 is used) can be cost effectively delivered at customer premises over a single fibre using NTEs (in some cases a significantly more expensive ADM is used, even today).

The LTE's functionality also includes the termination of the regenerator and multiplexer overheads and monitoring of line performance:

- Termination of the RSOH and MSOH for each STM-1 interface.
- Gathering performance monitoring data and events for each optical section (retrieved by the CU controller unit).

Table 16.4   Using a 3 port LTE on the ASDH CU

Base NTE supported per interface (either single or dual fibre version)	Number/type of expansion NTEs supported for each base NTE		
	4 × 2 expansion	16 × 2 expansion	34/45 expansion
4 × 2	2	–	–
16 × 2	–	–	–
34/45	–	–	–

Table 16.5   Using a single port LTE on the ASDH CU

Base NTE supported per interface (either single or dual fibre version)	Number/type of expansion NTEs supported for each base NTE		
	4 × 2 expansion	16 × 2 expansion	34/45 expansion
4 × 2	2	Plus one of either	
16 × 2 or 34/45	–	2	–
	–	1	1
	–	–	2

- Gathering performance monitoring data and events for the NTEs via the MS-DCC (retrieved by the CU controller unit).
- Passing configuration data from the CU controller to the NTEs (via the MS-DCC).

16.3.1.5 Element management

The access SDH element manager is comprised from two basic component elements:

1. The actual access SDH network components element management system (FENS-AN).
2. An extension to the element manager that provides a single interface (for multiple access SDH EMSs, up to a maximum of 10) to the BT's operational support systems and users. It offers a single point of user login and GUI navigation and provides a single interface to PACS and GTFM (FENS-CEM).

FENS-AN provides the necessary functionality to manage all connected ASDH elements and components for provision, configuration and maintenance. Table 16.6 provides an overview of its functionality.

16.3.1.5.1 Craft terminal

All access SDH equipment supports the connection of a craft terminal (or laptop), via an RJ45 interface supporting a VT100 terminal emulation. The features accessible via this interface, relate only to the equipment to which it is connected, and are:

- display of configuration data and active alarms;
- display of performance monitoring data;
- control of loopbacks (NTE only);
- configuration of monitor points (NTE only);
- configuration of comms (DCC, NSAP address on CU only).

Table 16.6   Overview of ASDH element manager features support

Configuration management:	Fault management:
• Build (create NE resources, create path connections).	• Retrieve, display, filter and co-ordinate alarm/event messages from all managed NEs.
• Remove (mark equipment for removal, remove path connections).	Test management:
• Modify (change resource and equipment configuration parameters).	• Control of loopbacks and PRBS
	• Path trace, Section trace
• Display (view resource and equipment details, view log files).	Performance management
	User and access security management
Software download (download embedded software images to shelf equipment).	Electronic interfaces, via FENS-CEM, for:
	• the forwarding of equipment alarms to GTFM
	• the receipt of commands (service provision only) from PACS

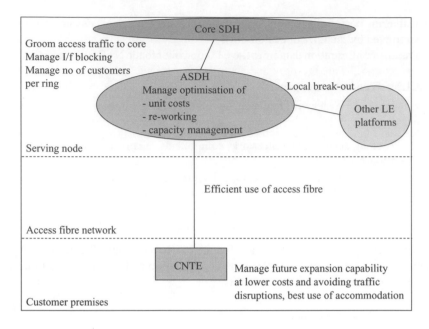

Figure 16.9   ASDH exchange architecture scope

16.3.2   Access SDH, the solution

16.3.2.1   Access SDH stage 1, wideband service support

As mentioned earlier, the introduction of ASDH technology offered a great opportunity to the BT team to affect the component designs in a manner which is exploited by both architecture and solution designs to improve both asset management and the customer experience. The diagram in Figure 16.9 presents an overview of how the technology is used in the access network to facilitate that.

1. Unit cost reductions – use of ASDH has facilitated cost reductions, against previous technology (especially where a three port LTE is used). There is also the added advantage that, like any new technology, there is increased scope for further reductions as deployment volumes increase and the technology matures. Additionally, there are 'indirect' economies of scale that can be achieved with ASDH due to the increased number of customers/traffic that can be supported per cubic metre in an exchange, compared with other/older systems.
2. Local break-out capability – ASDH provides break-out capability which avoids the use of core SDH capacity and components (at reduced cost).
3. Manage traffic into the core SDH – access traffic is combined into STM-1 interfaces before connection into the core SDH network. The ASDH technology, combined with both exchange architecture designs and relevant planning rules, grooms traffic into the core network offering increased potential to avoid

interface blocking on core ring ADMs and also utilise more efficiently the levels of traffic and number of customers a ring of given capacity can accommodate.

4. Improved capacity management – the CU, HDSL shelf and the links to the DDF and core SDH are pre-provided and capacity managed; only the LTEs/NTEs are planned and installed reactively. This means there should always be capacity available to provide a new access tail circuit.
5. More efficient use of access fibre – a single fibre, without the need for external splitters, is used to provide the connectivity between LTEs and NTEs. expansion ports on the NTEs can be used to increase the amount of traffic that can be delivered on a customer site over that single fibre (this again helps reduce costs for delivery of a number of services at a site which would otherwise require the use of a significantly more expensive ADM at the customer premises).
6. Customer benefits – the afforded unit cost reductions are passed to the customers. Improved service management from use of SDH technology. Increased reliability of components. Reduced lead times for successive orders from the use of expansion NTEs. Reduced accommodation requirements.

16.3.2.1.1 Access SDH BT node architectures

There are a number of different BT node architectures that have been designed, depending upon the forecast wideband customer growth at each serving exchange and the recommended use of components for optimising capacity management. As such, exchanges were classified as, large, medium, small and very small. A separate architecture, referred to as high band CU (HBCU), has also been introduced to support the increase on some of the limits imposed on the other architectures, on the number of 16×2 NTEs that each customer serving CU can accommodate (this implies that certain BT nodes will have both an HBCU architecture AND a standard architecture based on the criteria for wide band growth).

The actual planning rules for component use was also driven by early studies on wide band (0–2 Mbit/s services) traffic patterns at certain exchanges which revealed that:

- 17% of traffic is switched locally;
- 15% of traffic breaks out to other local platforms (like PSTN switches);
- 68% of traffic is routed through to the core network.

Large. A large BT node is one identified as requiring more than two CUs over a three-year planning period. Such a node architecture employs a two-tier CU structure (Figure 16.10).

One CU becomes a hub and is used for supporting up to a maximum of three collocated customer serving CUs; it has no direct connections to either customers or local platforms, only connections to the core SDH. The use of a hub reduces the number of connections required to the core SDH network and reduces the need for the core SDH network to support its own BT node circuits.

The connection between the hub CU and the core SDH network is via (up to six) unprotected STM-1 links (electrical or optical depending on inter-exchange distance).

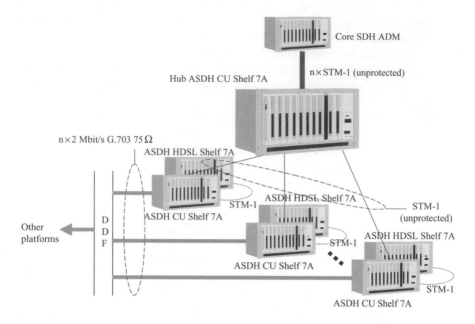

Figure 16.10   Large ASDH exchange architecture

The connection between the hub and its CUs is via STM-1 links (electrical or optical depending on distance, with electrical preferred). The connection to other platforms at 2 Mbit/s G.703 is via the non-hub CUs only. An HDSL shelf is used for copper delivery and is connected to the non-hub CUs via an STM-1o link.

As traffic requirements in these BT nodes grow, this architecture will be replicated. Where this occurs, any circuits between the hub CUs will be routed via the core SDH network.

Medium. A medium BT node is classed as having a customer growth not exceeding the need for two CUs over a three year planning period. In this architecture there is no hub CU. Each CU connects directly to core SDH and other local platforms and utilises an HDSL shelf for copper delivery (the connectivity guidelines remain the same as described for a large node).

Small. A small BT node is classed has having a customer growth over a three year period that can be accommodated by a single CU which uses direct connections to core SDH and local platforms. This architecture still uses an HDSL shelf for copper delivery (the connectivity guidelines remain the same as described for a large node).

Very small. This is the same as the small node with the only difference that copper delivery does not deploy a separate HDSL shelf but uses instead the 2-port HDSL CU line card.

HBCU. Appropriate traffic management measures have been put in place to ensure that ASDH node architectures can flex and adapt to changing customer demands and traffic pattern characteristics. As a result, it was early realised that in some nodes traffic patterns deviated from the expected resulting in some cases in poor equipment utilisation. This architecture was therefore introduced to overcome some of the limitations imposed by the other architectures at certain nodes, and hence improve equipment utilisation and capacity management. This was due to the following reasons:

- A higher than average deployment of 16×2 NTEs per CU or customer.
- A higher than average percentage of that traffic routed to the core SDH.

As such, each CU is used to only serve fibre customer delivery where each customer is going to be served by a 16×2 NTE. As a result, the number of STM-1 connections to the core can vary between 3 and 7, depending on the local traffic patterns.

16.3.2.1.2 Access SDH operational framework

The solution design for supporting the introduction of ASDH in the BT network aimed to minimise operational and business impact. Provision control and responsibility of ASDH tail end circuits has been therefore retained with the same operational teams that deal with DWSS and CWSS; the delivery of SDH all the way to the customer has necessitated the involvement of the SDH network operations unit. The required enhancements were introduced to systems involved with the support of SDH technology in the BT network and to systems that support operational activities in the access network.

- Wide band planners plan ASDH local ends and record them on the PACS database using the 'Planner' system.
- Wide band controls perform the job control function using activities generated from a COSMOSS template. Circuit design is done using SSD which can design a circuit onto a planned ASDH local end.
- The SDH NOU integration duty build and integrate the local end on PACS and the ASDH element manager.
- The SDH NOU CP and access SDH duties do the assignment and configuration of circuits on ASDH.
- Internal and external wide band works duties install the local end equipment (LTE and NTE) and connect to transmission medium (fibre or copper).

In a similar manner the same people that repair CWSS and DWSS circuits deal with access SDH. If a fault is proved onto the CU or HDSL shelf, i.e. at an LTE port or PDH break-out on the DDF, the SDH NMC will take responsibility for clearing the fault.

As described earlier, a CU is cabled to a DDF which provides a number of 2 Mbit/s PDH appearances. Also cabled from the CU are STM-1 links to either an ASDH hub CU or a core ring ADM directly, depending on the architecture deployed at a particular exchange. The CU, HDSL shelf and the links to the DDF and ring-ADM or hub-CU are pre-provided and capacity managed. This ensures there should always be capacity available to provide a new access tail. The access SDH LTE cards and NTEs are

Table 16.7 Plan, build and provision systems used with ASDH

PACS	Graphical Browser (GB)	Discoverer	Planner
This is used by the SDH NOUs for planning, assigning and configuring the SDH/ASDH network.	This provides a read-only graphical view of the SDH/ASDH network.	This system is used by wideband planners. It provides a read-only graphical view of the SDH/ASDH network. Basically it is a cut down version of GB.	This system is used by wideband planners for building new ASDH tails. Basically it is a cut down version of PACS.

Access media details
Optical fibre records are held on **INS**.
Copper pair records are held on **CSS**.

planned and installed reactively when an order is received that requires a wideband access section to a customer premises.

When the local end has been installed and connected up, the information is then entered on PACS and the ASDH element manager. It is then that PACS issues a configuration command to activate the cross-connection and the required circuit connectivity through the CU. Circuits that need to be routed onto a local service platform (e.g. ISND30 or KiloStream) or routed over the PDH network, are cross-connected through to a DDF appearance (i.e. PDH break-out). Circuits that are to be routed through the core SDH network will be cross-connected through to the core ring ADM or hub CU instead of a PDH break-out.

Circuits are recorded on INS along with the PDH routing details. ASDH equipment and the ASDH/SDH portion of circuit routings are recorded on the SDH network database. Table 16.7 provides an overview and brief description of the systems involved with plan, build and provision for ASDH whilst Figure 16.11 depicts how the technology was incorporated in the OSS SDH model described earlier.

16.3.2.1.3 Service support

The wide band service support capability of ASDH has already started field deployment using the components and exchange architectures described earlier. Figure 16.12 depicts how the technology is used to deliver either copper or fibre bearer support for 2 Mbit/s services. The following list offers an overview of the services that can already be supported (note that ASDH is used to provide the layer 1 private-circuit access tail to the relevant switch platform for the delivery of layer 2/3 data services):

- MegaStream 2.
- KiloStream N (data) and (voice) – ASDH currently used as the 2 Mbit/s access bearer to feed an NTU8 at the customer premises.

- KiloStream aggregate bearer.
- MultiStream bearer.
- ISDN30 – DASS2 PBX.
- ISDN30 – I.421.
- Data services like CellStream 2 Mbit/s and frame relay, $0 \sim 2$ Mbit/s.

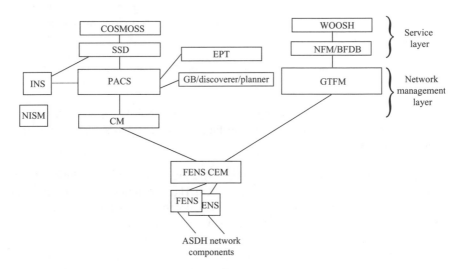

Figure 16.11   OSS for ASDH

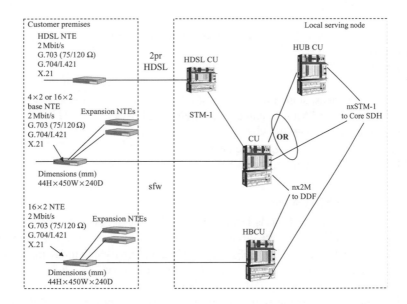

Figure 16.12 $n \times 2$ Mbit/s delivery using HDSL and fibre NTEs

16.3.2.2   Access SDH stage 1, high band service support

The primary drivers for utilising ASDH technology to support high band services (34/45/140/155 Mbit/s access bearers) and customer designs that require an ADM located at the customer premises are:

- Unit cost reductions compared with existing technology.
- Delivery of X21-based services with higher resilience requirements (steel/silver).
- Support of all the customer SDH access configurations options presented earlier (see Section 16.2.3).

However, use of ASDH technology to support high band customer solutions is currently limited by:

- the line rate back to the parenting node can either be STM-1 or STM-4, and
- the maximum bandwidth customer interface can be 155 Mbit/s, STM-1 electrical or optical presentation.

Development work is ongoing looking to culminate in a suitable solution design framework that will accommodate the use of ASDH as a complementary proposition to satisfy service demand that drives the installation of an SDH ADM at customer premises. It is a complementary proposition since it is acknowledged that a number of parameters, including current and future service mix and total bandwidth requirements at a customer site can exclude the choice for ASDH as being the more appropriate.

The list (non-exhaustive) that follows presents some of the services that can be supported using ASDH. Figure 16.13 depicts the access configurations that can be supported with ASDH.

- MegaStream Genus 2~155 Mbit/s (note that for 2 Mbit/s, if an NTE it must be the dual fibre type).
- MetroStream Premium/Standard 2 Mbit/s.
- NTEstream.
- 2 Mbit/s broadcast standard.
- Cellstream 2~155 Mbit/s standard/secure/secure+.
- MegaStream 2~155 Mbit/s.
- MegaStream long line aggregate bearer.
- Midband STM-1o connection to SP.

16.3.2.3   Access SDH stage 2

The current BT data portfolio is largely provided in the access network by the use of fixed rate, permanent private circuits. By and large, products in this sector either comprise a layer 1 circuit, e.g. KiloStream or MegaStream (which today make up the majority of private circuit volumes and sales), or a data service such as frame relay or CellStream, which rely on PDH or SDH bearers in the access network.

This requirement profile will initially be well served by access ASDH stage 1 and BT's other existing access transmission systems such as low cost KiloStream.

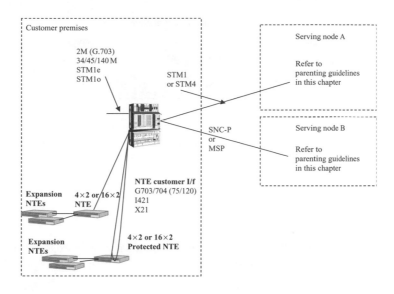

Figure 16.13   Service delivery using dual parented ASDH CU-Shelf 7A

The requirement to explore potential solutions, like ASDH stage 2, comes from an assumption that increasing demand for data services will be seen throughout all segments of the business market.

Access SDH stage 2 is an evolution of the equipment deployed for ASDH stage 1, and essentially comprises the addition of an ATM layer to the stage 1 SDH-only capability. The objective of this is to allow cost effective consolidation and concentration of data services, as such services begin to move into volume demand across geographies and the customer base.

The core component of ASDH, the consolidation unit (an SDH ADM) has been designed with a dual bus serving the majority of its tributary and line slots. The (TDM) circuit-based bus connects traffic (max 1 AU4 per slot) to the high and low order SDH cross connect switches (Figure 16.14). The ATM cell-bus offers similar traffic connectivity, only this time it directs traffic from the tributary slots to the ATM switch which then interconnects to the high order SDH switch for access to the line slots (the CU has 2 (worker/protector) switch slot positions dedicated for the use of an ATM switch fabric). There are two ATM switch cards with different throughput capacity; the choice of card also determines the actual number of slots with ATM connectivity.

Using ATM-enabled cards, a tributary slot can offer connectivity and delivery of either circuit-based or cell-based only traffic, or a mixture of both, depending on the choice of customer NTEs (note that the total traffic capacity a tributary slot can accommodate is 1 × AU4 of ATM and/or 1 × AU4 of TDM). Equally, an STM-1 card with cell-bus access needs to be used in a line slot position to provide the ATM loaded VC-4 connectivity to other platforms.

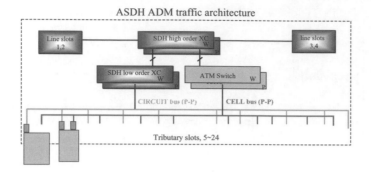

Figure 16.14   ASDH dual-bus architecture

A range of ATM fibre NTEs currently appear on the suppliers' roadmap with the following common design concept:

A base component which comes in three variants:

- single fibre with 2 TUG3 expansion ports and one PIM (plug in module) slot
- dual (MSP) protected fibre with 2 TUG3 expansion ports and one PIM slot
- expansion with no fibre connections, 1 TUG3 port and one PIM slot

A set of PIMs with a range of customer interfaces (e.g. 4 × E1, ATM25, E3 and STM-1 'native' ATM UNI, 10/100 Base-T).

The TUG3 expansion ports allow the capability to mix and match ATM with TDM fibre NTEs at the customer end depending on traffic requirements.

There is an additional ATM-enabled card being designed from the supplier which can occupy any of the tributary slots and connects to both the ATM and the low order switch fabric. SDH VC-12 tributaries (max 63 per card since each slot has a limit of 1 AU4) carrying cell traffic can then be directed to such a card which can 'extract' the ATM payload and map it to the AU4 connection to the ATM switch. Additional functionality for circuit emulation and IMA groupings on the card could then be offered. The main benefit of such a component is the ability to interconnect cell-based customer traffic to the CU ATM switch without the need for ATM-NTEs and ATM-enabled tributary cards; a component that could be used to support a more cost-efficient asset management given the uncertainty of potential traffic mix variations at different geographies. Further options and design propositions are being considered for ASDH stage 2, such as the use of APON, and potential integration with the broadband xDSL platform.

ASDH stage 2 is still in the definition and design phase, and as such, a date for implementation and subsequent field deployment is not yet agreed. Work is ongoing to further validate the business benefits that can be accrued from its deployment looking more specifically at component design propositions, suitable network architectures, OSS systems integration and operational fit. The work forms part of a wider strategic initiative within BT looking at the benefits and impact of access network evolution propositions.

16.4   Conclusions

As stated earlier in this chapter, BT has, and continues to evolve, SDH transmission equipment in the core network. At the same time, deployment of access SDH (stage 1) equipment is well under way both in access and the outer core; discussions as to how and when to introduce the stage 2 component of access SDH are ongoing. An investigation is under way to explore the options open for both core and access technologies to understand how best to meet the needs of both customers (for interface flexibility and cost) and operator networks (primarily for reducing cost and managing capacity) with an agreed portfolio of equipment. This will undoubtedly involve deployment of Access SDH equipment in parts of the core network and within some customer segments, whereas other segments will require and could only be served by deployment of core SDH equipment. Resolution of this area is critical and proposals should be forthcoming in the very near future, but it is clear that there is a role for both technology options within the UK network.

Chapter 17

Managing access networks

S. J. Rees

17.1 Introduction

This chapter considers the deployment of network management systems to support the access network. It considers some of the key drivers for the integration of network management systems (NMS) and discusses the technology and standards that can be used to achieve this integration. A brief tutorial on the Telecommunication Management Network (TMN) standards is given. The chapter goes on to consider an example of a managed network and some of the implementation aspects of introducing TMN systems.

In today's competitive telecoms environment operators are increasingly trying to find ways to reduce the costs of running their networks. Additionally there is a need to differentiate from competitors by introducing new services to attract new customers and generate additional revenue. This has led to requirements to deploy increasingly sophisticated network management systems and the key requirement to be able to integrate them. With the expected migration of millions of traditional residential customers to new DSL-based broadband networks, providing fast Internet access, a non-integrated NMS no longer supports effective or efficient deployment. By automating the operation of the network, operators achieve greater customer focus and provide a better service tailored towards the needs of their customers. The drivers for integrated network management are summarised in Table 17.1.

Providing new services faster than the competitors allows an operator to increase their market share. A flexible integrated NMS allows capital investment in new technologies to begin generating revenue quickly. Many operators now have the vision of 'single touch provisioning' where service is provided to the customer from a single entry into the network management system from the front office. In fact, some operators even take this further with 'zero touch provisioning' where the service is activated automatically via an e-Commerce transaction over the Internet. Removing the human interaction from the provisioning process reduces the chances of costly

Table 17.1   Drivers for integrated network management

Increase revenue	Reduce operating costs
New value added services	Reduce costly human error
Increased customer satisfaction	Greater automation
Increase speed of provisioning	Reduce customer downtime

errors being made. These errors can result in the service being delayed or even the loss of the customer and associated revenue.

Many operators provide customers with service level agreements (SLAs). It is critical that the operator can diagnose and repair faults in the network quickly to avoid expensive rebates to customers. Offering SLAs increases customer's confidence and makes customer retention more likely. With large volumes of broadband customers, where margins will be low, reducing the cost of fault diagnosis and repair is key.

17.2   Operational processes

Traditionally, requirements for network management systems have been governed by the capabilities of the equipment being managed; however, recently requirements have been defined using a top-down approach, considering first the service aspects and then the requirements of the network management system in order to integrate with these service processes. A typical example of this is shown in Figure 17.1.

With many different technologies and vendors in the access network and different systems available for service management, standardisation of the management interfaces becomes essential. The availability of common standards and off-the-shelf NMS products increases an operator's purchasing options and reduces the lock-in and reliance on existing suppliers or internal developments. Products are differentiated by aspects such as price, quality, availability, support and management features.

The key functions required for managing an access network are illustrated by Figure 17.1. The process starts with the operator planning the network. This is supported by systems that hold records of equipment already installed, available space in exchanges and external plant such as copper or fibre pairs. Market forecasts will dictate what equipment is required in which locations. Providing sufficient equipment to meet demand is key to meeting customer expectations. If customers have to wait too long for their service to be provisioned then they may move to a different operator.

Once capacity has been built then orders can be accepted from customers. An order management system will capture details of the order and track progress of fulfilment. It is necessary to convert the order into a service request to the network management systems including details such as which exchange and which copper pair the customer's service is terminated on. Notification also needs to be sent to installation management systems and on to systems that schedule and manage the

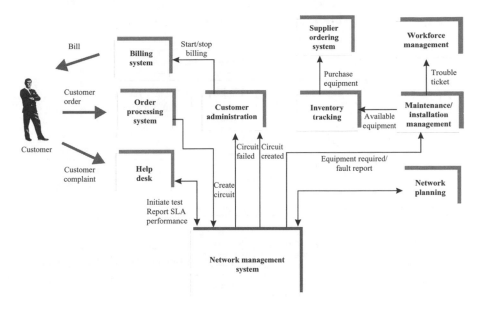

*Figure 17.1   Integration of a network management system with service management
processes*

workforce. It may also be necessary to interrogate inventory management systems in
order to determine what equipment is available to meet the requirements.

Once the service is provisioned, a customer administration system can be updated
in order to show the services the customer is receiving and notification can be sent
on to systems to start billing for the service. Should a fault occur it is preferable for
network management to notify maintenance systems and dispatch field engineers to
fix the fault before the customer complains. However, if a complaint is received,
this needs to be tracked by the help desk system against progress of fault resolution.
Network management must then support the testing of the customer's service and
use sophisticated fault analysis to find the root cause. One single problem within the
network can result in many different alarms being generated. Accurate correlation of
this fault information is key. Systems may need to be provided to track the performance
of the customer's service against his contracted service level agreement.

A detailed analysis of the process of managing a network is considered by the
TeleManagement Forum. They have performed a detailed analysis of the processes
involved in operating a network to define interfaces that will allow applications to
interoperate. More details about this can be found in the Telecom Operations Map
[1] which is illustrated in Figure 17.2.

Where possible, interface definitions are taken from industry standard bodies
such as the ITU-T and ETSI. They have defined the principles and specifications that
allow network management systems to integrate with other applications. This is by
an initiative known as the TMN or Telecommunication Management Network.

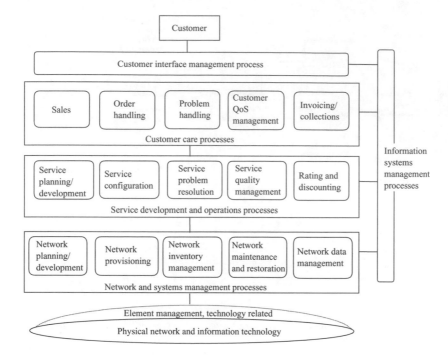

Figure 17.2   The telecom operations map

17.3   The Telecommunication Management Network (TMN)

In 1985 an initiative was launched in order to develop standards for the management of telecommunications networks. The work has been carried out under the heading of the TMN.

TMN is the term used to describe an area of standardisation that is being undertaken in order to harmonise and integrate the network management of telecommunications networks. An outline of the TMN is given in ITU-T Recommendation M.3010 [2]. The objective of the TMN is to provide a framework that will achieve interconnection of various types of operations systems (OSs), which perform the management, and telecommunications equipment within the managed network. This will have the advantage of easing the operator's task in planning, installing, provisioning and maintaining his networks and services. It will also make systems easier to specify for the vendor and increase the size of the market he has access to.

The TMN essentially defines three architectures and these will be described in turn.

17.3.1   Functional architecture

The TMN functional architecture divides a management service into a number of functional blocks. Reference points are then defined between functional blocks that

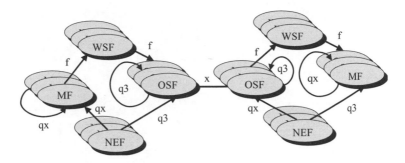

Figure 17.3   Interconnection of functional blocks in the TMN

exchange information. Each functional block is made up of functional components. The following functional blocks are defined:

Operation system function	–	OSF
Network element function	–	NEF
Workstation function	–	WSF
Mediation function	–	MF
Q Adaptor function	–	QAF

Figure 17.3 shows how functional blocks can be connected together and the reference points between them. The reference points provide a point where standardised interfaces may be defined.

17.3.2   Physical architecture

The TMN functional blocks can be implemented in a variety of physical configurations. The TMN defines a mapping of functional blocks onto physical platforms. For instance an operation system may consist of an OSF, MF, WSF and QAF. When reference points are implemented between physical platforms they form interfaces. These are indicated by the appropriate capital letter. An example of a TMN physical architecture is shown in Figure 17.4.

The following physical blocks are defined:

Operation system	–	OS
Data communications network	–	DCN
Work station	–	WS
Mediation device	–	MD
Q adaptor	–	QA
Network element	–	NE

17.3.3   Information architecture

Where physical interfaces are implemented, information is exchanged between functional blocks using the ISO layer seven service CMISE (common management

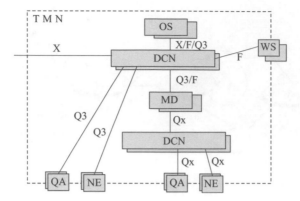

Figure 17.4   Example physical architecture for a TMN

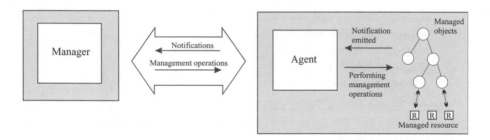

Figure 17.5   Manager/agent interaction

information service element). CMISE has a number of primitives that perform operations on what are known as managed objects. An abstract 'information model' of a system is created by defining a set of managed objects and the relationships between them. This is a formal representation of the management capabilities of a system. Across an interface there is a manager/agent relationship with the manager invoking operations on the agent and the agent sending replies and notifications. This is illustrated in Figure 17.5.

The protocol allows a manager to create and delete managed objects, get and set attribute values and also to invoke some predefined actions. CMISE also allows an agent to send an event report containing detailed information concerning an event (e.g. alarms) that occurs within the network. Typical resources modelled by managed objects include fault logs, network elements, field replaceable units, etc. Managed objects are defined using a language called GDMO (guidelines for the definition of managed objects) and the syntax is defined in ASN.1 (Abstract Syntax Notation 1).

17.3.4   The logical layered architecture

The TMN allows for a layered approach to be taken to the management of telecommunications networks where each layer presents a different abstraction of the

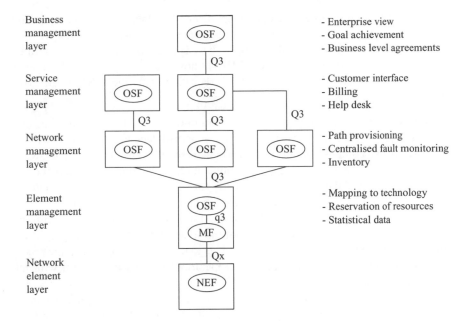

Figure 17.6   TMN layered architecture

management information. An example of this is shown in Figure 17.6. It should be noted that this layered approach is intended to help requirements capture by dividing up the problem space but should not constrain any physical realisation that an operator wishes to deploy.

17.4   Interfacing technologies

Although the TMN originally defined CMISE as the mechanism for interfacing in the TMN, relatively few implementations of this have been achieved. Using the full OSI 7 layer stack has provided performance challenges and the flexibility provided by the models is difficult to implement. Also the absolute dominance of IP networks has limited scope for its use. The work though has proved valuable in defining the information to be passed and a framework around which systems can be implemented. A number of other technologies are in reality being used to transport information between systems.

Transaction Language 1 (TL1). TL1 is an ASCII or man-machine management protocol defined by Bellcore. It is particularly used by North American vendors and provides a simple easy to understand way of managing network elements.

Simple network management protocol (SNMP). SNMP originated from the IP community and is defined by the Internet Engineering Task Force (IETF).

It is commonly implemented on equipment from the Datacoms community and is now the dominant protocol on access equipment such as DSLAMs. It uses similar concepts to CMISE and is object based with interfaces being defined using management information bases (MIBs); however, it uses a much simpler protocol for transport. This protocol uses IP rather than the OSI seven layer stack.

CORBA. CORBA is the acronym for common object request broker architecture, OMG's (Object Management Group's) open, vendor-independent architecture and infrastructure that computer applications use to work together over networks. CORBA applications are composed of objects, individual units of running software that combine functionality and data, and that frequently (but not always) represent something in the real world. This is a similar concept to managed objects in the TMN.

For each object type, an interface is defined in OMG IDL (interface definition language). The interface is the syntax part of the contract that the server object offers to the clients that invoke it. Any client that wants to invoke an operation on the object must use this IDL interface to specify the operation it wants to perform with the parameters it wants to send.

The IDL interface definition is independent of programming language, but maps to all of the popular programming languages via OMG standards: OMG has standardised mappings from IDL to C, C++, Java, COBOL, Smalltalk, Ada, Lisp, Python, and IDLscript.

XML. The extensible markup language (XML) is a universal format for structured documents and data on the Web. Although defined to provide a standard format for computer documents, its syntax is now commonly used to specify interface definitions between network management applications. Data is included in XML documents as strings of text and the data is surrounded by text markup that describes the data. The use if XML has the advantage of being easily human-readable and usable from a variety of Web based applications.

17.5   Example of managed networks – broadband

The deployment of broadband has represented a significant challenge to operators. Mature operations systems for the delivery of POTS have had to be combined with new systems for managing the high speed data aspects. Figure 17.7 shows a typical broadband network based on ADSL.

The operations that need to take place to provision a service across it, outlined in Figure 17.7, are as follows:

1. *Allocate and configure DSLAM Resources*: based on a forecast for a particular exchange, suitable resources should be installed and configured.
2. *Pre-qualify line*: when a customer requests ADSL, the first step involves determining if the line is suitable to carry ADSL. This may be done manually by determining the loop length or may be the result of a line test.

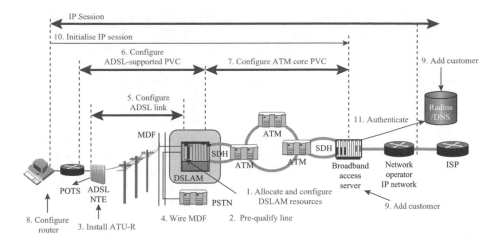

Figure 17.7   Provisioning process for a broadband network

3. *Build ATU-R*: once a customer has bought ADSL, a command is sent to OSS to
    build an ATU-R. This command can specify the ATU-C port to use or it can be
    auto allocated.
4. *Wire MDF*: The installation engineer then needs to connect the POTS line to the
    splitter via the MDF. Potentially the information on the jumpering could be stored
    in the management system at this stage.
5. *Configure ADSL Link*: Having built an ATU-R, the bandwidth of the ADSL link
    can be configured.
6. *Configure PVC*: Once an ADSL link is in place it is possible to build an end-to-end
    PVC.
7. *Configure core PVC*: The configuration of the access PVC needs to be
    co-ordinated with configuration of a PVC across the core network.
8. *Configure router*: If the routing functionality of the NTE is used, this then needs
    to be configured on the ATU-R.

Once these processes have been carried out and a user has been added to the
access server, a user session can be initiated.

9. *Configure broadband access server*: A user name and password need to be added
    to the RADIUS database allowing the customer to be authenticated when initiat-
    ing a user session. Additional parameters may be set on the BAS to control the
    bandwidth and quality of service that the customer receives.

Once a service is provisioned, then diagnosing problems is also a challenging
process. Common problems include: loss of syncronisation, where the DSL modem
drops the connection (commonly caused by noise on the line on long loops or faulty
DSLAM equipment); loss of connection to RAS, where the PPP session to the

broadband RAS has failed (commonly caused by problems in the ATM layer or faulty BAS equipment); CPE problems, CPE misconfigured or driver problems.

Diagnosing the fault incorrectly can be costly. For instance, a hardware failure on a RAS may be mis-diagnosed as a problem on the DSLAM and several engineers tasked to the same exchange to resolve the problem. Accurate fault correlation is key.

17.5.1   Copper loop

A key problem in the provision of ADSL services is the quality of the copper loop. Many loops were installed decades ago and over the years may have undergone changes that may or may not have been documented. The copper pairs may be incorrectly jumpered to the main distribution frame (MDF), may be degraded or may be subject to noise from other pairs in the same bundle. There are two approaches being taken to address this problem.

17.5.1.1   Test access matrices

In order to save costly visits to exchanges, operators are installing test access matrices. These matrices allow an operator to switch the line to a test head to provide various testing capabilities. New broadband test heads are being provided that test the higher frequency levels that DSL occupies compared with POTs.

These frequency levels are more susceptible to interference from other data services and so a clear analysis of the signals present is key to fault diagnosis. The wiring in the home can also impact service. Poorly filtered telephones, and noisy appliances can degrade the DSL performance. Test access matrices are being provided to analyse the following:

- Loop length.
- POTS detection and status.
- Wideband noise.
- DSL signal type.
- Spectrum analysis.
- Time domain reflectometry.
- Modem connection.
- Customer equipment emulation.
- IP connectivity throughput.
- Loop impedance.
- Loop capacitance.

17.5.1.2   Single ended line testing

Loop testing through the use of dedicated hardware can, however, be expensive so technology is now being developed to utilise the power of the DSPs in the ADSL equipment to provide line test facilities. This is known as single ended line testing or SELT. At the time of writing an ITU standard known as G.selt is being drafted to cover this area. This technology will provide facilities such as:

- Time-domain reflectometry.
- Frequency domain reflectometery.

- Loop impedance.
- Loop capacitance.
- Power spectrum of noise on the line.

17.6   NMS implementation

The implementation of a network management system will typically involve the following phases.

Phase	Description
Requirements	Determine the requirements for the application
Conceptual design	Model the underlying business that the application will support
Logical design	Design in general terms how the application will operate
Physical design	Design in specific terms how the application will be constructed
Construction	Construct the application
Usability testing	Test the usability of the user interface

17.6.1   Requirements capture

Understanding the requirements and the environment in which a network management system will be used is vital to successful NMS implementation. A number of tools are available for capturing requirements, and traceablility of how the requirements are implemented through the subsequent phases of development is important to implementing a successful system. Involving a wide range of people in the project from marketing, development engineers' to testers and end-users in reviewing the requirements is key.

17.6.2   Conceptual design

The next step is to produce a model of the system. This typically involves a data model, i.e. the information that will need to be stored, and a business process model describing the functions that need to take place. This is typically described by defining cases using languages such as the Unified Modelling Language (UML^{TM}) as defined by the Object Management Group^{TM} (OMG) and a communications model describing how information flows around the key parts of the system. The conceptual stage should be independent of any implementation.

17.6.3   Logical design

The logical design gets more into specifics about how the application will meet the business processes described in the conceptual design. This will include activities such as prototyping user interfaces and determining which technology will be used for implementation. This will typically involve definitions of managed objects in which interfacing or implementation technology is chosen, for instance GDMO or IML.

17.6.4   Physical design

The physical design is concerned with determining how the logical design will be implemented, for instance choice of hardware platform, database, and communication protocols. This has to take into consideration aspects such as scalability, i.e. how many users and how many lines will be managed. A typical system may need to handle several million events a day. Defining an implementation that can be distributed and grow as the network grows is key. Also resilience needs to be considered. Where an NMS is critical to an operator's business it has to be fault tolerant using stratagies such as disk mirroring, regular backups and distributed processors. This phase will also choose what user interface technology to use, for instance a windowing technology provides exciting graphics and a rich user experience whereas Web technologies provide easier remote access and lower cost terminals. An example of a Web-based user interface is shown in Figure 17.8.

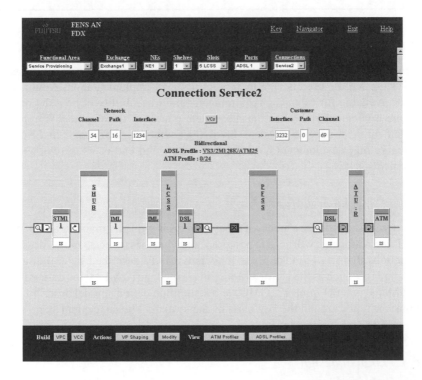

Figure 17.8   Service provisioning screen from Fujitsu's FENS AN access network management system

17.6.5 Testing

Testing of the system will check that the requirements of the system have been met. A large number of test cases will be defined to validate all aspects of the system. The testing stage usually goes in a number of phases with defects being fixed as the testing progresses. Regression testing is performed when the software is upgraded to check that functions previously tested still work with the new release. Provocative testing can often be carried out where the testers try everything they can to break the system.

17.7 References

1 Tele Management Forum: Network Management Detailed Operations Map GB 908 v1.0
2 ITU-T Recommendation M.3010 (1992): 'Principles for a telecommunications management network'
3 ITU-T Recommendation M.3100 (1992): 'Generic network information model'
4 ITU-T Recommendation Q.811 (1993): 'Lower layer protocol profiles for the Q3 interface'
5 ITU-T Recommendation Q.812 (1993): 'Upper layer protocol profiles for the Q3 interface'
6 ITU-T Recommendation X.720 (1992): 'Management Information Model'
7 ITU-T Recommendation X.721 (1992): 'Definition of management'
8 O'Reilly: XML In a Nutshell: A Desktop Quick Reference

Glossary

2B + D	Two Bearer plus Data
2B1Q	Two Binary, One Quaternary
2G	Second Generation
3B2T	3 Binary, 2 Ternary line code
3G	Third Generation
4B3T	4 Binary, 3 Ternary line code
A1141	A1141 form
AC	Alternating Current
ACE	Automatic Cross-connection Equipment [kilostream]
ACK	Acknowledgement
ADC	Analogue to Digital Converter
ADM	Add-drop Multiplexer [SDH]
ADPCM	Adaptive Differential Pulse Code Modulation
ADSL	Asymmetric Digital Subscriber Line
ALC	Analogue Line Card
AM	Amplitude Modulation
AM-VSB	Amplitude Modulation Vestigial Sideband [modulation]
AMI	Alternate Mark Inversion
ANFP	Access Network Frequency Plan
ANSI	American National Standards Institute
AoD	Analogue over Digital
AP	Access Point
API	Application Programming Interface
APON	ATM over Passive Optical Networks [BTL]
ASCII	American Standard Code for Information Interchange
ASDH	Access Synchronous Digital Hierarchy [BT]
ASN.1	Abstract Syntax Notation One
AT&T	American Telephone and Telegraph
ATM	Asynchronous Transfer Mode
ATM25	Asynchronous Transfer Mode at 25 Mbit/s [ATM Forum]
ATU-C	ADSL Transceiver Unit – Central office end
ATU-R	ADSL Transceiver Unit – Remote terminal end

AU4	Administrative Unit 4 [SDH]
AUC	Authentication Centre [GSM]
B-ISDN	Broadband Integrated Service Digital Network [ITU-T]
BAS	Broadband Access Server
BE	Best Effort service [IETF]
BellCoRe	Bell Company Research
BER	Bit Error Rate
BFWA	Broadband Fixed Wireless Access
BIDS	Broadband Integrated Distributed Star
BPON	Broadband Passive Optical Network
BPSK	Binary Phase Shift Keying
BSC	Base-Station Controller
BSS	Base-Station Subsystem
BT	British Telecommunications plc
BTexaCT	BTexact [BT]
BTS	Base Transceiver Station
BTTJ	BT Technology Journal [BT]
CA	Collision Avoidance
CAA	Civil Aviation Authority [UK]
CAC	Connection Admission Control
CAP	Carrierless Amplitude & Phase modulation
Cat5	Category 5 cabling
CBL	Common Business Library [CommerceOne]
CBR	Constant Bit Rate
CCA	Clear Channel Assessment
CD	Collision Detection
CDMA	Code Division Multiple Access
Cdma2000	Code Division Multiple Access for the year 2000
cdmaOne	Code Division Multiple Access One
CellStream	BT's ATM service to customers
CEPT	Conference of European Post & Telecommunications administrations
CLASS	Custom-calling Local Area Signalling System
CLEC	Competitive Local Exchange Carrier [USA]
CLIP	Calling Line Identification Presentation [ITU-T]
CLIR	Calling Line Identification Restriction
CM	Configuaration Manager
CMISE	Common Management Information Service Element [ISO]
CMS	Call Management Server
CMTS	Cable Modem Termination System
COBOL	COmmon Business Oriented Language
CODEC	Coder/Decoder
CORBA	Common Object Request Brokerage Architecture
COSMOSS	Customer Oriented System for the Management Of Special Services [BT]

CP	Cplan re Capacity Planning
CPE	Customers' Premises Equipment
CS	Carrier Sense
CSMA	Carrier Sense Multiple Access
CSS	Customer Services System [BT]
CTS	Clear To Send
CU	Consolidation Unit [BT]
CWSS	Copper Wide Band Serving Section
DAB	Digital Audio Broadcasting
DACS	Digital Access Carrier System
DACS1	Digital Access Carrier System No. 1
DACS2	Digital Access Carrier System No. 2
DASS2	Digital Access Signalling System v.2
dB	deciBel
DBA	Dynamic Bandwidth Assignment
dBW	deciBel Watt
DC	Direct Current
DCC	Data Communications Channel
DCN	Data Communications Network
DDF	Digital Distribution Frame
DDI	Direct Dialling In [BT]
DDS	Digital Data Service
DECT	Digital Enhanced Cordless Telecommunications
DECTForum	Digital Enhanced Cordless Telecommunications Forum
DEL	Direct Exchange Line [BT]
DFS	Dynamic Frequency Selection
DHCP	Dynamic Host Configuration Protocol
DiffServ	Differentiated Services [IETF]
DigiMedia	DigiMedia Vision Ltd
DLC	Digital Line Card [Telspec]
DLC	Data Link Connection
DLE	Digital Local Exchange [BT]
DLS	Digital Line System
DMSU	Digital Main Switching Unit
DMT	Discrete Multi-Tone
DN	Directory Number [BT]
DoCoMo	DoCo? Mobile [NTT]
DOCSIS	Data Over Cable Service Interface Specification
DP	Distribution Point
DPCN	Digital Private Circuit Network
DPRS	DECT Packet Radio Service
DQoS	Dynamic Quality of Service
DS	Direct Sequence
DSL	Digital Subscriber Line
DSLAM	Digital Subscriber Line Access Multiplexer

DSP	Digital Signal Processor/Processing
DTI	Department of Trade and Industry [UK]
DTV	Digital Television
DVB	Digital Video Broadcast
DVB-RCS	Digital Video Broadcasting group – Return Channel over Satellite
DVB-S	Digital Video Broadcasting group
DVD	Digital Versatile Disc
DWDM	Dense Wavelength Division Multiplexing
DWSS	Digital Wideband Serving System
E1	{Europe} 1 [ITU-T]
E3	{Europe} 3 [ITU-T]
EC	European Community
ECAT	Event Collection and Translation [BT]
EDM6003	Equipment Digital Muldex [GPT]
EEC	European Economic Community
EFM	Ethernet in the First Mile [IEEE]
EIR	Equipment Identity Register [GSM]
EIRP	Effective Isotropic Radiated Power
EM	Equipment Modem
EMC	Electro-Magnetic Compatibility
EMS	Element Management System
EN	European Norm (Norme Europeene)
ENA	Equipment Network Access
EPT	Equipment Planning Tool [BT]
ERC	European Radiocommunications Committee
ESCOM	A lower rate interface
ETR	European Telecommunications Requirement [ETSI]
ETS	European Telecommunications Standard [ETSI]
ETSI	European Telecommunications Standards Institute
EU	European Union
EURESCOM	European institute for REsearch & Strategic studies in teleCOMmunications
FBS	Flexible Bandwidth Service
FDD	Frequency Division Duplex
FDM	Frequency Division Multiplexing
FDMA	Frequency Division Multiple Access
FEC	Forward Error Correction
FEXT	Far-End Cross-Talk
FM	Frequency Modulation
FS	Fixed Service
FS-VDSL	Full Service Access Network – Very high speed Digital Subscriber Line
FSAN	Full Service Access Network [BT, NTT]
FSS	Fixed-Satellite Services

FT	Frequency Time
FTP	File Transfer Protocol [IETF]
FTTB	Fibre To The Business/Building
FTTC	Fibre To The Cabinet
FTTCab	Fibre to the Cabinet
FTTE	Fibre To The Exchange
FTTH	Fibre To The Home
FTTK	Fibre to the Kerb
FTTO	Fibre To The Office
G.703	Physical/electrical characteristics of hierarchical digital interfaces [ITU-T]
G7	Group of Seven
GDMO	Guidelines for Definition of Managed Objects [ISO]
GEO	Geostationary Orbit
GFR	Guaranteed Frame Rate
GFSK	Gaussian Frequancy Shift Keying
GGSN	Gateway GPRS (General Packet Radio Service) Support/Service Node
GHz	Gigahertz
GPRS	General Packet Radio Service
GPS	Global Positioning System
GPT	GEC-Plessey Telecommunications
GRE	Generic Routing Encapsulation [IETF]
GSM	Global System for Mobile (was Group Speciale Mobile)
GSO	Geo-Stationary Orbit
GTE	GTE Corporation
GTFM	Generic Technology Fault Management [BT]
GUI	Graphical User Interface
HAP	High Altitude Platform
HAPS	High Altitude Platform Station
HBCU	High Band Consolidation Unit
HDF	Handover Distribution Frame [BT]
HDFSS	High Density Fixed Satellite Service
HDSL	High-speed Digital Subscriber Line
HDSL2	High Bit Rate Digital Subscriber Line 2
HDT	Host Digital Terminal [BBT]
HDTV	High Definition Television
HFC	Hybrid Fibre-Coax
HIPERLAN	High Performance European Radio Local Area Network
HLR	Home Location Register [TACS]
HPF	High Pass Filter
HTTP	HyperText Transfer Protocol
HTTPS	HTTP over SSL
I.421	Primary rate user-network interface [ITU-T]
IBTE	Institution of British Telecommunications Engineers

ICO	ICO Global Communications
ID	Identity
IDL	Interface Definition Language
IDLscript	Interface Definition Language script
IEE	Institution of Electrical Engineers [UK]
IEEE	Institute of Electronic and Electrical Engineers [USA]
IETF	Internet Engineering Task Force
ILEC	Incumbent Local Exchange Carrier [USA]
IMA	Inverse Multiplexing for ATM
IMT	International Mobile Telecommunications [ITU-T]
IMT2000	International Mobile Telecommunications for the year 2000
IN	Intelligent Network
INS	Integrated Network Systems
IP	Internet Protocol [IETF]
IPSec	IP Security protocols [IETF]
IPStream	BT IPStream
IRE	Institute of Radio Engineers
ISA	Industry Standard Architecture
ISBN	International Standard Book Number
ISDH	Initial Synchronous Digital Hierarchy
ISDN	Integrated Services Digital Network
ISDN2	Integrated Services Digital Network (Basic Rate, 2B + D)
ISDN2e	Integrated Services Digital Network (Basic Rate, 2B + D) european
ISDN30	Integrated Services Digital Network (Primary Rate, 30B + D)
ISI	Inter-Symbol Interference
ISM	Industrial, Scientific and Medical
ISO	International Organisation for Standardization
ISP	Internet Service Provider
IT	Information Technology
ITU	International Telecommunications Union
ITU-T	International Telecommunication Union – Telecommunications Standardization Sector
JANET	Joint Academic NETwork
kBaud	kilo Baud
kHz	Kilohertz – 1,000 Hz
KPN	Royal Dutch Telecom
kW	kilo Watt
L2TP	Layer 2 Tunnelling Protocol [IETF]
LAN	Local Area Network
LE	Local Exchange
LEO	Low Earth Orbit
LLC	Logical Link Control
LLFN	London Local Fibre Network

LLU	Local loop unbundling
LLUAF	Local Loop Unbundling Automation Forum [UK]
LMDS	Local Multipoint Distribution System
LNEL	Local Network Evaluation Laboratory [BT]
LPF	Low Pass Filter
LTE	Line Termination Equipment
M.3010	ITU-T Recommendation M.3010 (1992), Principles for a telecommunications management network
M.3100	ITU-T Recommendation M.3100 (1992), Generic network information model
MAC	Media Access Control
MAN	Metropolitan Area Network
MBit	Mega Bit/s Second
MC	Multicarrier
MCID	Malicious Call Identification
MCL	Mercury Communications Ltd [C&W]
MD	Mediation Device
MDF	Main Distribution Frame
MEO	Medium-Earth Orbit
MF	Multi-Frequency
MF	Mediation Function
MHz	Megahertz – 1,000,000 Hz
MIB	Management Information Base [IETF]
MMAC	MultiMedia A? C?
MOD	Ministry Of Defence
MOLO	Mobile Other Licensed Operator
MP3	MP3
MPE	Multiprotocol Encapsulation
MPEG	Moving Picture Experts Group
MPEG1	Moving Picture Experts Group [ISO/ITU-T compression algorithm]
MPEG2	Moving Picture Experts Group [ISO/ITU-T compression algorithm]
MSC	Mobile Switching Centre
MSc	Master of Science
MSH64	Marconi STM-64 ADM
MSN	Multiple Subscriber Number [ISDN]
MSO	Multiple System Operator
MSOH	Mux Section OverHead
MSP	Multiplex Section Protection
MTA	Multimedia Terminal Adapter
MTU	Multi-Tenanted Unit
mW	milli Watt
NAT	Network Address Translation [IETF]

NB	Narrowband
NE	Network Element
NEF	Network Element Function
NetMod	Network Model [BT]
NEXT	Near-End Cross-Talk
NFAP	National Frequency Assignment Panel [UK]
NFM	Network Fault Manager
NGSO	Non Geo-Stationary Orbit
NICC	Network Interoperability Consultative Committee
NISM	Network Inventory and Spares Management [BT]
NIU	Network Interface Unit
NMC	Network Management Centre
NMS	Network Management System
NOU	Network Operations Unit [BT]
NRIC	Network Interoperability and Reliability Council
NSAP	Network Service Access Point
NT	Network Termination
NTE	Network Termination Equipment
NTE5	Network Terminating Equipment No. 5
NTP	Network Terminating Point
NTT	Nippon Telephone and Telegraph
NTU	Network Terminating Unit
OAM	Operations Administration and Maintenance
OAN	Optical Access Network
OFDM	Orthogonal Frequency Division Multiplexing
OFTEL	OFfice of TELecommunications
OLO	Other Licensed Operator
OLT	Optical Line Termination
OMG	Object Management Group
ONT	Optical Network Termination
ONU	Optical Network Unit
OPF	Operator Policy Forum [Oftel]
OS	Operations System [ITU-T]
OSF	Operations System Function
OSI	Open Systems Interconnection
OSS	Operational Support System
PacketCable	Packet over Cable
PACS	Planning Assignment and Configuration System
PAN	Personal Access Network
PBO	Power Back-Off
PBX	Private Branch Exchange
PC	Personal/Portable Computer
PCI	Peripheral Component Interconnect/Interface
PCM	Pulse Code Modulation
PCN	Personal Communications Network

PCP	Primary Cross-connect Point
PCS	Personal Communications Services
PDA	Personal Digital Assistant
PDC	Personal Digital Cellular
PDH	Plesiochronous Digital Hierarchy
PEP	Performance Enhancing Proxy
PES	Permanent Earth Station
PHY	Physical Layer
PIC	Polythylene Insulated Cable
PID	Packet IDentifier
PIM	Plug In Module
PLOAM	Physical Layer Operations and Maintenance
PON	Passive Optical Network
POTS	Plain Old Telephony Service
PPP	Point-to-Point Protocol [IETF]
PPTP	Point-to-Point Tunneling Protocol [IETF]
PSD	Power Spectral Density
PSTN	Public Switched Telephone Network
PTT	Post, Telegraph & Telephones [Organisation]
PVC	Permanent Virtual Connection
Q3	Management communication protocol [ETSI]
QA	Q Adaptor
QAF	Q Adaptor Function
QAM	Quadrature Amplitude Modulation
QoS	Quality of Service
QPSK	Quadrature Phase-Shift Keying
R&TTE	Radio & Telecommunications Terminal Equipment
RA	DTI Radiocommunications Agency
RADIUS	Remote Authentication Dial In User Service [IETF]
RAS	Remote Access Server
RBOC	Regional Bell Operating Company
RBW	Resolution BandWidth
RCS	Return Channel via Satellite
RCST	Return Channel Satellite Terminal
RCU	Remote Concentrator Unit
RedABC	RedCare Alarms By Carrier [BT]
RedCARE	RedCare Alarm service [BT]
REN	Ring-Equivalent Number [BABT]
RENACE	Remote Network Automatic Cross-Connect Equipment
RF	Radio Frequency
RFC	Request for Comment [IETF]
RFI	Radio Frequency Interference
RJ45	Registered Jack 45
RKS	Record Keeping Server
RNC	Radio Network Controllers

RNS	Radio Network S?
RR	Radio Regulations [ITU]
RS232	Interface Standard
RSOH	Regenerator Section OverHead
RSVP	Resource ReSerVation Protocol [IETF]
RTS	Ready To Send
RU	Remote Unit
SBC	SBC Communications Inc.
SBR	Statistical Bit Rate
SC	Single Carrier
SCM	Sub Carrier Modulation
SDH	Synchronous Digital Hierarchy [ITU-T]
SDSL	Symmetric Digital Subscriber Line
SELT	Single Ended Line Testing
SES	Satellite Earth Stations & Systems (ETSI)
SFW	Single Fibre Working
SGSN	Serving GPRS Support Node
SIT	Satellite Interactive Terminal
SMDS	Switched Multimegabit Data Service
SME	Small Medium Enterprise
SMS	Short Message Service
SNCP	Switched Network Circuit Protection [BT]
SNMP	Simple Network Management Protocol [IETF]
SNR	Signal/Noise Ratio
SOHO	Small Office Home Office
SONET	Synchronous Optical Network
SP	Service Provider
SPM	Subscriber Private Meter
SPRing	Shared Protection Ring
SSD	Service Solution Design
SSL	Secure Sockets Layer
SSU	Synchronisation Supply Unit
STB	Set Top Box
STM	Synchronous Transport Module
STM-1	Synchronous Transport Module Level 1 (155 Mbit/s)
STM-4	Synchronous Transport Module Level 4 (622 Mbit/s)
SuperJANET	Super Joint Academic NETwork [UK]
SuperPON	Super Passive Optical Network
SUT	Satellite User Terminal
T1	T-Carrier 1 [USA]
TC	Time And Code (Division Multiple Access)
TCP	Transmission Control Protocol [IETF]
TD	Time Division
TDD	Time Division Duplex

TDM	Time Division Multiplexing
TDMA	Time Division Multiple-Access
TEM	TPON Element Manager
TEP	Telecommunications Equipment Practice
TL1	Transaction Language 1
TM	Transport Module [SDH]
TMN	Telecommunications Management Network [ITU-T]
TNS	Transmission Network Surveillance [BT]
TOS	Type-Of-Service
TPON	Telecommunications over Passive Optical Network
TR	Technical Recommendation
TS	Transport System [SDH]
TS0	Time Slot 0
TV	Television
TVRO	Television Receive Only
UCL	University College London
UDP	User Datagram Protocol [IETF]
UGS	Unsolicited Grant Service
UK	United Kingdom
UML	Unified Modeling Language
UMTS	Universal Mobile Telecommunications System [ITU-T]
UNI	User-to-Network Interface
URL	Uniform Resource Locators [IETF]
US	United States
USA	United States of America
USB	Universal Serial Bus
UWB	Ultra Wide Band
V.32	ITU-T Modem standard
V.34	ITU-T Modem standard
VB.net	Visual Basic Internet [Microsoft]
VBR	Variable Bit Rate
VC	Virtual Connection [ATM]
VC-4	Virtual Container Level 4 concatenated
VDSL	Very high speed Digital Subscriber Line
VLAN	Virtual Local Area Network
VLR	Visitor Location Register [TACS]
VoATM	Voice over Asynchronous Transfer Mode
VOD	Video on Demand
VoDSL	Voice over Digital Subscriber Line
VoIP	Voice over Internet Protocol
VP	Virtual Path
VPN	Virtual Private Network
VSAT	Very Small Aperture Terminal [satellite]
VT100	Virtual Terminal 100

W-CDMA	Wideband Code Division Multiple Access
WAL2	{a line code}
WAN	Wide Area Network
WAP	Wireless Application Protocol
WARC	World Administrative Radio Conference
WB900	Model no. of an analogue pair-gain system
WDM	Wavelength Division Multiplexing
WECA	Wireless Ethernet Compatibility Alliance
WiFi	Wireless Fidelity
WildBlue	{a Geostationary Orbit (GEO) satellite system}
WLL	Wireless Local Loop
WRC	World Radiocommunications Conference
WS	WorkStation
WSF	WorkStation Function
WT	Wireless Telegraphy
WWW	World Wide Web [CERN]
X.21	Interface between data terminal equipment and data circuit-terminating equipment for synchronous operation on public data networks [ITU-T]
xDSL	ADSL, HDSL, SDSL, VDSL, etc.
XML	eXtensible Markup Language

Index